D0457498

FREQUENCY DOMAIN CRITERIA FOR ABSOLUTE STABILITY

ELECTRICAL SCIENCE
A Series of Monographs and Texts

Editors: Henry G. Booker and Nicholas DeClaris
A complete list of titles in this series appears at the end of this volume

Frequency Domain Criteria for Absolute Stability

KUMPATI S. NARENDRA

Yale University
New Haven, Connecticut

JAMES H. TAYLOR

The Analytic Sciences Corporation
Reading, Massachusetts

ACADEMIC PRESS New York and London 1973

A Subsidiary of Harcourt Brace Jovanovich, Publishers

ACADEMIC PRESS, INC.
111 Fifth Avenue, New York, New York 10003

United Kingdom Edition published by
ACADEMIC PRESS, INC. (LONDON) LTD.
24/28 Oval Road, London NW1

LIBRARY OF CONGRESS CATALOG CARD NUMBER: 72-82641

PRINTED IN THE UNITED STATES OF AMERICA

To

BARBARA AND ANNE-MARIE

CONTENTS

Foreword xi
Preface xiii
Acknowledgments xv
Special Notation xvii

I. Introduction

1. The System 1
2. Stability of Motion 6
3. Lyapunov's Direct Method 7
4. The Quadratic Lyapunov Function 10
5. Some Problems in Stability 11
6. The Conjectures of Aizerman and Kalman 13
7. The Absolute Stability Problem 14
8. The Criterion of Popov 15
9. Synopsis 16

II. Problem Statement

1. System Definition 18
2. Definitions of Stability 33
3. Formal Problem Statement 38

III. Mathematical Preliminaries

1. Sufficiency Theorems 40
2. The Absolute Lyapunov Function Candidates 44

3. Restated Stability Theorems 48
4. The Kalman–Yakubovich Lemma 48
5. Positive Real Functions 57
6. Existence Theorems 61

IV. Linear Time-Invariant Systems and Absolute Stability

1. Relations between Linear Time-Invariant and Nonlinear Time-Varying Systems 67
2. The Existence of the Quadratic Lyapunov Function x^TPx and the Hurwitz Condition 81
3. The Existence of the Quadratic Lyapunov Function $x^TPx + \kappa x^TMx$ and the Nyquist Criterion 84

V. Stability of Nonlinear Systems

1. The Popov Stability Criterion 91
2. Stability Criteria for Monotonic Nonlinearities 99
3. Linear Systems 113
4. Odd Monotonic Gains 116
5. The General Finite Sector Problem 118

VI. Stability of Nonlinear Time-Varying Systems

1. The Circle Criterion 123
2. An Extension of the Popov Criterion—Point Conditions 126
3. Stability Criteria for Restricted Nonlinear Behavior—Point Conditions 131
4. The General Finite Sector Problem 135
5. Periodic Nonlinear Time-Varying Gains 137
6. Extension of the Popov Criterion—Integral Conditions 138
7. Integral Conditions for Restricted Nonlinear and Linear Gains 141
8. Integral Conditions for Linear Time-Varying Systems 142

VII. Geometric Stability Criteria

1. Linear Time-Invariant Systems 149
2. The Circle Criterion 155
3. The Popov Criterion 158
4. Monotonic Nonlinearities: An Off-Axis Circle Criterion 166
5. Further Geometric Interpretations for Time-Varying Systems 175

VIII. The Mathieu Equation: An Example

1. Solutions of the Mathieu Equation 186
2. Linear Case ($a \approx 1$): A Perturbation Analysis 187
3. Linear Case ($a \approx 4$): A Floquet Analysis 189

4. Application of Stability Criteria to the Linear Case 190
5. Application of Stability Criteria to the Nonlinear Case 198

IX. **Absolute Stability of Systems with Multiple Nonlinear Time-Varying Gains**

1. Introduction 202
2. Problem Statement 203
3. Mathematical Preliminaries 209
4. Linear Time-Invariant Systems and Absolute Stability 213
5. Stability of Nonlinear Systems 216
6. Stability of Nonlinear Time-Varying Systems 225

APPENDIX. **Matrix Version of the Kalman–Yakubovich Lemma** 230

References 235

INDEX 243

FOREWORD

To write a highly interesting book in a field in which there exists a rich literature is certainly a difficult task. The present book fully realizes this performance. Its success is due mainly to the great care with which the authors selected the material included in the book, with the obvious aim of giving the reader a deep and broad understanding of the subject.

While the main theorems are recent ones, the authors show very clearly the strong connections of the theory with the classical results of Hurwitz, Lyapunov, Nyquist, etc. Their reappraisal of the traditional engineering methods of control theory—including the daring but often successful technique of "describing functions"—provides an opportunity to point out another important source of the ideas that generated the contemporary view of the problem. While developing the theory with care for rigor and generality, the authors also show much concern for concrete examples and often illustrate the general theory by significant and illuminative applications, treated in detail.

These qualities make the book very useful even for persons who have little or no previous knowledge of the subject. These people will find this book an excellent introduction to the field. On the other hand, those already familiar with the subject will find a detailed exposition of some of the most advanced results which are harder to find elsewhere and which are due mainly to the outstanding research done in the field by the authors themselves.

Books which successfully cover such a broad range of interests are rare. They are also very much needed, because they are bound to produce a favorable influence upon research.

V. M. Popov

PREFACE

This book presents some recent generalizations of the well-known Popov solution to the absolute stability problem proposed by Lur'e and Postnikov in 1944. The Popov frequency domain stability criterion and the results of several earlier approaches to the Lur'e–Postnikov problem are presented in detail in the excellent books of Lefschetz and of Aizerman and Gantmacher; the work that led to the formulation of the absolute stability problem and the first solutions to it are not considered here.

The success of Popov's elegant criterion inspired many extensions of the basic Lur'e–Postnikov problem. Studies of these related questions gave rise to a great number of stability criteria, derived using both the direct method of Lyapunov and the positive operator concept of functional analysis. The great interest in this area has resulted in a continuing state of rapid development. The generation of this type of frequency domain stability criteria has now reached a relative state of completeness. It is also notable that the two seemingly disparate analytic approaches have led to stability criteria that are equivalent in most respects, and thus it is possible to present a unified picture of the recent research in this area using only Lyapunov's direct method. In each of the two fundamental approaches there are several points of view which have been used to good effect by various groups of researchers. It should thus be noted that this book is founded on a single set of techniques based on the direct method of Lyapunov and developed first at Harvard University and then at Yale University and the Indian Institute of Science (Bangalore, India). This makes the book rather specialized in its overall scope, but the techniques are found to be applicable to a wide range of important questions regarding the stability of nonlinear systems.

In view of the approach taken, several important results derived using a functional analysis viewpoint have either been omitted entirely or only mentioned in passing. Since the emphasis is on the application of Lyapunov's direct method to generate frequency domain criteria for stability, many fine results related to other aspects of the stability problem are omitted. The bibliography is by no means complete for this reason and contains only works directly related to the problems discussed.

Continuous-time systems are considered here, although many similar results already exist for discrete systems. In the first eight chapters, systems with a single nonlinear function or time-varying parameter are treated. Systems with multiple nonlinearities or time-varying gains are considered in Chapter IX; some criteria are derived in detail while others are presented in outline form as an indication of the state of current research.

This book can serve very well as a reference for research courses concerning stability problems related to the absolute stability problem of Lur'e and Postnikov. Engineers and applied mathematicians should also find the results contained herein, particularly the geometric stability criteria, of use in practical applications. Because of the diversity of the audience being addressed, rigorous theory is developed with what we hope can be considered a minimum of mathematical formalism. Certain sections contain some quite condensed technical material required as a foundation for the derivations; these may be omitted by those whose interest is limited to applications.

It is assumed that the reader is familiar with matrix operations that are utilized in dealing with the state vector representation of dynamic systems. All definitions and theorems are developed as needed so that the derivations are independent of other works; some acquaintance with the basic concepts of stability and Lyapunov's direct method would be helpful. The historical development of the work associated with the Lur'e–Postnikov problem has been strongly linked to the theory of automatic control, so control systems terminology is used sparingly wherever it is reasonable to expect that the meaning is clear to all readers.

ACKNOWLEDGMENTS

The authors are indebted to Roger M. Goldwyn, Charles P. Neuman, and Yo-Sung Cho (who were graduate students under the guidance of the first-named author), and M. A. L. Thathachar and M. D. Srinath, all of whose work forms an important basis for the material presented here. An early review of this effort by V. M. Popov was also significant; his suggestions and generous comments provided both encouragement and improvement in completing the final version of this work.

It is a pleasure for the authors to acknowledge colleagues and graduate students who have generously given their assistance both in general discussions and in recommending specific changes in the manuscript. The efforts of R. Viswanathan, M. D. Srinath, N. Viswanadham, and S. Rajaram are especially appreciated in this regard. Finally, an important factor in the completion of this effort was the able assistance provided by Mrs. Coralie Wilson, Mrs. Jean Gemmell, and Mrs. Anne-Marie Taylor, who typed many drafts and corrections.

The support of various institutions has also been invaluable. The second-named author would like to acknowledge the support received from Yale University while a graduate student and also the generosity of the Indian Institute of Science where he was recently a Visiting Assistant Professor.

One of the authors was in Bangalore, India while the other was in New Haven, Connecticut during the entire period of preparation of this book. That the sequence of corrections, additions, and revisions carried across several continents finally converged is in itself an achievement in the eyes of the authors. It may be safely said that without the patience, understanding, and encouragement of our wives, Barbara Narendra and Anne-Marie Taylor, this book would not have been completed.

SPECIAL NOTATION

Throughout this book the following symbol conventions are generally adhered to:

(i) Scalars are denoted by lower case Greek characters (ρ, τ, $\sigma_0 = h^T x + \rho\tau$ etc.). The principal exception is the independent variable t (time).

(ii) Column vectors and explicit functions are denoted by lower case Latin characters (x, h; $f(\sigma_0)$, etc.). The notation $x = 0$ signifies that all elements x_i of the vector are zero.

(iii) Matrices, transfer functions, function classes, function bounds, and n-dimensional Euclidean spaces are denoted by capital Latin characters ($A = [a_{ij}]$; $G(s)$; $f(\sigma_0) \in \{F\}$; $\underline{K} \leqslant k(t) \leqslant \bar{K}$; $x \in X$). Upper and lower bounds are distinguished by bars above and below, respectively. $A = \underline{0}$ denotes the null matrix ($a_{ij} = 0$, all i and j) and I is the unit matrix ($I = \text{diag}\,(1, 1, \ldots, 1)$).

(iv) Function ranges (as in (iii) above) may be specified by the notation $k(t) \in [\underline{K}, \bar{K}]$. A bracket indicates a closed interval, whereas a parenthesis indicates an open interval. Thus $k(t) \in (0, \bar{K}]$ signifies that $0 < k(t) \leqslant \bar{K}$.

(v) The transpose of a vector or matrix is designated by a superscript T ($x^T = [x_1, x_2, \ldots, x_n]$; $A^T = [a_{ji}]$), and the inverse of a non-singular matrix is denoted by a superscript -1 [$(sI - A)^{-1}$].

(vi) The notation $G(s) \in \{Z_N\}$ denotes the membership of $G(s)$ in some class $\{Z_N\}$ of transfer functions. In particular, {PR} is the class of all positive real functions and {SPR} the class of all strictly positive real functions (Chapter III, Section 5).

(vii) Systems may be classified as being
LTI (linear time-invariant)
LTV (linear time-varying)
NLTI (nonlinear time-invariant)
or NLTV (nonlinear time-varying).

(viii) The complete notational designation of nonlinear time-varying gain functions ($g(\sigma_0, t) \in \{G_i[N, T]\}$) is detailed in Chapter II, Section 1.

(ix) The classes of matrices $\{A_1\}$ and $\{A_0\}$ are defined in Chapter II, Section 1.

(x) The superscript * denotes the complex conjugate of a scalar or the complex conjugate transpose of a vector or matrix.

(xi) An open square (□) indicates the end of a theorem, lemma, definition, or proof.

(xii) In all theorems and lemmas, $Z(s)^{\pm 1}$ implies that either $Z(s)$ or $Z^{-1}(s)$ can be used in satisfying the indicated condition.

FREQUENCY DOMAIN CRITERIA FOR ABSOLUTE STABILITY

I

INTRODUCTION

In this study we restrict our attention to real dynamic systems which are governed by ordinary differential equations containing a finite number of parameters; our principal aim is the development of conditions that are sufficient to ensure certain stability properties of such systems. In the introductory comments that follow, an attempt is made to provide a brief outline of the necessary theoretical background and to establish a context for the derivations of succeeding chapters. First, however, it is necessary to consider the class of dynamic systems under study in some detail.

1. The System

If $x_1, x_2, \ldots, x_n$ represent n coordinates in a Euclidean n-space X, and t the time, the behavior of a typical dynamic system is described by the differential equations

$$dx_i/dt \triangleq \dot{x}_i = f_i(x_1, x_2, \ldots, x_n, t), \qquad i = 1, 2, \ldots, n, \tag{1-1}$$

or, if x_i and f_i are considered to be elements of column vectors x and f, by the vector differential equation

$$\dot{x} = f(x, t). \tag{1-2}$$

The variables x_i are referred to as the state variables, and these constitute the state vector x. X is said to be the state space.

If a solution exists in some neighborhood of a given initial condition (x_0, t_0) where $x_0 \in X$, $t_0 \in (-\infty, +\infty)$, then it is denoted by $x(t; x_0, t_0)$. This solution may be represented as a curve in the Euclidean $(n + 1)$-space M with coordinates $x_1, x_2, \ldots, x_n, t$; points on this curve are specified by the independent variable t. M is said to be the motion space; the curve $x(t; x_0, t_0) \in M$ specifies a motion in a general sense, that is, the state variables may represent voltages, temperatures, or pressures as well as spacial coordinates. The projection of the motion onto the state space X is called a trajectory. A typical solution to a second order system $\dot{x}_1 = x_2, \dot{x}_2 = -x_1$ [an oscillation with period $T = 2\pi$ and initial conditions $x_1(0) = A$, $x_2(0) = 0$] is depicted both as a trajectory and a motion in Fig. 1-1.

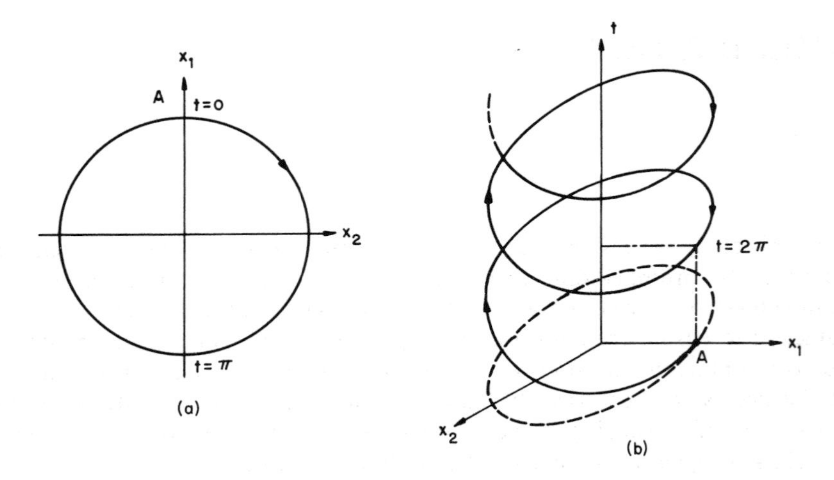

Fig. 1-1. The solution $x(t; x_0, 0)$ in the state and motion spaces: (a) state space X; (b) motion space M.

The point x_e which satisfies the equation $f(x_e, t) \equiv 0$ for all t is called a singular point of the differential equation. To a singular point corresponds a constant solution which can be thought of as an equilibrium for a physical system. Throughout this book, only systems with a single equilibrium at the origin $(x \equiv 0)$ will be considered so that $f(0, t) = 0$.

It is crucial in the application of Lyapunov's direct method that the system be sufficiently well behaved so that a solution (not necessarily unique) $x(t; x_0, t_0)$ exists in some neighborhood of any initial condition $(x_0, t_0) \in M$. The following well-known theorem provides elementary conditions that suffice to guarantee the existence of solutions in one specific sense (see Coddington and Levinson [1]).

THEOREM 1 (A vector form of the Cauchy–Peano existence theorem). If $f(x, t)$ is continuous with respect to x and t in a region

$$R \triangleq \{(x, t): \ |t - t_0| \leqslant \alpha, \ \ \|x - x_0\| \leqslant \beta; \ \ \alpha, \beta > 0\},$$

then for some $\hat{\alpha} > 0$ there exists a solution $x(t; x_0, t_0)$ to Eq. (1-2) for $|t - t_0| \leqslant \hat{\alpha}$ which has a continuous time derivative. □

Additional assumptions must be made about $f(x, t)$ if the solutions to the differential equation (1-2) are required to have other more stringent properties.

(1) If $x(t; x_0, t_0)$ must be continuous with respect to all of its arguments, then solutions for any $(x_0, t_0) \in M$ must be unique [see (2)].

(2) If for every $(x_0, t_0) \in M$ there must exist a unique solution, the following conditions are sufficient.

THEOREM 2 (see Coddington and Levinson [1, Theorem 2.3]). If $f(x, t)$ is continuous with respect to x and t in R (Theorem 1) and in addition $f(x, t)$ satisfies a local Lipschitz condition (LLC), that is, some $K(\alpha, \beta)$ exists such that

$$\|f(x, t) - f(x', t)\| \leqslant K \|x - x'\|$$

for all (x, t), $(x', t) \in R$, then for some $\hat{\alpha} > 0$ there exists a unique solution $x(t; x_0, t_0)$ to Eq. (1-2) for all $|t - t_0| \leqslant \hat{\alpha}$ which has a continuous time derivative. □

(3) If a constant $\gamma > 0$ is to exist such that

$$(d/dt) \|x(t; x_0, t_0)\| \leqslant \gamma \|x(t; x_0, t_0)\|,$$

then it is sufficient that $f(x, t)$ satisfy a uniform Lipschitz condition (ULC) with respect to t, that is, $K = K(\beta)$ (see Hahn [1, §38]).

(4) If it must be possible to continue $x(t; x_0, t_0)$ for all $t \geqslant t_0$, it is sufficient that $f(x, t)$ satisfy a global uniform Lipschitz condition (GULC), that is, that a constant K exists such that

$$\|f(x, t) - f(x', t)\| \leqslant K \|x - x'\|$$

for all (x, t), $(x', t) \in M$ (see Kalman and Bertram [1]).

Whereas statements (1) to (4) are not directly germane to the derivation of absolute stability criteria using the second method of Lyapunov, they are mentioned here because they shed a certain amount of light on the properties of solutions to differential equations. In applying the direct method of Lyapunov to determine stability of solutions for all $(x_0, t_0) \in M$, however, it is found that the solutions are not required to be unique or to have a continuous derivative with respect to time; hence the above restrictions on $f(x, t)$, even those of Theorem 1, are unnecessarily stringent.

The following theorem provides conditions that are more nearly in consonance with the requirements of this study.

THEOREM 3 (The existence theorem of Caratheodory; see Coddington and Levinson [1]). If for all $(x, t) \in R$ (Theorem 1), $f(x, t)$ is defined, measurable in t for all fixed x, continuous in x for all fixed t, and

$$\| f(x, t) \| \leqslant \mu(t)$$

is satisfied where $\mu(t)$ is Lebesque integrable over $|t - t_0| \leqslant \alpha$, then for some $\hat{\alpha} > 0$ there exists a solution $x(t; x_0, t_0)$ for $|t - t_0| \leqslant \hat{\alpha}$ that is absolutely continuous with respect to time and which satisfies $(d/dt)\, x(t; x_0, t_0) = f(x(t; x_0, t_0), t)$ for almost all t, $|t - t_0| \leqslant \hat{\alpha}$. □

The minimal conditions on $f(x, t)$ that are assumed hereafter are summarized by defining the class $\{S\}$ of acceptable functions $f(x, t)$.

FUNCTION CLASS $\{S\}$. A function $f(x, t) \in \{S\}$ if

(a) $f(x, t)$ is defined for all x, t;
(b) $f(x, t)$ is continuous with respect to x for all fixed t;
(c) $f(x, t)$ has at most an enumerable set of isolated discontinuities with respect to t for all fixed x;
(d) $\| f(x, t) \|$ is bounded for all finite x and for all t as in Theorem 3;
(e) $f(0, t) \equiv 0$ for all t, and there exists no $x_e \neq 0$ such that $f(x_e, t) \equiv 0$; hence the origin is the sole singular (equilibrium) point. □

Three important special cases arise naturally when dealing with the properties of systems of the form

$$\dot{x} = f(x, t), \qquad f(x, t) \in \{S\}. \tag{1-3}$$

(i) *Linear time-invariant (LTI) systems*

$$\dot{x} = A_0 x, \tag{1-3a}$$

where A_0 is an $(n \times n)$ matrix of constants.

(ii) *Linear time-varying (LTV) systems*

$$\dot{x} = A_0(t) x, \tag{1-3b}$$

where $A_0(t)$ is an $(n \times n)$ matrix any of the elements of which may be functions of time.

(iii) *Nonlinear time-invariant (NLTI) systems*

$$\dot{x} = f_0(x), \tag{1-3c}$$

where $f_0(x)$ explicitly does not vary with time. Such systems are often referred to as autonomous systems. For the sake of completeness, systems represented by the general equation (1-2) are considered to be a fourth class.

(iv) *Nonlinear time-varying (NLTV) systems*

$$\dot{x} = f_0(x, t). \tag{1-3d}$$

If $f_0(x, t)$ satisfies $f_0(0, t) = 0$ for all t, the origin $x = 0$ is by definition an equilibrium point of the system. For cases (i) and (ii) this condition is automatically satisfied, while for NLTI systems the origin is an equilibrium state if $f_0(0) = 0$. It is the behavior of the solutions of Eq. (1-3) with respect to this equilibrium state that is of particular interest to us.

Exhaustive studies of the properties of LTI systems have been undertaken, and the stability problem with respect to Eq. (1-3a) has been completely resolved. Chapter IV outlines the two basic stability criteria for such systems and relates these to the developments of succeeding chapters. The question of the stability of LTV systems [Eq. (1-3b)] has been investigated extensively, but the great complexity introduced by assuming the existence of time-varying system parameters thus far has precluded any general solution. Many techniques for treating specific forms have been developed, however, so that numerous classes of systems may be successfully analyzed.

The stability properties of NLTI or NLTV systems [Eqs. (1-3c) or (1-3d)] present even more formidable obstacles. Whereas Lyapunov's direct method provides a powerful general tool for use in such situations, the application of this theory often is not practical. Until comparatively recently the stability criteria derived in this area have been restricted to quite specific system models that are generally of a low order and possess a given structure. A widespread practice in dealing with such systems has been, in fact, to attempt to relate a problem that is not amenable to fairly simple analysis to the more tractable form of Eqs. (1-3a) or (1-3b).

When total linearization is not possible (for example, the physical system contains a nonlinear device that is not sufficiently linear over the normal operating range to be approximated by a linear element), partial linearization is often profitable. Such systems become easier to deal with analytically while they continue to reflect the effect of the nonlinearity on the nature of the solutions. Partial linearization of Eq. (1-2) leads to the system of equations (1-4) where the single nonlinear time-varying element $g(\cdot, t)$ is explicitly separated from the linear part:

$$\dot{x} = Ax + b\tau, \qquad \sigma_0 = h^T x + \rho\tau, \qquad \tau = -g(\sigma_0, t). \tag{1-4}$$

The vectors b and h are of dimension n and ρ is a scalar constant, σ_0 is a linear combination of the state variables x_i and τ, while τ is a nonlinear and/or time-varying function of σ_0. The time functions $\tau(t)$ and $\sigma_0(t)$ may be thought of as the input and output of an LTI plant while the relation $\tau = -g(\sigma_0, t)$ may be considered to be the mathematical representation of an NLTV controller. This control theory terminology is an interpretation of the diagram of Fig. 1-2 which represents the dynamics of Eq. (1-4).

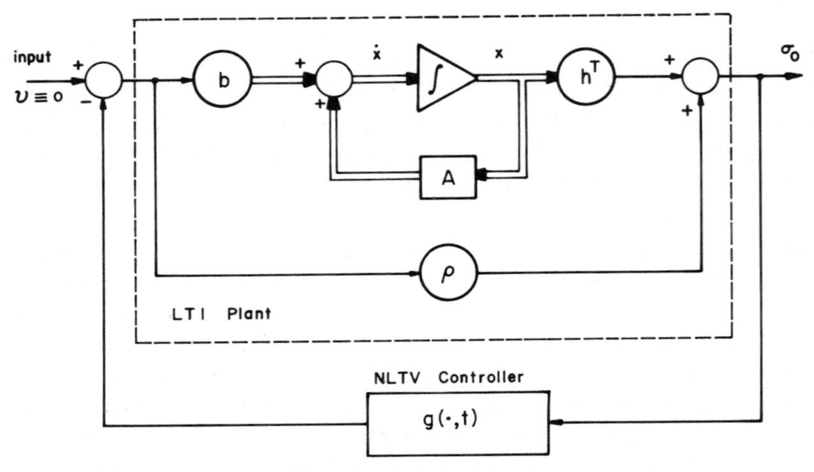

Fig. 1-2. System dynamics.

The four cases (i) to (iv) correspond to

(i′) *LTI systems:* $\tau = -\kappa\sigma_0$; $A_0 = A - \dfrac{\kappa}{1+\rho\kappa} bh^{\mathrm{T}}$.

(ii′) *LTV systems:* $\tau = -k(t)\sigma_0$; $A_0(t) = A - \dfrac{k(t)}{1+\rho k(t)} bh^{\mathrm{T}}$.

(iii′) *NLTI systems:* $\tau = -f(\sigma_0)$.

(iv′) *NLTV systems:* $\tau = -g(\sigma_0, t)$.

2. Stability of Motion

If for the system described by Eq. (1-3), $\hat{x}(t) = x(t; \hat{x}_0, t_0)$ represents a solution of particular interest arising from initial conditions $(\hat{x}_0, t_0)$ and $x(t) = x(t; x_0, t_0)$ corresponds to a solution that is initially close to $\hat{x}_0$ in the sense that $\| \hat{x}_0 - x_0 \| \leqslant \mu$, then

$$\dot{\hat{x}} = f_0(\hat{x}, t), \qquad \dot{x} = f_0(x, t). \tag{1-5}$$

If the difference $\varepsilon(t) \triangleq x(t) - \hat{x}(t)$ is studied, $\varepsilon(t)$ satisfies the differential equation

$$\dot{\varepsilon} = f_0(\hat{x} + \varepsilon, t) - f_0(\hat{x}, t) \triangleq f_1(\varepsilon, t). \tag{1-6}$$

As $f_1(0, t) = 0$, the solution $\varepsilon \equiv 0$ corresponds to the equilibrium of (1-6); thus, the behavior of the solutions of Eq. (1-3) in the neighborhood of $\hat{x}(t)$ corresponds to the behavior of the solutions of Eq. (1-6) in the neighborhood of the origin. This implies that the general problem of determining the stability of a motion $\hat{x}(t)$ of a system can be transformed into an equivalent stability problem where the origin is the equilibrium point under investigation. Hence, without any loss of generality we can deal solely with the equilibrium state $x(t) \equiv 0$.

This same approach also allows the study of the stability of a motion with respect to initial conditions even when specified input functions are present, namely,

$$\dot{x} = f(x, u, t). \tag{1-7}$$

3. Lyapunov's Direct Method

The direct method of Lyapunov, which is the most general method of stability analysis available, is used exclusively in this book for all of the problems considered. The system described by Eq. (1-3) is assumed to have no input function so that the stability of the null solution $x(t) \equiv 0$ is investigated in all cases.

A. Concepts of Stability

The most fundamental definitions of stability considered in using this approach were originally proposed by Lyapunov.

DEFINITION 1 STABILITY. The equilibrium $x \equiv 0$ of the differential equation (1-3) is stable if for every real $\varepsilon > 0$ and t_0, there exists a real $\delta(\varepsilon, t_0) > 0$ such that

$$\|x_0\| \leqslant \delta \longrightarrow \|x(t; x_0, t_0)\| \leqslant \varepsilon$$

for all $t \geqslant t_0$. □

DEFINITION 2 ATTRACTIVITY. The equilibrium $x \equiv 0$ of the differential equation (1-3) is attractive if for some $\rho > 0$ and for every $\eta > 0$ there exists a number $T(\eta, x_0, t_0)$ such that

$$\|x(t; x_0, t_0)\| \leqslant \eta \qquad \text{for all} \quad t - t_0 \geqslant T$$

for all $\|x_0\| \leqslant \rho$. □

DEFINITION 3 ASYMPTOTIC STABILITY. The equilibrium $x \equiv 0$ of the differential equation (1-3) is asymptotically stable if it is both stable and attractive. □

In this case, for any given $\varepsilon > 0$ and t_0, there exists a constant $\delta(\varepsilon, t_0)$ such that

$$\|x(t; x_0, t_0)\| \leqslant \varepsilon \qquad \text{for all} \quad \|x_0\| \leqslant \delta \tag{1-8a}$$

and

$$\lim_{t \to \infty} x(t; x_0, t_0) = 0. \tag{1-8b}$$

Definition 1 ensures that all solutions can be made to lie within an arbitrary neighborhood of the origin by choosing the initial state within a suitable neighborhood of $x = 0$. If ε is decreased, a correspondingly smaller value of δ can be found such that relation (1-8a) holds.

Definition 2 assures that condition (1-8b) holds for all trajectories starting at time t_0 from a closed ball of radius ρ but does not indicate how the trajectories behave for finite values of t. The concept of attractivity is independent of the concept of stability; a system can be attractive while being unstable. Hahn [1] has discussed several systems exhibiting such behavior, the most interesting of which is an autonomous system of second order originally suggested by Vinograd.

Hahn [1] has also suggested alternative definitions in terms of comparison functions for the three concepts defined in Definitions 1 to 3, and has established the equivalence of the two sets of definitions. The comparison functions provide a convenient and intuitively satisfactory method of describing the various concepts. We merely present the alternate definitions without proving the equivalence of the two definitions. The comparison functions are found to be particularly useful while considering further refinements of Definitions 1 to 3 (see Chapter II).

FUNCTION CLASS $\{K\}$. A real valued function $\phi(\rho, t_0)$ belongs to class $\{K\}$ if for the specified t_0

(a) it is defined, continuous and strictly increasing for all $\rho, 0 \leqslant \rho \leqslant \rho_1$, where $\rho_1 > 0$ is arbitrary;
(b) $\phi(0, t_0) = 0$. □

FUNCTION CLASS $\{L\}$. A real valued function $\psi(t; x_0, t_0)$ belongs to class $\{L\}$ if for the specified x_0 and t_0

(a) it is defined, continuous and strictly decreasing for all $t \geqslant t_0$;
(b) $\lim_{t \to \infty} \psi(t; x_0, t_0) = 0$. □

Using these function classes, the alternative stability definitions of Hahn are:

DEFINITION 1′ STABILITY. The equilibrium $x \equiv 0$ of the differential equation (1-3) is stable if there exists a function $\phi \in \{K\}$ such that

$$\|x(t; x_0, t_0)\| \leqslant \phi(\|x_0\|, t_0)$$

for all t_0 and for all $t \geqslant t_0$. □

DEFINITION 2′ ATTRACTIVITY. The equilibrium $x \equiv 0$ of the differential equation (1-3) is attractive if for some $\rho(t_0) > 0$ there exists a function $\psi \in \{L\}$ such that

$$\|x(t; x_0, t_0)\| \leqslant \psi(t - t_0; x_0, t_0)$$

for all $t \geqslant t_0$ and $\|x_0\| \leqslant \rho$. □

DEFINITION 3′ ASYMPTOTIC STABILITY. The equilibrium $x \equiv 0$ of the differential equation (1-3) is asymptotically stable if there exists a function $\phi \in \{K\}$, and for some $\rho(t_0) > 0$ there exists a function $\psi \in \{L\}$ such that

$$\|x(t; x_0, t_0)\| \leqslant \phi(\|x_0\|, t_0)\psi(t - t_0; x_0, t_0)$$

for all $t \geqslant t_0$ and $\|x_0\| \leqslant \rho$. □

For autonomous systems either set of definitions is considerably simpler both in concept and in application. The parameters δ, ρ, and T are not functions of the initial time in Definitions 1 to 3, and the comparison functions of Definitions 1′ to 3′ are likewise unaffected by t_0.

In recent years several other types of stability have been defined in terms of the inputs and the outputs of the system. The functional analytic techniques used in such cases when applied to the problems considered in this book yield similar stability criteria. While no attempt is made to present these techniques, references are provided wherever relevant since many of the contributions to the field have been made using such an approach.

B. Lyapunov Functions

The most attractive feature of both the direct method of Lyapunov and the more recent functional analytic approach is the fact that they enable the stability of a system to be ascertained directly from the equations describing the system without recourse to the explicit form of the solutions.

In the former theory a suitable function is defined in the state space or motion space and the sign of the function and the sign of its time derivative are examined. For autonomous systems a positive definite scalar function $v(x)$ is chosen and the total time derivative of v along the system trajectories is investigated. If $v(x) = \beta$ for positive constant values of β defines a family of closed nested hypersurfaces [the surface $v = \beta_1$ encloses all surfaces $v = \beta$ for $0 \leqslant \beta < \beta_1$ for each $\beta_1 > 0$] in some neighborhood of the origin, and $v(x(t))$ decreases monotonically along any trajectory $x(t)$ for sufficiently small $\|x\|$, the trajectory crosses each curve $v(x) = \beta$ from the exterior to the interior for increasing values of the parameter t. If the origin is an interior point of each hypersurface, the trajectory approaches the origin arbitrarily closely and the equilibrium is asymptotically stable. For asymptotic stability in the whole we require that $v(x) = \beta$ remain closed for arbitrarily large values of β, that $v(x) \longrightarrow \infty$ as $\|x\| \longrightarrow \infty$, and that the time derivative $\dot{v}$ be negative for all $x \neq 0$.

For the system described in Eq. (1-3c),

$$dv/dt \triangleq \dot{v} \triangleq (\nabla v)^{\mathrm{T}}\dot{x} = (\nabla v)^{\mathrm{T}} f_0(x) \tag{1-9}$$

where ∇v is the gradient:

$$(\nabla v)^T \triangleq [\partial v/\partial x_1, \partial v/\partial x_2, \ldots, \partial v/\partial x_n].$$

If $\dot{v}$ satisfies the condition $\dot{v} \leqslant 0$, then $v(x)$ is said to exist as a Lyapunov function for the specific system (1-3c) and the system is stable. If $\dot{v}(x) \leqslant 0$ but not identically zero for any solution $x(t) \not\equiv 0$, then (see Chapter III, Section 1) the equilibrium is also asymptotically stable. For nonautonomous systems of the form (1-3d) a similar approach is used but the positive definite function v is also generally a function of time. In this case the total derivative $\dot{v}(x, t)$ taken along the trajectories is given by

$$\dot{v}(x, t) \triangleq \partial v/\partial t + (\nabla v)^T f_0(x, t). \quad (1\text{-}10)$$

If $v(x, t)$ is positive definite and $\dot{v} \leqslant 0$, v is a Lyapunov function for the nonautonomous system and the system is stable as before. For asymptotic stability somewhat stronger conditions have to be imposed on the Lyapunov function. These conditions are discussed in detail in Chapter III.

For NLTV systems, treated in Chapter VI, the theorem of Corduneanu [1] is applied in the development of one of the criteria for stability. In this case, $\dot{v}(x, t)$ has to satisfy the less restrictive condition that

$$\dot{v}(x, t) \leqslant m[v(x, t), t] \quad (1\text{-}11)$$

for stability where $\dot{y} = m(y, t)$ is an asymptotically stable scalar differential equation. While $v(x, t)$ defined in this fashion is strictly speaking not a Lyapunov function, the Corduneanu theorem is based on concepts derived from Lyapunov's method.

4. The Quadratic Lyapunov Function

For all of the generality of Lyapunov's direct method, it has one important weakness when used in specific situations. There is no way of determining the definitive Lyapunov function candidate that would yield necessary and sufficient conditions for stability except for the case of LTI systems. In general, when a candidate $v(x, t)$ is chosen, the conditions that have to be imposed on the system (1-3), in order to guarantee that $\dot{v}(x, t)$ satisfies the inequality $\dot{v} < 0$ or that (1-11) is satisfied, are conservative. For the example of the damped Mathieu equation in Chapter VIII, for instance, extensive effort has been made to determine a suitable form of $v(x, t)$ which yields necessary and sufficient conditions for stability. The failure to determine such a function for a seemingly elementary second order LTV system demonstrates the magnitude of this difficulty.

One of the principal modes of attack used in this study is the development of Lyapunov functions from the quadratic form $x^T Px + \kappa x^T Mx$ that yields necessary and sufficient conditions for the stability of LTI systems (1-3a).

This form is suitably modified to yield sufficient conditions for stability of the many classes of nonlinear and/or time-varying systems considered. The modifications in the form of the Lyapunov function for each specific class of systems, the motivation behind each modification, and the resulting sufficient conditions for stability form the essence of this book. To the authors, this unified approach in studying the stability of various classes of systems using a single form of the Lyapunov function is most appealing.

5. Some Problems in Stability

Within the framework provided by the system model given by Eq. (1-4) a great variety of stability problems may be posed. Among these, three important classes arise quite frequently and thus call for some discussion. These may be categorized as questions concerning stability regions in state space, in parameter space and in function space.

A. Stability Regions in State Space

The problem of determining the stability behavior of the equilibrium state and the regions R in the state space such that $x_0 \in R$ implies that the solution $x(t; x_0, t_0)$ is stable (or is asymptotically stable) constitutes one of the most significant questions of practical and mathematical interest. In general only estimates $\hat{R}$ of the region R can be made; for instance, it is often only possible to estimate ρ in Definition 2. If the relation (1-8) holds for all initial values (that is, if R is the entire state space X), then the system (1-4) is said to be asymptotically stable in the whole. Whereas local stability in the state space is a very important concept, we are concerned in the following chapters only with systems which are asymptotically stable in the whole.

B. Stability Regions in Parameter Space

If the dynamic system is represented by the differential equation

$$\dot{x} = f(x, a, t), \tag{1-12}$$

where a is an explicit $(r \times 1)$ vector of parameters the elements of which are a_i $(i = 1, \ldots, r)$, the properties of the family of motions defined by Eq. (1-12) depends on the values of each parameter a_i. The constraints on the parameters a_i which assure some degree of stability of the equilibrium state are of considerable interest. In particular, we are concerned with conditions on a_i which guarantee the asymptotic stability of the system in the whole.

For linear time-invariant systems, definitive methods such as the Hurwitz conditions and the Nyquist criterion exist for the determination of the conditions that must be satisfied by the parameters. By definitive, we mean that the entire region A_s in the r-dimensional Euclidean space A of the parameters a_i $(i = 1, \ldots, r)$ within which a must lie for asymptotic stability can be determined using such methods; the conditions are both necessary and sufficient. For the LTV, NLTI, and NLTV cases it is difficult if not impossible in most general cases to find such definitive regions of stability in parameter space. Again in such instances we must be satisfied with estimates of the region A_s, so that $a \in \hat{A}_s$ is a sufficient condition on the parameters of the system for asymptotic stability in the whole.

C. *Stability Regions in Function Space*

Given an NLTI system of the form described by Eq. (1-4) with $\tau = -f(\sigma_0)$, we have the equations

$$\dot{x} = Ax - bf(\sigma_0), \qquad \sigma_0 = h^{\mathrm{T}}x - \rho f(\sigma_0). \tag{1-13}$$

Frequently very little is known about the nonlinear function $f(\sigma_0)$ aside from the fact that it belongs to a specified function class $\{N\}$ $[f(\sigma_0) \in \{N\}]$. Examples of such classes of nonlinear functions include continuous functions which lie in the first and third quadrants of the σ, $f(\sigma)$ plane, functions which are monotonically nondecreasing,

$$(f(\sigma_1) - f(\sigma_2))/(\sigma_1 - \sigma_2) \geqslant 0 \qquad \text{for all} \quad \sigma_1, \sigma_2; \quad \sigma_1 \neq \sigma_2,$$

and those with a specified range, for example those that satisfy the inequality $0 \leqslant f(\sigma)/\sigma < \bar{F}$, denoted by $f(\sigma)/\sigma \in [0, \bar{F})$.

The problem, then, given specific system parameters a_{ij}, b_i, h_i, and ρ in Eq. (1-13), is to determine the constraints on the function $f(\sigma_0)$ as to its behavior and range that are sufficient to guarantee the asymptotic stability in the whole of the system.

A more general form of Eq. (1-13) is

$$\dot{x} = Ax + F(x), \tag{1-14}$$

where $F(x)$ is a vector, every element $f_i(x)$ of which must be constrained to belong to some class $\{N_i\}$ in order for the system to be asymptotically stable in the whole for some given A matrix. Some problems of this type are considered in Chapter IX.

In general, many of the stability problems that arise in practice represent a fusion of several of the above questions. R in Section 5A evidently is determined by the system parameters and functions that are specified. The class $\{N\}$ in Section 5C may be determined by the values of the parameters included in A, h, b, and ρ. The Aizerman and Kalman conjectures and the absolute stability problem mentioned in Sections 6 and 7, for instance, represent natural combinations of questions 5B and 5C.

6. The Conjectures of Aizerman and Kalman

For a linear time-invariant system described by the equation

$$\dot{x} = Ax,$$

applying the Hurwitz conditions to the parameters of the characteristic equation

$$\det[\lambda I - A] = 0$$

determines the domain of asymptotic stability in the space of parameters a_{ij}. When $a_{ij}(x_j)$ is a nonlinear function of the variable x_j, the question arises as to whether the system is equiasymptotically stable in the whole (Chapter II, Section 2) for all functions $a_{ij}(x_j)$ satisfying

$$\underline{a}_{ij} \leqslant a_{ij}(x_j) \leqslant \bar{a}_{ij}$$

where the LTI system is asymptotically stable for all constant gains a_{ij} in the same range. This represents a procedure of linearization that is implemented by replacing each nonlinearity $f_{ij}(x_j) \triangleq a_{ij}(x_j)x_j$ with a linear gain $a_{ij}x_j$, where the constant a_{ij} takes on all values in the range

$$\underline{a}_{ij} \triangleq \min_{x_j}\{f_{ij}(x_j)/x_j\}, \qquad \bar{a}_{ij} \triangleq \max_{x_j}\{f_{ij}(x_j)/x_j\}.$$

Aizerman [1] conjectured in 1949 that this would indeed be the case for a system with a single nonlinear element. This conjecture stimulated considerable research in the area; it was ultimately shown by Pliss [1] and Krasovskii [1] that the Hurwitz inequalities applied to the range of $f_{ij}(x_j)/x_j$ are not sufficient to ensure stability in general.

A second conjecture along this same line by Kalman [1] in 1957 was also prompted by the tantalizing prospect that some method of replacing a nonlinearity with a linear gain or gains might be meaningfully and rigorously applied to NLTI systems. In this case it was proposed that $f_{ij}(x_j)$ might be replaced by $a_{ij}x_j$, where the constant a_{ij} takes on values in the range

$$\underline{a}'_{ij} \triangleq \min_{x_j}\{df_{ij}/dx_j\}, \qquad \bar{a}'_{ij} \triangleq \max_{x_j}\{df_{ij}/dx_j\};$$

if the Hurwitz technique guarantees asymptotic stability for $a_{ij} \in [\underline{a}'_{ij}, \bar{a}'_{ij}]$, then it was hoped that the NLTI system could be guaranteed to be equiasymptotically stable in the whole. This conjecture was disproved as was that of Aizerman by the generation of counterexamples, that is, NLTI systems with a single nonlinearity $f(\sigma_0)$ were found exhibiting oscillatory behavior, although $df(\sigma_0)/d\sigma_0$ lies in the Hurwitz range (Fitts [1]).

Had these appealing conjectures been shown to be true, the problem of determining stability conditions for nonlinear systems would have been resolved to a great extent. As this was not the case, however, attention has shifted to such related problems as the determination of conditions that are sufficient to guarantee the absolute stability of a system.

7. The Absolute Stability Problem

A very important contribution to the area of nonlinear system stability is the absolute stability problem of Lur'e and Postnikov [1]. In 1944 they proposed the study of the stability of systems of the form depicted in Fig. 1-3 which are assumed to be completely linearizable with the exception of a single NLTI element denoted by $f(\cdot)$. The operator W is assumed to be LTI, while the nonlinear characteristic $f(\cdot)$ satisfies the following conditions.

FUNCTION CLASS $\{F\}$. A function $f(\sigma_0) \in \{F\}$ if it is a real continuous single-valued scalar function that satisfies

$$\sigma_0 f(\sigma_0) \geqslant 0. \quad \square \tag{1-15}$$

This system is equivalent to the time-invariant form of Eq. (1-4) or of Fig. 1-2, that is, to a system described by the differential equation (1-4) with $\tau = -f(\sigma_0)$, where $f(\sigma_0)$ is constrained to lie in the first and third quadrants of the σ_0, $f(\sigma_0)$ plane.

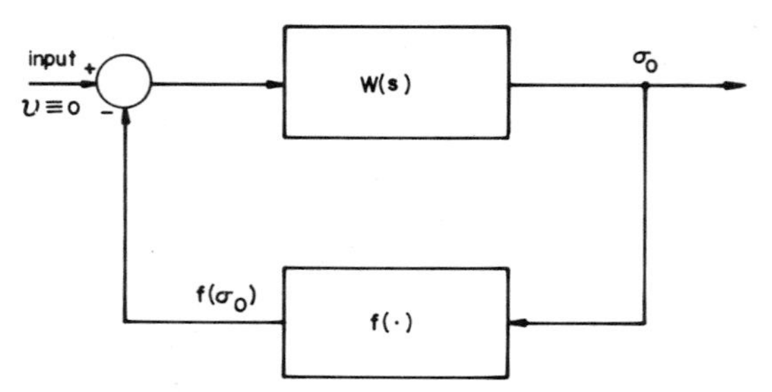

Fig. 1-3. The problem of Lur'e and Postnikov.

Lur'e and Postnikov proposed to determine conditions that would serve as constraints on the LTI plant that would be sufficient to ensure the stability of the system for any nonlinear gain $f(\cdot) \in \{F\}$. The resulting criterion then could be applied to any specific problem, where the nonlinear function is any member of the class $\{F\}$. This concept is an important philosophical basis for many recent developments in stability theory.

Another important contribution of Lur'e and Postnikov was the introduction of the Lyapunov function candidate

$$v(x) = \tfrac{1}{2}x^{\mathrm{T}}Px + \beta \int_0^{\sigma_0} f(\zeta)\, d\zeta, \tag{1-16}$$

where $P = P^T > 0$ [P is a symmetric positive definite matrix] and generally $\beta \geqslant 0$. For $f(\sigma) = \kappa\sigma$ this becomes a member of the fundamental form $x^T Px + \kappa x^T Mx$ mentioned in Section 4.

The precise reason for choosing the form (1-16) is not known, though heuristic arguments can be presented. Using an argument of Pliss (see Aizerman and Gantmacher [1]), for instance, it is possible to substitute a search for $v = x^T Px + \kappa x^T hh^T x$ for an attempt to find a Lyapunov function of the Lur'e–Postnikov form. As pointed out by Lefschetz [1], the great merit of this form when applied to the Lur'e problem is that $\dot{v}$ is a quadratic form in x and τ, and thus the problem of manipulating $\dot{v}$ into a negative semidefinite form is greatly simplified.

For over a decade after the problem was posed the significant results in this area were obtained by research workers in the Soviet Union. The names of V. A. Yakubovich, I. G. Malkin, M. A. Aizerman, V. A. Pliss, and E. N. Rozenvasser are associated with some of the major contributions. More recently, however, the works of Kalman, LaSalle, and Lefschetz in the United States have also added considerably to our knowledge of this important problem.

8. The Criterion of Popov

Although numerous workers continued to deal with the absolute stability problem, it was only in 1961 when the Rumanian applied mathematician Popov [1] presented an elegant criterion based on the frequency response of the linear part of the system that there was a resurgence of interest in the field. Popov also demonstrated that the conditions of his theorem are satisfied if a Lyapunov function derived from the form (1-16) existed for the given problem. Further work by Yakubovich [1], and more recently by Kalman [2], Meyer [1], and Lefschetz [1], showed that the frequency condition of Popov was also sufficient for the existence of a Lyapunov function of the modified Lur'e–Postnikov form. The many special cases of the Popov criterion, as well as the intimate relationship that exists between the method of Lur'e resolving equations and Popov's method, have been treated in detail in the monograph of Aizerman and Gantmacher [1].

The results of Popov are particularly attractive in that they are easy to apply. A large body of similar results has accumulated over the past years since Popov's famous result and are termed "frequency domain stability criteria." This book is chiefly devoted to this problem of generating frequency domain stability criteria for systems of the form depicted in Fig. 1-2 having a single nonlinear and/or time-varying gain.

9. Synopsis

The recent monographs of Aizerman and Gantmacher [1] and Lefschetz [1] have considered in detail the work done on the problem of Lur'e and Postnikov up to and including the work of Popov. The scope of this book consequently is restricted to the result of Popov and the many subsequent criteria inspired by his contribution. The material presented has been organized to provide insight into the relation between the class of nonlinear or time-varying function in the system and the corresponding extension of the Lur'e–Postnikov form of the Lyapunov function used for proving stability.

The system description and the basic stability definitions and theorems are presented in Chapters II and III. The stability of linear time-invariant systems is presented in Chapter IV. In particular, it is shown that (1) the Hurwitz conditions for determining stability boundaries in parameter space correspond to the necessary and sufficient conditions for the existence of a quadratic Lyapunov function $x^{\mathrm{T}}Px$, and (2) the Nyquist criterion for determining the stability range $0 \leqslant \kappa < \bar{K}$ of a single constant parameter κ is equivalent to the conditions for the existence of a quadratic Lyapunov function of the form $x^{\mathrm{T}}[P + \kappa M]x$ ($P > 0$, $M \geqslant 0$) over the same range, provided that the LTI plant has more poles than zeros. It is this form of the Lyapunov function that is subsequently modified suitably in all succeeding chapters for different classes of nonlinear and/or time-varying systems.

The stability of nonlinear time-invariant systems is treated next in Chapter V. The frequency domain criterion of Popov for systems with a single first and third quadrant nonlinear function is derived. By restricting the class of allowable nonlinear functions, less restrictive frequency domain conditions are derived by suitably modifying the Lur'e–Postnikov Lyapunov function used for the Popov problem. The classes of nonlinear functions for which stability conditions are derived include monotonic and odd monotonic functions.

The ensuing chapter deals with two generalizations of the problem treated in Chapter V which render the results applicable to time-varying situations as well. In the first instance, the application of Lyapunov's stability theorem to the system containing a nonlinear time-varying gain of the form $k(t)f(\cdot)$ yields an upper bound on $1/k(t)\ dk/dt$. The use of a stability theorem due to Corduneanu in lieu of that of Lyapunov yields a relaxation of the restriction of dk/dt by substituting an integral or time averaged constraint on dk/dt. The stability of linear time-varying systems is discussed at the end of the chapter as a special case of the problems considered.

Several useful geometric criteria for determining the stability of the overall system from the behavior of the linear part are discussed in Chapter VII.

These criteria are expressed in terms of the Nyquist plot [Im $W(i\omega)$ versus Re $W(i\omega)$], the modified Nyquist plot [ω Im $W(i\omega)$ versus Re $W(i\omega)$], or the root locus plot of the LTI plant. Since the geometric criteria assure the existence of a specific Lyapunov function, the stability properties of the system can be determined directly from the characteristics of the linear part of the system, which can be obtained experimentally.

The stability analysis of a damped nonlinear Mathieu equation is undertaken in the next chapter. The study is intended to demonstrate the application of every relevant stability criterion considered in the book and provide a basis for the comparison of the different results with the necessary and sufficient conditions obtained using classical perturbation techniques.

The final chapter indicates the generalizations of the various results to systems with multiple nonlinear and/or time-varying gains. It is demonstrated that many of the criteria of Chapters IV to VI may be extended in a straightforward manner to such systems.

II

PROBLEM STATEMENT

In this chapter, we give a complete formal definition of the stability problems that are treated in all subsequent chapters save Chapter IX. This is possible due to the basic unity of this subject: all systems whose stability properties are to be investigated may be described by one general system model (Section 1), and the concept of absolute stability introduced by Lur'e and Postnikov may be directly generalized to apply to all of these situations (Section 2). The detailed specification of system properties and the extended definition of absolute stability thus form the essential foundation for the stability problem stated in Section 3, to which the remainder of this study is directed.

1. System Definition

The most basic system model is described by the state-vector formulation of Chapter I:

$$\dot{x} = Ax + b\tau, \qquad \sigma_0 = h^{\mathrm{T}}x + \rho\tau, \qquad \tau = -g(\sigma_0, t). \tag{1-4}$$

The structure of a system whose behavior is represented by this first-order ordinary vector differential equation is depicted in Fig. 1-2. In the introductory comments we establish the somewhat arbitrary but convenient division of the system into an LTI plant and an NLTV controller. Two alternative representations of such a system prove to be useful in dealing with them.

The first formulation equivalent to Eq. (1-4) is defined with the aid of the Laplace transform pair:

$$H(s) \triangleq \int_0^\infty h(t)e^{-st}\,dt \triangleq L[h(t)],$$

and the inverse transformation

$$h(t) \triangleq (2\pi i)^{-1} \int_{\mathrm{Br}} H(s)e^{+st}\,ds \triangleq L^{-1}[H(s)],$$

where Br denotes the Bromwich contour.

Formally substituting X for $L[x(t)]$ (a vector of the Laplace transforms of the state variables), sX for $L[\dot{x}]$, Σ_0 for $L[\sigma_0(t)]$, and T for $L[\tau(t)]$, the LTI part of Eq. (1-4) is reduced to transfer function form:

$$W(s) \triangleq \Sigma_0/T = h^{\mathrm{T}}(sI - A)^{-1}b + \rho. \tag{2-1}$$

If (h, A, b) are of a particular form (the phase variable canonical form), that is,

$$h = \begin{bmatrix} h_1 \\ h_2 \\ \cdot \\ \cdot \\ \cdot \\ h_{n-1} \\ h_n \end{bmatrix}, \quad A = \left[\begin{array}{c|c} 0 & \\ 0 & \\ \cdot & \\ \cdot & I \\ \cdot & \\ 0 & \\ \hline -a_1 & -a_2, \ldots, -a_n \end{array}\right], \quad b = \begin{bmatrix} 0 \\ 0 \\ \cdot \\ \cdot \\ \cdot \\ 0 \\ 1 \end{bmatrix}, \tag{2-2}$$

then by substitution,

$$W(s) = \rho + \frac{h_n s^{n-1} + \cdots + h_2 s + h_1}{s^n + a_n s^{n-1} + \cdots + a_2 s + a_1}. \tag{2-3}$$

The parameters ρ, h_i and a_i correspond to those of Eq. (1-4) with (h, A, b) specified by Eq. (2-2). A second definition of parameters yields

$$W(s) = \frac{\rho s^n + p_n s^{n-1} + \cdots + p_2 s + p_1}{s^n + a_n s^{n-1} + \cdots + a_2 s + a_1}, \tag{2-4}$$

where $p_i \triangleq (h_i + \rho a_i)$ allows a more compact notation. If $\rho = 0$, then we have the special case where the number of zeros of $W(s)$ is less than the number of poles, or, in terms of frequency response, $W_\infty \triangleq \lim_{\omega\to\infty} W(i\omega) = 0$. The representation of the system in transfer function form is portrayed in Fig. 2-1a. The nomenclature plant for $W(s)$ and controller for $g(\cdot, t)$ is purely a matter of convenience; as is demonstrated in Fig. 2-1b, the controller dynamics may also be described by an NLTV differential equation. By defining $W(s) \triangleq H_1(s)W_1(s)H_2(s)$, we have a system of the form of Fig. 2-1a that is equivalent to that of Fig. 2-1b for the purposes of this study.

The second alternative representation of the system takes the form of an

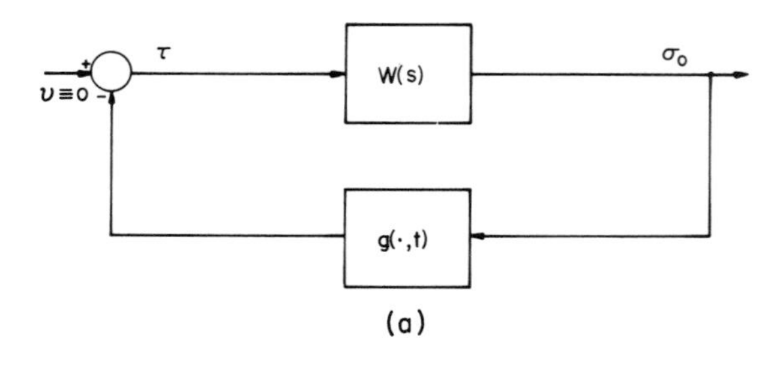

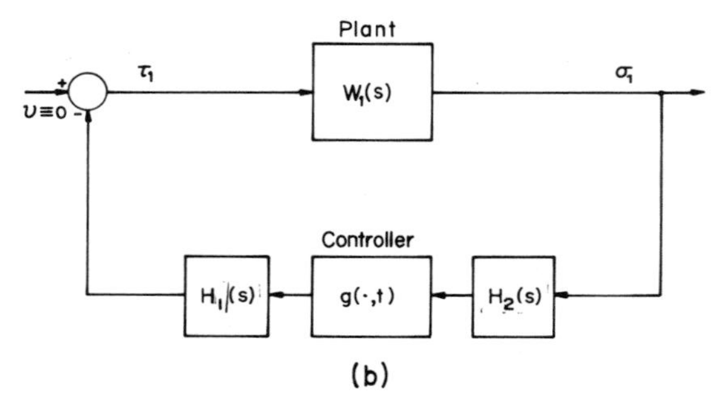

Fig. 2-1 (a) The general system model. (b) An alternative form.

nth order ordinary scalar differential equation. The transfer function $W(s)$ [Eq. (2-4)] plus the controller equation $\tau = -g(\sigma_0, t)$ corresponds to

$$[D^n + a_n D^{n-1} + \cdots + a_2 D + a_1]\xi + g[(\rho D^n + p_n D^{n-1} + \cdots + p_2 D + p_1)\xi, t] = 0, \tag{2-5}$$

where in differential operator notation $D^m \triangleq d^m/dt^m$, and again the parameters a_i, p_i and ρ are the same as in previous formulations. The variable ξ is related to the state vector by

$$x^{\mathrm{T}} = [\xi, D\xi, D^2\xi, \ldots, D^{n-1}\xi].$$

This description is of less utility than Eqs. (1-4) or (2-1) for our purposes; however, it is useful in proving certain subsidiary results.

In completing the formal definition of the system, various other properties of the plant and controller must be specified. In the sequel, the triple (h, A, b) is in the phase variable canonical form [Eq. (2-2)], the A matrix possesses certain stability properties, denoted by $A \in \{A_i\}$, and the nonlinear time-varying controller is specified as to its separability, nonlinear behavior and time variation (denoted in the aggregate by $g(\sigma_0, t) \in \{G_i[N, T]\}$) and range (denoted by $g(\sigma_0, t)/\sigma_0 \in [\underline{G}_N, \bar{G}_N]$, for example). Each of these properties is discussed at greater length below. Before we concern ourselves with these details, however, we formally state the complete state vector differential equation description of the system to be dealt with in Chapters IV through VIII:

$$\begin{aligned}&\dot{x} = Ax + b\tau,\ A \in \{A_i\}, \qquad \sigma_0 = h^{\mathrm{T}}x + \rho\tau, \qquad \tau = -g(\sigma_0, t),\\ &\quad (h, A, b) \quad \text{of the form (2-2)},\\ &\quad g(\sigma_0, t) \in \{G_i[N, T]\}, \qquad g(\sigma_0, t)/\sigma_0 \in [\underline{G}_N, \bar{G}_N].\end{aligned} \tag{2-6}$$

A. *Properties of the LTI Plant*

Since the state vector description

$$\dot{x} = Ax + b\tau, \qquad \sigma_0 = h^{\mathrm{T}}x + \rho\tau \tag{2-7}$$

of the single input $[\tau(t)]$ single output $[\sigma_0(t)]$ plant has been analyzed more fully and systematically from a systems theoretic viewpoint than has the preceding transfer function formulation, the properties of this portion of the overall system are best set forth in terms of (h, A, b) of Eq. (2-7).

In writing Eq. (2-7) we assume that the variables of the state vector x have been chosen in such a manner that a knowledge of $x(t)$ is sufficient to completely specify the behavior of the plant. In an electrical system, for instance, $x(t)$ might be made up of the node voltages or loop currents, while in mechanics, the Lagrangian coordinates $(x_i, \dot{x}_i)$ serve this purpose. Knowing the interrelations between these variables [the A matrix of Eq. (2-7)], the input or control function $\tau(t)$, and the vector b which defines the distribution of τ to the differential equations of the state variables, it is possible to obtain $x(t; x_0, t_0)$ or the state vector solution as a function of time for the initial condition (x_0, t_0) as

$$x(t; x_0, t_0) = \Phi(t - t_0)x_0 + \int_{t_0}^{t} \Phi(t - \xi)b\tau(\xi)d\xi, \qquad t \geqslant t_0,$$

where

$$\Phi(t) \triangleq \exp(At) \triangleq L^{-1}[(sI - A)^{-1}]$$

is the transition matrix of the dynamic system. The output $\sigma_0(t)$ is then specified to be a linear combination of the state variables and the input, $\sigma_0 = h^{\mathrm{T}}x + \rho\tau$. In discussing the properties of the LTI plant, the feedback relation $\tau = -g(\sigma_0, t)$ is completely disregarded, that is, $\tau(t)$ is considered to be completely independent of σ_0.

Looking at the system dynamics from the viewpoint provided by the concept of state, it may be appreciated that the following two questions are of fundamental significance.

(i) Given an arbitrary initial condition (x_0, t_0) and a second arbitrary final condition (x_f, t_f) where $t_f > t_0$ and finite, is it possible to find a suitable control function $\tau(t)$ such that $x(t_f; x_0, t_0) = x_f$? If this may be answered affirmatively, the system is said to be *completely controllable*. The simple example

$$\dot{x} = \begin{bmatrix} -\alpha & \beta \\ \beta & -\alpha \end{bmatrix} x + \begin{bmatrix} 1 \\ 1 \end{bmatrix} \tau(t)$$

demonstrates that it is not always possible to do this; although we may take one state variable or the other from any value to any other value, it is not possible to change them both arbitrarily, as the difference $\delta \triangleq x_1 - x_2$ satisfies $\dot{\delta} = -\alpha\delta$ or $\delta(t_f) = \delta(t_0)\exp[-\alpha(t_f - t_0)]$, irrespective of τ.

(ii) Given a knowledge of $\tau(t)$ and $\sigma_0(t)$ over some finite interval of time, is it possible to determine the state vector $x(t)$? This question is concerned with the mapping of the state into the output and the conditions under which a unique state vector can be associated with every output that can occur. If the system defined by Eq. (2-7) permits the determination of the value of each state variable, it is said to be *completely observable*. Again, it is a simple matter to construct a system that is not completely observable: given

$$\dot{x} = \begin{bmatrix} 0 & 1 \\ -\alpha\beta & -(\alpha + \beta) \end{bmatrix} x + \begin{bmatrix} b_1 \\ b_2 \end{bmatrix} \tau,$$

$$\sigma_0 = [\alpha \quad 1] x + \rho\tau,$$

we can successively differentiate $(\sigma_0 - \rho\tau)$ to obtain

$$\begin{aligned} \alpha x_1 + x_2 &= \sigma_0(t) - \rho\tau(t), \\ -\beta(\alpha x_1 + x_2) &= (d/dt)(\sigma_0 - \rho\tau) - (\alpha b_1 + b_2)\tau, \\ \beta^2(\alpha x_1 + x_2) &= (d^2/dt^2)(\sigma_0 - \rho\tau) - (\alpha b_1 + b_2)(d\tau/dt - \beta\tau), \\ &\vdots \end{aligned}$$

where by assumption the right hand side of each equation is known. This procedure always gives us the form $(-\beta)^m(\alpha x_1 + x_2)$, so only this particular linear combination of x_1 and x_2 may be determined.

In the statement of controllability we note that the output is of no relevance; hence h and ρ do not come into consideration. Similarly in the case of observability, the input enters into the determination of the state as a known function, so that a formal definition can be given entirely in terms of the parameters of A and h.

In the following chapters we are concerned only with feedback systems whose linear parts are described by the linear time-invariant differential equations (2-7). For such systems the conditions for complete controllability and complete observability are well established in terms of the triple (h, A, b). These conditions play an important role in considering the relation between the frequency domain criteria and the existence of Lyapunov functions.

This dependence is noted in considering the various forms of the Kalman–Yakubovich lemma (Chapter III, Section 4).

Another result of Kalman [3] further clarifies the properties of controllability and observability: any dynamic system may be transformed by a nonsingular transformation $y = Tx$ into the canonical form

$$\dot{y} = \begin{bmatrix} A_{11} & A_{12} & A_{13} & A_{14} \\ 0 & A_{22} & 0 & A_{24} \\ 0 & 0 & A_{33} & A_{34} \\ 0 & 0 & 0 & A_{44} \end{bmatrix} \begin{bmatrix} y_1 \\ y_2 \\ y_3 \\ y_4 \end{bmatrix} + \begin{bmatrix} b_1 \\ 0 \\ b_3 \\ 0 \end{bmatrix} \tau$$

$$\sigma_0 = [0 \quad 0 \quad h_3 \quad h_4] y + \rho\tau.$$

Subsystem Class:

1: cc, no
2: nc, no
3: cc, co
4: nc, co

The subsystems characterized by the states that make up y_1 and y_3 are completely controllable (cc). The input τ cannot affect the state variables of y_2 and y_4 directly, as b_2 and b_4 are null vectors, and since A_{21}, A_{23}, A_{41}, and A_{43} are also composed of zero elements, the influence of τ is not brought to bear indirectly upon y_2 and y_4 through its effect upon y_1 and y_3 either, so the states of y_2 and y_4 are strictly noncontrollable (nc). Similarly the states of y_1 and y_2 are strictly nonobservable (no), as h_1 and h_2 are null vectors which precludes direct observation, and as the matrices A_{31}, A_{32}, A_{41}, and A_{42} are composed solely of zeros, y_1 and y_2 do not interact with y_3 and y_4 in any way and hence they cannot be observed indirectly by their effect upon y_3 and y_4. This canonical decomposition is shown in Fig. 2-2.

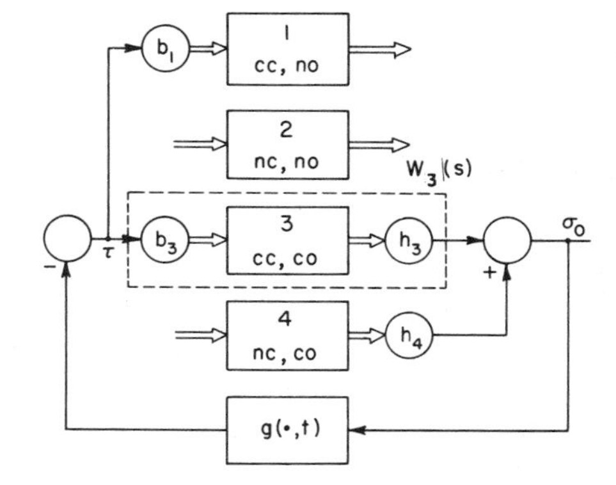

Fig. 2-2. Canonical decomposition of linear systems.

In the first example of a system that is not cc, the nonsingular transformation $y = T_1 x$ yields

$$y = \begin{bmatrix} t_{11} & (1 - t_{11}) \\ -t_{22} & t_{22} \end{bmatrix} x \longrightarrow \dot{y} = \begin{bmatrix} -(\alpha - \beta) & d_{12} \\ 0 & -(\alpha + \beta) \end{bmatrix} y + \begin{bmatrix} 1 \\ 0 \end{bmatrix} \tau$$

(where $t_{22} \neq 0$ ensures the nonsingularity of T_1), so the second state $y_2 = t_{22}(x_2 - x_1) \triangleq -t_{22}\delta$ is evidently uncontrollable. The system in which it is not possible to observe both states is reduced by $y = T_2 x$ to

$$y = \begin{bmatrix} t_{11} & t_{12} \\ \alpha & 1 \end{bmatrix} x \longrightarrow \dot{y} = \begin{bmatrix} -\alpha & (t_{11} - \beta t_{12}) \\ 0 & -\beta \end{bmatrix} y + c'\tau$$

$$\sigma_0 = [0 \quad 1] y + \rho\tau;$$

this transformation is nonsingular if $t_{11} \neq \alpha t_{12}$. Only the state variable $y_2 = \alpha x_1 + x_2$ may be observed.

If a system cannot be transformed into one in which the control affects only certain states (that is, if none of the system state variables fall into categories 2 and 4 above) then the total system is completely controllable, and similarly if no states of a system belong to categories 1 and 2, the entire system is completely observable.

The formal conditions that must be satisfied by (A, b) for complete controllability and by (h^{T}, A) for complete observability and certain other ramifications of these properties are given as follows.

(1) *Controllability*

(a) The necessary and sufficient condition that Eq. (2-7) is completely controllable is that the matrix B,

$$B \triangleq [A^{n-1}b \mid A^{n-2}b \mid \cdots \mid Ab \mid b], \tag{2-8}$$

must be nonsingular, that is, $|B| \neq 0$ (Kalman, Ho, and Narendra [1]).

(b) If A and b are in the phase variable canonical form (Eq. (2-2)], then Eq. (2-7) are completely controllable. To demonstrate that this is so, determine the matrix B which is triangular with diagonal terms equal to unity. For such a matrix, $|B| = 1$, which by (a) establishes the result.

(c) Any system

$$\dot{y} = Dy + c\psi, \qquad \phi_0 = p^{\mathrm{T}}y + \rho\psi$$

that is completely controllable may be transformed using the transformation $x = Ty$ into the phase variable canonical form set forth in Eq. (2-2), that is, $TDT^{-1} = A$ and $Tc = b$, where A and b have the forms indicated in Eq. (2-2) (Johnson and Wonham [1]), and T is a constant nonsingular matrix. From this viewpoint, then, this useful form may be considered with no loss of generality.

(2) *Observability*

(a) The necessary and sufficient condition that Eq. (2-7) is completely observable is that

$$C \triangleq [(A^{\mathrm{T}})^{n-1}h \mid (A^{\mathrm{T}})^{n-2}h \mid \cdots \mid A^{\mathrm{T}}h \mid h] \tag{2-9}$$

is a nonsingular matrix. The pair (h^T, A) is completely observable if and only if (A^T, h) is completely controllable (Kalman [3]).

(b) Since the observability of Eq. (2-7) is determined by the h vector and the A matrix, and the elements of h are not specified in the phase-variable canonical form (2-2), the use of this form does not guarantee complete observability.

A lack of either complete observability, or complete controllability, or both results in a degeneracy of the impulse response of the system. This response,

$$w(t) \triangleq L^{-1}[W(s)],$$

depends only on the controllable and observable states of the system (Kalman [3]). For the examples of systems that are not completely controllable and observable given earlier, we have

$$\text{ncc:}\quad W(s) = \frac{(h_1 + h_2)(s + \alpha + \beta)}{(s + \alpha - \beta)(s + \alpha + \beta)}, \qquad w(t) = (h_1 + h_2)e^{-(\alpha-\beta)t},$$

$$\text{nco:}\quad W(s) = \frac{s + \alpha}{(s + \alpha)(s + \beta)}, \qquad w(t) = e^{-\beta t}.$$

The transfer function of the completely controllable and completely observable part and the corresponding impulse response are those of a first-order system. In general, if $W_3(s) \triangleq h_3^T(sI - A_{33})^{-1}b_3$ (Fig. 2-2), then $w(t) = L^{-1}[W_3(s)]$.

Any function $W(s)$ which is degenerate, that is, that has pole-zero cancellations, may be described by a state vector differential equation that is either not completely controllable, or not completely observable, or neither. Thus, from our viewpoint, we may consider only systems that possess one or the other of these forms of degeneracy, and hence the phase variable canonical form is used throughout this book to guarantee complete controllability with no loss in generality.

The stability properties of the total or closed-loop system (Fig. 2-2) are always specified for at least one value of LTI feedback gain. In terms of the lower bound $\underline{G}_N$ of the NLTV function $g(\sigma_0, t)$ (usually this is $\underline{G}_N = 0$; see Section 1B), we take $\tau = -\underline{G}_N\sigma_0$ in Eq. (2-6); eliminating σ_0 and τ yields

$$\dot{x} = [A - (\underline{G}_N/(1 + \rho\underline{G}_N))bh^T]x \triangleq A_{\underline{G}N}x. \qquad (2\text{-}10)$$

The elements of $A_{\underline{G}N}$ are always assumed to be bounded (that is, $\rho\underline{G}_N \neq -1$) and the roots of the characteristic equation

$$|\lambda I - A_{\underline{G}N}| = 0$$

must have nonpositive real parts, Re $\lambda_i \leqslant 0$. The case most frequently treated (the principal case; see Section 1C) is denoted and defined by

$$A_{\underline{G}N} \in \{A_1\}: \quad \text{Re}\,\lambda_i < 0, \quad i = 1, 2, \ldots, n.$$

Only one particular case is treated:

$$A_{GN} \in \{A_0\}: \quad \lambda_1 = 0, \quad \operatorname{Re} \lambda_i < 0, \qquad i = 2, 3, \ldots, n.$$

If the stability properties of the matrix A_{GN} are specified by $A_{GN} \in \{A_1\}$, this condition justifies not requiring that $W(s)$ be the ratio of relatively prime polynomials. By assuming the asymptotic stability of the overall system for any value of κ, $A_\kappa \in \{A_1\}$, it is guaranteed that the eigenvalues of subsystems 1, 2, and 4 have negative real parts for all values of κ. This point is demonstrated in Fig. 2-2; it is also evident from the earlier analytic formulation:

$$|\lambda I - A_\kappa|$$
$$= |\lambda I - A_{11}||\lambda I - A_{22}||\lambda I - (A_{33} - (\kappa/(1 + \rho\kappa))b_3 h_3^{\mathrm{T}})||\lambda I - A_{44}|.$$

The value of κ affects only the eigenvalues of subsystem 3.

Essentially, the direct method of Lyapunov applied to the absolute stability problem using the techniques given in subsequent chapters provides a generalization of this result to the nonlinear time-varying case. If $A_{ii} \in \{A_1\}$, $i = 1, 2, 4$, then the absolute stability of the closed loop subsystem $\{W_3(s), g(\sigma_0, t)\}$ (Fig. 2-2) guarantees the absolute stability of the total system.

In the particular case $A_{GN} \in \{A_0\}$, we note that $\lambda_1 = 0$ must be an eigenvalue of subsystem 3 from this same line of reasoning; this means that $s = 0$ must not also be a zero of $W_3(s)$. In phase-variable canonical form this reduces to the requirement that $h_1 \neq 0$; see also Section 1D for a second interpretation of this condition.

B. Properties of the NLTV Controller

A primary consideration in defining the properties of the nonlinear time-varying function $g(\sigma_0, t)$ is its separability. It is said that g is separable (denoted $g \in \{G_1\}$) if it can be expressed as the product of a memoryless nonlinear function $f(\cdot)$ and a time-varying gain $k(t)$; thus

$$g(\sigma_0, t) \in \{G_1\} \longrightarrow g(\sigma_0, t) = k(t)f(\sigma_0).$$

If g is not separable in this manner, then $g \in \{G_0\}$.

The behavior of $g(\sigma_0, t)$ with respect to σ_0 and t is restricted by defining classes of nonlinear functions $\{N\}$ and classes of time-varying gains $\{T\}$. Thus, if $g(\sigma_0, t) \in \{G_1[N, T]\}$, it is understood that g is separable, $f(\sigma_0)$ belongs to class $\{N\}$ and $k(t) \in \{T\}$. These classes are defined in terms of $f(\sigma_0)$ and $k(t)$, although the properties so defined apply equally directly to the more general case of $g(\sigma_0, t) \in \{G_0\}$. The meaning of the range of a nonlinear time-varying gain is discussed simultaneously, first over the semiclosed interval $[0, \bar{G}_N)$.

(1) The gain $k(t)$ is a real single-valued nonnegative bounded function of time, $0 \leqslant k(t) < \bar{K} < \infty$. Three classes of such functions are considered.

(a) $\{K_0\}$; *Discontinuous functions:* $k(t) \in \{K_0\}$ if it is discontinuous only at an enumerable number of distinct instants of time. □

(b) $\{K_1\}$; *Continuous functions:* $k(t) \in \{K_1\}$ if it is continuous everywhere. □

(c) $\{K_2\}$; *Continuously differentiable functions:* $k(t) \in \{K_2\}$ if its derivative exists everywhere and is continuous. □

The class $\{K_0\}$ is the most general considered in this study under the constraint that solutions to the system differential equation (2-6) must exist for every initial condition (x_0, t_0), as discussed in Chapter I, Section 1. Further constraints are often necessitated in the application of Lyapunov's and Corduneanu's stability theorems.

(2) The nonlinearity $f(\sigma)$ is a real, continuous, single-valued, scalar function of the real scalar argument σ, which again is a requirement for solutions to the differential equation (2-6) to exist. It is guaranteed that $x \equiv 0$ is an equilibrium of the system by requiring that $f(0) = 0$. It is further assumed that $f(\sigma)$ is a member of one of the following classes.

(a) $\{F\}$; *First and third quadrant functions:*

(i) The infinite sector case: $f(\sigma) \in \{F\}$, $f(\sigma)/\sigma \in [0, \infty)$ if

$$0 \leqslant f(\sigma)/\sigma < \infty \qquad \text{for all finite } \sigma \neq 0. \ \square$$

This condition constrains the function to lie within the sector corresponding to linear gains $\kappa\sigma$ for $\kappa \in [0, \infty)$, that is, the entire first and third quadrants of the σ, $f(\sigma)$ plane (Fig. 2-3a). The function may tend to infinity more rapidly than $|\sigma|$, for example $f(\sigma) = \sigma^3$ is allowed, but $f(\sigma)$ is bounded for finite σ.

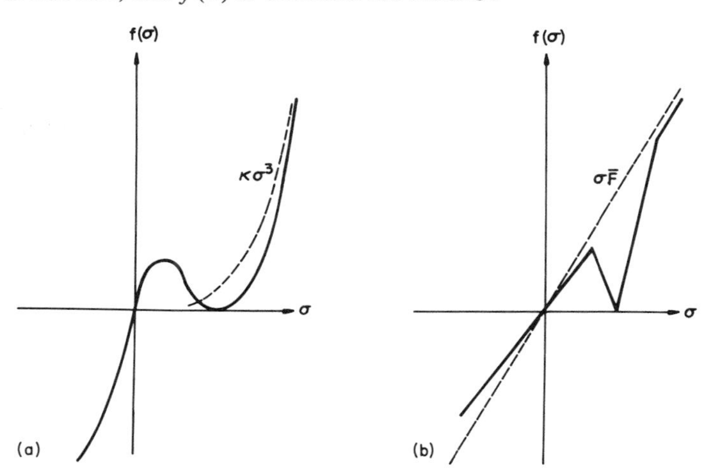

Fig. 2-3. Class $\{F\}$: First and third quadrant functions: (a) $f(\sigma) \in \{F\}$, $f(\sigma)/\sigma \in [0, \infty)$; (b) $f(\sigma) \in \{F\}$, $f(\sigma)/\sigma \in [0, \bar{F})$.

(ii) The finite sector case: $f(\sigma) \in \{F\}$, $f(\sigma)/\sigma \in [0, \bar{F})$, if

$$0 \leqslant f(\sigma)/\sigma < \bar{F} < \infty \qquad \text{for all finite } \sigma \neq 0. \quad \square$$

In this case the function must lie in the sector corresponding to $f(\sigma) = \kappa\sigma$, $\kappa \in [0, F)$, as shown in Fig. 2-3b.

This type of nonlinearity and the concomitant concept of range are directly related to the Aizerman conjecture discussed in Chapter I, Section 6. Given a system with a single nonlinear gain $f(\sigma)$, it was proposed that $\underline{F} \triangleq \min_\sigma\{f(\sigma)/\sigma\}$ and $\bar{F} \triangleq \max_\sigma\{f(\sigma)/\sigma\}$ be determined; if all linear time-invariant systems corresponding to $\tau = -\kappa\sigma$, $\kappa \in [\underline{F}, \bar{F}]$ prove to be asymptotically stable, then it was incorrectly inferred that the nonlinear system is equiasymptotically stable in the whole (Chapter IV, Section 1B).

(b) $\{F_m\}$; *First and third quadrant monotonic functions:*

(i) The infinite sector case: $f(\sigma) \in \{F_m\}$, $\Delta f/\Delta\sigma \in [0, \infty)$ if

$$0 \leqslant \frac{f(\sigma_1) - f(\sigma_2)}{\sigma_1 - \sigma_2} < \infty$$

for all finite σ_1 and $\sigma_2 \neq \sigma_1$. $\square$

(ii) The finite sector case: $f(\sigma) \in \{F_m\}$, $\Delta f/\Delta\sigma \in [0, \bar{M})$ if

$$0 \leqslant \frac{f(\sigma_1) - f(\sigma_2)}{\sigma_1 - \sigma_2} < \bar{M} < \infty$$

for all finite σ_1 and $\sigma_2 \neq \sigma_1$. $\square$

This class of functions and the related idea of range are related to the conjecture of Kalman (Chapter I, Section 6). Rather than finding $\underline{F}$ and $\bar{F}$ as defined previously, it was surmised that the bounds $\underline{M} \triangleq \min_\sigma\{df/d\sigma\}$ and $\bar{M} \triangleq \max_\sigma\{df/d\sigma\}$ should be obtained and the LTI system $\tau = -\kappa\sigma$, $\kappa \in [\underline{M}, \bar{M}]$ should be analyzed. The corresponding inference that the asymptotic stability of the LTI system for $\kappa \in [\underline{M}, \bar{M}]$ implies the equiasymptotic stability in the whole of the NLTI system has also been disproved (Chapter IV, Section 1B).

Typical monotonic nonlinearities are depicted in Fig. 2-4. The upper bound $\bar{M}$ is equal to $\bar{F}$, the upper bound on $f(\sigma)/\sigma$, if $f(\sigma)$ achieves its maximum slope at $\sigma = 0$ or if $df/d\sigma \longrightarrow \bar{M}$ as $|\sigma| \longrightarrow \infty$. Further, $f(\sigma)$ need not be differentiable everywhere in order for it to belong to this class of nonlinear gains; where $df/d\sigma$ exists it must lie in the range $[0, \bar{M})$.

(c) $\{F_{mo}\}$; *First and third quadrant monotonic odd functions:*

(i) and (ii) In both the infinite and finite sector case, the only requirement over and above those of (b) is that $f(\sigma)$ be an odd function, that is, $f(-\sigma) = -f(\sigma)$ for all σ. $\square$

The definitions used so far to specify range for various classes of NLTI functions $f(\sigma)$ are generalized readily to the NLTV case. If $g(\sigma, t)$ is separable and nonmonotonic, $g(\sigma, t) \in \{G_1[F, T]\}$, then $g(\sigma, t)/\sigma \in [0, \bar{G}_F)$, where

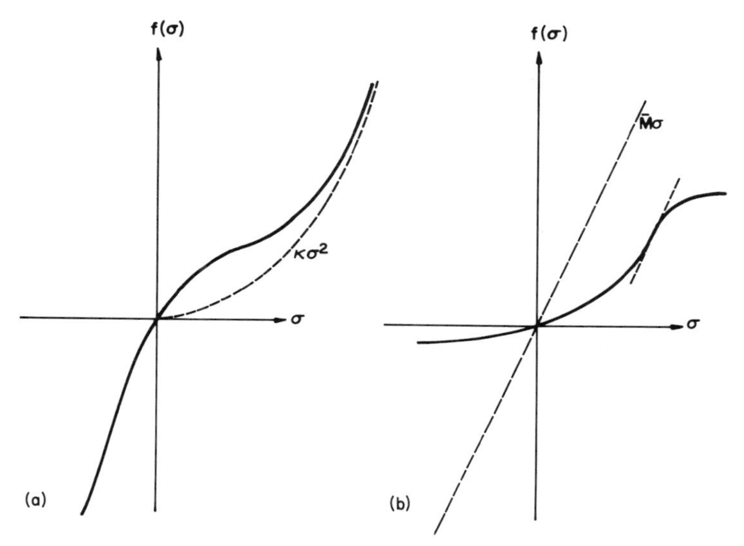

Fig. 2-4. Class $\{F_m\}$: First and third quadrant monotonic functions: (a) $f(\sigma) \in \{F_m\}$, $\Delta f(\sigma)/\Delta\sigma \in [0, \infty)$; (b) $f(\sigma) \in \{F_m\}$, $\Delta f(\sigma)/\Delta\sigma \in [0, \bar{M})$.

$f(\sigma)/\sigma \in [0, \bar{F})$, $k(t) \in [0, \bar{K})$ and $\bar{G}_F \triangleq \bar{F}\bar{K}$. If $f(\sigma) \in \{F_m\}$ or $\{F_{mo}\}$ then $\Delta g(\sigma, t)/\Delta\sigma \in [0, \bar{G}_M)$ where $\bar{G}_M \triangleq \bar{M}\bar{K}$. Similarly for the case of nonseparable gains, $g(\sigma, t) \in \{G_0[N, T]\}$, we have

$$\bar{G}_F \triangleq \max_{\sigma, t}\{g(\sigma, t)/\sigma\}, \qquad \{N\} = \{F\};$$

$$\bar{G}_M \triangleq \max_{\sigma, \hat{\sigma} \neq \sigma, t} \{(g(\sigma, t) - g(\hat{\sigma}, t))/(\sigma - \hat{\sigma})\}, \qquad \{N\} = \{F_m\} \quad \text{or} \quad \{F_{mo}\}.$$

For the majority of the systems analyzed subsequently, the range specified is of the form $[0, \bar{G}_N]$. In treating systems with a marginally stable plant (in particular, A having a single zero eigenvalue, $A \in \{A_0\}$) it is necessary to exclude zero gains: either $g(\sigma, t)/\sigma > 0$ or $g(\sigma, t)/\sigma \geqslant \varepsilon > 0$. At the end of most derivations we consider the general finite sector case, $g(\sigma, t)/\sigma \in [\underline{G}_N, \bar{G}_N]$.

If a system is linear but time-varying (LTV), it is denoted by $g(\sigma_0, t)/\sigma_0 = k(t) \in \{T\}$, and if it is nonlinear but time-invariant (NLTI) by $g(\sigma_0, t) = f(\sigma_0) \in \{N\}$. If the system is LTI $[g(\sigma_0, t) = \kappa\sigma_0]$, this notation is dispensed with entirely.

The parameter $\underline{\Phi}$ defined by

$$\underline{\Phi} \triangleq \min_{\sigma} \phi(\sigma) \triangleq \min_{\sigma} \left\{ \frac{\sigma f(\sigma)}{\int_0^\sigma f(\zeta)\, d\zeta} \right\} \tag{2-11}$$

is found to arise in the stability criteria for separable NLTV systems, $g(\sigma, t) \in \{G_1[N, T]\}$; it provides an effective index of nonlinearity for $f(\sigma)$. The bound on the rate of time variation of the gain for stability, that is, the upper bound on dk/dt, is shown to be proportional to $\underline{\Phi}$ (see Chapter *VI*). Thus it is important to be able to estimate $\underline{\Phi}$ as liberally as possible for a specific nonlinearity.

As a preliminary step, it is possible to establish ranges of $\underline{\Phi}$ for each of the nonlinearity classes defined previously. The following points and examples should be noted in obtaining the results tabulated in Table 2-1.

TABLE 2-1 MAXIMUM RANGES OF $\underline{\Phi}$ FOR NONLINEARITY CLASSES

Nonlinearity class $\{N\}$	Range of $\underline{\Phi}$
$\{F\}$	$[0, \infty)$
$\{F_m\}, \{F_{mo}\}$	$[1, \infty)$
$f(\sigma) = \kappa\sigma$	2

(i) Only for $f(\sigma) \in \{F\}$ can $f(\sigma) = 0$ for some $\sigma \neq 0$ when $\int_0^\sigma f(\zeta)\, d\zeta > 0$, so $\underline{\Phi} = 0$ may occur only for this class.
(ii) For monotonic gains we have $\underline{\Phi} \geqslant 1$, since for any finite value of $\sigma \neq 0$ we see that necessarily $\int_0^\sigma f(\zeta)\, d\zeta < \sigma f(\sigma)$. The strict inequality follows from the continuity of $f(\sigma)$ at $\sigma = 0$. If the case $\lim_{|\sigma| \to \infty} |f(\sigma)| = \psi$ (ψ being any finite constant) is considered we obtain $\underline{\Phi} = 1$.
(iii) If $f(\sigma)$ is a power-law function,

$$f(\sigma) = \kappa |\sigma|^\alpha \operatorname{sgn} \sigma, \qquad \alpha \geqslant 1,$$

then $f(\sigma)$ belongs to all of the above classes of nonlinearities. By substitution, $\phi(\sigma) = \alpha + 1$ for all σ, so $\underline{\Phi} = \alpha + 1$. For a linear system, $\alpha = 1$ and $\underline{\Phi} = 2$.

For NLTV systems with separable monotonic gains it is noted in Chapter VI that the rate of time variation that can be tolerated is at least one-half that allowed for a comparable LTV system that satisfies the same frequency domain condition. This is based on the magnitude of $\underline{\Phi}$ which for monotonic gains is never less than unity as compared with $\underline{\Phi} = 2$ for linear systems.

The above comments indicate that in some simple cases when the asymptotic behavior of $f(\sigma)$ is known an upper bound of $\underline{\Phi}$ can be inferred directly. For example, if $f(\sigma)$ approaches a power-law function as $\sigma \to \pm\infty$, then $\underline{\Phi} \leqslant \alpha + 1$, while $\underline{\Phi} \leqslant 2$ if it approaches a linear gain. If $f(\sigma)$ approaches some constant value γ^+ or γ^- as $\sigma \to \pm\infty$, then $\underline{\Phi} \leqslant 1$.

C. *Principal and Particular Cases*

It has been a general practice in the past to consider a division of control systems into equations of direct control and equations of indirect control as being fundamental and to treat each case separately. The system described by (2-6) is said to be direct, and

$$\dot{x} = Ax + b\tau, \qquad \sigma_0 = h^{\mathrm{T}}x + \rho\tau, \qquad d\tau/dt = -g(\sigma_0, t) \tag{2-12}$$

to represent the structure of an indirect control system. The feedback path of such a system may be seen to include an integration. It is simply demonstrated, however, that the two formulations are equivalent and thus one may consider (2-6) with no loss in generality: Define

$$z \triangleq \begin{bmatrix} x \\ \hline \tau \end{bmatrix}, \qquad b_0 \triangleq \begin{bmatrix} 0 \\ \hline b \end{bmatrix}, \qquad A_0 \triangleq \left[\begin{array}{c|c} A & b \\ \hline 0, \quad 0, \ldots, 0 & 0 \end{array}\right]$$

and $h_0^{\mathrm{T}} = [h^{\mathrm{T}}, \rho]$; then Eq. (2-12) is transformed into the form of Eq. (2-6) where by inspection A_0 has a zero eigenvalue, and $\rho_0 = 0$.

This transformed system still possesses the property that (A_0, b_0) is completely controllable; the B matrix of Eq. (2-8) is again triangular with diagonal terms of unity. Hence as previously stated a second nonsingular transformation T may be applied to bring the indirect control system of Eq. (2-12) into the phase variable canonical form. It is easily shown that if A is in the phase variable canonical form then

$$T = \left[\begin{array}{c|c} & 0 \\ & 0 \\ I & \cdot \\ & \cdot \\ & \cdot \\ & 0 \\ \hline -a_1, \quad -a_2, \ldots, -a_n & 1 \end{array}\right]$$

yields

$$TA_0T^{-1} = \left[\begin{array}{c|c} 0 & \\ 0 & \\ \cdot & \\ \cdot & I \\ \cdot & \\ 0 & \\ \hline 0 & -a_1, \quad -a_2, \ldots, -a_n \end{array}\right], \qquad Tb_0 = b_0. \tag{2-13}$$

Since a control system may have a zero eigenvalue irrespective of the actual physical nature of the controller (direct or indirect), the authors share the belief of Aizerman and Gantmacher [1] that this classification is artificial and unproductive. Thus, the principal case (when A is a stable matrix, $A \in \{A_1\}$) and particular cases (when A is marginally stable) are considered as fundamental classes of control systems. Various particular cases include A having a single eigenvalue of zero ($A \in \{A_0\}$), having multiple eigenvalues of zero, and having distinct eigenvalues on the imaginary axis. The analysis of the Lur'e–Postnikov problem for all of these cases has been treated by Yakubovich.

The treatment of particular cases in this book is confined to a consideration of those systems that possess a single zero eigenvalue, that is, $A \in \{A_0\}$. The original system under consideration may have been an indirect control system, or it may have possessed an inherently marginally stable linear part (plant).

D. *The Origin as the Sole Equilibrium Point*

Now that an investigation of the system properties has been completed, it is possible to show that $x \equiv 0$ is the only equilibrium state of Eq. (2-6), provided that the corresponding LTI system ($\tau = -\kappa\sigma_0$) is asymptotically stable for κ in the same range as the NLTV gain. The principal case and the particular case corresponding to a single zero eigenvalue must be considered separately. The case $\underline{G}_N = 0$ is considered with no loss in generality.

(1) *The Principal Case*

Since A is an asymptotically stable matrix, we may assume that $0 \leqslant g(\sigma_0, t)/\sigma_0 \leqslant \bar{G}$ (or $< \bar{G}$; the two cases are considered simultaneously). Then clearly a necessary condition for absolute stability is that the linear system corresponding to $g(\sigma_0, t) = \kappa\sigma_0$ be asymptotically stable for all $\kappa \in [0, \bar{G}]$ or $\kappa \in [0, \bar{G})$; this assumption is made prior to every analysis. From the Nyquist criterion (see Chapter IV) this is equivalent to the statement that $W(i\omega)$ does not intersect the negative real axis to the left of $(-1/\bar{G})$, that is, that if $\operatorname{Im} W(i\omega) = 0$, then $\operatorname{Re} W(i\omega) > -1/\bar{G}$ [or $\geqslant 1/\bar{G}$]. Since $W(0)$ is real for the principal case, we require $W(0) > -1/\bar{G}$ [or $\geqslant 1/\bar{G}$], or

$$-1/(\rho - h^{\mathrm{T}}A^{-1}b) = -1/W(0) > \bar{G} \quad [\text{or} \geqslant \bar{G}].$$

Assume that $x_e \neq 0$ exists: then the first relation in Eq. (2-6) yields $\dot{x}_e \equiv 0$, or

$$Ax_e \equiv -b\tau_e,$$

and since A^{-1} exists (because asymptotic stability of A ensures that $|A| \neq 0$)

$$x_e \equiv -A^{-1}b\tau_e.$$

The next two relations in Eq. (2-6) subject to this condition give us

$$g(\sigma_{0e}, t)/\sigma_{0e} \equiv -1/W(0)$$

which implies that $g(\sigma_{0e}, t)/\sigma_{0e}$ must be greater that $\bar{G}$ [or $\geqslant \bar{G}$]. This contradiction to the assumed upper bound on $g(\sigma_0, t)/\sigma_0$ guarantees that $x_e \neq 0$ does not exist.

(2) *The Particular Case of One Zero Eigenvalue*

Since A is only marginally stable, the range of $g(\sigma_0, t)/\sigma_0$ for absolute stability must be either $(0, \bar{G}]$ or $(0, \bar{G})$; again both are treated simultaneously. If the lower bound 0 is not precluded then $g(\sigma_0, t) \equiv 0$ is possible, and asymptotic stability cannot ensue.

By inspection, $a_1 = 0$ yields the requisite characteristic equation

$$|\lambda I - A| = \lambda^n + a_n\lambda^{n-1} + \cdots + a_3\lambda^2 + a_2\lambda = 0$$

with one zero root. Again consider Eq. (2-6) under the assumption that $x_e \neq 0$ exists: setting $\dot{x}_e \equiv 0$ yields

$$\left[\begin{array}{c|c} \begin{matrix} 0 \\ 0 \\ \cdot \\ \cdot \\ \cdot \\ 0 \end{matrix} & I \\ \hline 0 & -a_2, \quad -a_3, \ldots, -a_n \end{array}\right] x_e = \begin{bmatrix} x_{2e} \\ x_{3e} \\ \cdot \\ \cdot \\ \cdot \\ x_{ne} \\ \left(-\sum_{i=2}^{n} a_i x_{ie}\right) \end{bmatrix} = \begin{bmatrix} 0 \\ 0 \\ \cdot \\ \cdot \\ \cdot \\ 0 \\ -\tau_e \end{bmatrix}.$$

The first $(n - 1)$ conditions require that $x_{2e} = x_{3e} = \cdots = x_{ne} \equiv 0$. Since the last row gives τ_e as a linear combination of just these state variables (not x_{1e}), $\tau_e \equiv 0$.

Since $\sigma_0 = 0$ is the only zero of $\tau = -g(\sigma_0, t)$, then $\tau_e \equiv 0$ requires that $\sigma_{0e} \equiv 0$. Finally the expression $\sigma_0 = h^{\mathrm{T}}x + \rho\tau$ yields

$$h_1 x_{1e} \equiv 0.$$

If $h_1 = 0$, then $W(s)$ is of the form

$$W(s) = \rho + \frac{s[h_n s^{n-2} + \cdots + h_3 s + h_2]}{s[s^{n-1} + a_n s^{n-2} + \cdots + a_3 s + a_2]},$$

hence, the possibility of the pole at $s = 0$ cancelling with a zero would allow $x_{1e} \neq 0$ to exist. If, however, this is precluded by assumption, then $x_{1e} \equiv 0$ or $x \equiv 0$ is the only equilibrium. It is worth noting, however, that other poles and zeros may cancel, that is, complete observability is not required.

2. Definitions of Stability

As indicated in Chapter I, the concept of stability with regard to general nonlinear time-varying systems is quite complex. The most fundamental definitions establishing the concepts of stability, attractivity and asymptotic stability are considered in Chapter I, Section 3. The great variety of possibilities that exist in the behavior of nonlinear systems has given rise to a multiplicity of definitions which are extensions and refinements of these concepts.

Although the definition of absolute $\{G_i[N, T]\}$ stability could be set forth directly, a discussion of the foundation of this concept is called for to clarify the significance of the various factors involved. This is accomplished most simply by giving certain basic definitions and their interrelations.

In the definition of stability (Definition 1), the initial point x_0 and the initial time t_0 occur as parameters. For a stable system the function $\|x(t; x_0, t_0)\|$ is bounded with respect to time by ε when $\|x_0\| \leqslant \delta(\varepsilon, t_0)$ and t_0 are specified, and this bound tends monotonically towards zero as x_0 tends to the origin, that is, $\lim_{\varepsilon\to 0} \delta(\varepsilon, t_0) = 0$. For attractivity (Definition 2), the time T, defined by

$$\|x(t; x_0, t_0)\| \leqslant \eta, \qquad t \geqslant t_0 + T \tag{2-14}$$

is a function of x_0, t_0 and η. Simple examples can be given where the bound on $\|x(t; x_0, t_0)\|$ for a given x_0 increases monotonically with t_0. Similarly in an attractive system the time $T(\eta, x_0, t_0)$ may increase arbitrarily with t_0. Such systems may exhibit modes of behavior that are quite pathological and not at all in accordance with the intuitive notions about stability that develop from a study of LTI systems. As an instance of this (Kalman and Bertram [1]) the first order system $\dot{x} + x/t = 0$ is asymptotically stable but not uniformly so with respect to time. As a consequence, it is apparent that we must take care to emphasize the dependence of the motion on initial conditions and to specify stability definitions that avoid such peculiarities. The significant question to be investigated is whether or not the function $\sup_{t\geqslant t_0}\{\|x(t; x_0, t_0)\|\}$ can be uniformly estimated with respect to x_0 and t_0. This has given rise to Definitions 4 to 7 that follow.

Definition 4 Uniform Stability (Persidskii [1]). The equilibrium $x \equiv 0$ of the differential equation (1-3) is uniformly stable if for each $\varepsilon > 0$, a number $\delta = \delta(\varepsilon) > 0$, independent of t_0, can be determined such that for all $\|x_0\| \leqslant \delta$

$$\|x(t; x_0, t_0)\| \leqslant \varepsilon, \qquad t \geqslant t_0. \ \square$$

Definition 5 Uniform Attractivity (Hahn [1], Antosiewicz [1]). The equilibrium $x \equiv 0$ of the differential equation (1-3) is uniformly attractive:

(a) with respect to x_0, if for some $\rho > 0$ and for every $\eta > 0$ there exists a $T = T(\eta, \rho, t_0)$ such that

$$\|x(t; x_0, t_0)\| \leqslant \eta \qquad \text{for} \quad t \geqslant t_0 + T, \|x_0\| \leqslant \rho;$$

(b) with respect to t_0, if for some $\rho > 0$ and for every $\eta > 0$ there exists a $T = T(\eta, x_0)$ such that

$$\|x(t; x_0, t_0)\| \leqslant \eta \qquad \text{for} \quad t \geqslant t_0 + T, \|x_0\| \leqslant \rho. \ \square$$

In case (i) T is independent of the initial state x_0 (but not of $\rho \geqslant \|x_0\|$) while in case (ii) T is independent of the initial time t_0.

DEFINITION 6 EQUIASYMPTOTIC STABILITY (Massera [1]). The equilibrium $x \equiv 0$ of the differential equation (1-3) is equiasymptotically stable if it is both stable and uniformly attractive with respect to x_0. □

Definition 6 obviously implies asymptotic stability as the system is both stable and attractive. By the second condition equiasymptotic stability can be interpreted as asymptotic stability which is uniform with respect to the initial space coordinates x_0. For LTI and LTV systems asymptotic stability implies equiasymptotic stability; Massera has given examples to show that this does not hold in the general case of nonlinear systems. It is shown by Kalman and Bertram [1] that asymptotic stability and equiasymptoic stability are equivalent if they are uniform in t_0.

DEFINITION 7 UNIFORM ASYMPTOTIC STABILITY (Malkin [1]). The equilibrium $x \equiv 0$ of the differential equation (1-3) is uniformly asymptotically stable if the equilibrium is both uniformly stable and uniformly attractive with respect to x_0 and t_0. □

For the attractivity specified in Definition 7, it is guaranteed that for every $\eta > 0$ there exists a number $T(\eta, \rho)$ such that the inequality

$$\|x(t; x_0, t_0)\| \leqslant \eta \qquad \text{for} \quad t \geqslant t_0 + T$$

is satisfied for all $\|x_0\| \leqslant \rho$, where ρ and η are independent. Uniform asymptotic stability is also seen to imply equiasymptotic stability.

From the above definitions it is clear that if an autonomous system is stable (equiasymptotically stable) it is also uniformly stable (uniformly asymptotically stable). The more significant result that asymptotic stability and uniform asymptotic stability are equivalent for differential equations that are time invariant or that have periodic coefficients was proven by Hahn [1].

In Definitions 1′ to 3′ it is indicated that stability, attractivity, and asymptotic stability can be expressed in terms of comparison functions $\phi(\cdot) \in \{K\}$ and $\psi(\cdot) \in \{L\}$. The various Definitions 4 to 7 can also be expressed conveniently in terms of these functions, which indicate the dependence of the motion in each case on the initial values x_0 and t_0. For asymptotic, equiasymptotic, and uniform asymptotic stability respectively, the inequalities

$$\|x(t; x_0, t_0)\| \leqslant \phi(\|x_0\|; t_0)\psi(t - t_0; x_0, t_0),$$
$$\|x(t; x_0, t_0)\| \leqslant \phi(\|x_0\|; t_0)\psi(t - t_0; t_0),$$
$$\|x(t; x_0, t_0)\| \leqslant \phi(\|x_0\|)\psi(t - t_0)$$

have to be satisfied for any t_0 and for all $t \geqslant t_0$ and for all x_0 in some fixed ball $\|x_0\| \leqslant \rho$. For further details on comparison functions the reader is referred to the recent book by Hahn [1]. The relationships between the stability concepts given by Definitions 1–7 are shown diagrammatically in Fig. 2-5.

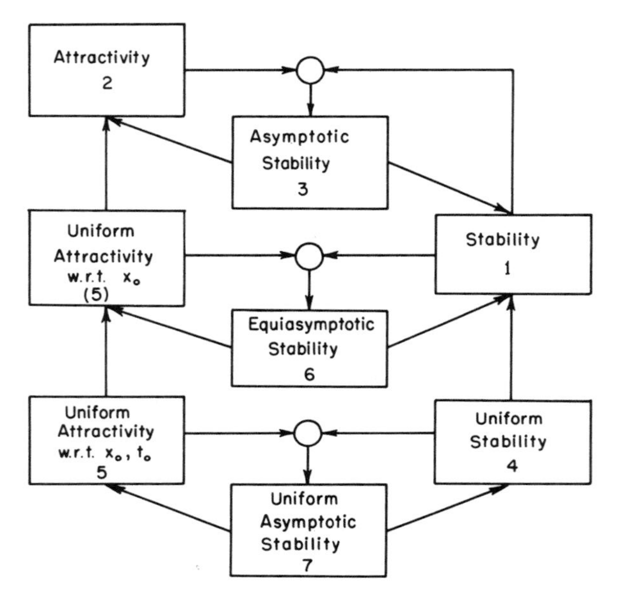

Fig. 2-5. Basic stability relationships (numbers refer to Definitions).

All Definitions 1 to 7 only imply certain system properties in some neighborhood of $x = 0$; for motions defined in the entire state space X further extensions are needed.

DEFINITION 8 UNIFORM STABILITY IN THE WHOLE. The equilibrium $x \equiv 0$ of the differential equation (1-3) is uniformly stable in the whole if in Definition 4 $\delta(\varepsilon)$ exists for all ε no matter how large, and

$$\lim_{\varepsilon \to \infty} \delta(\varepsilon) = \infty. \quad \square$$

The latter condition added to the definition of uniform stability implies that solutions to the differential equation (1-3) are uniformly bounded for all x_0, which is required for stability in the whole. This concept, like attractivity, is independent of stability.

DEFINITION 9 UNIFORM ATTRACTIVITY IN THE WHOLE. The equilibrium $x \equiv 0$ of the differential equation (1-3) is uniformly attractive in the whole (i) with respect to x_0 or (ii) with respect to t_0 if the provisions of Definition 5 hold for any $\rho > 0$. $\square$

DEFINITION 10 UNIFORM ASYMPTOTIC STABILITY IN THE WHOLE. The equilibrium $x \equiv 0$ of the differential equation (1-3) is uniformly asymptotically stable in the whole if it is both uniformly stable in the whole and uniformly attractive in the whole with respect to x_0 and t_0. □

The equivalence of Definitions 9 and 10 under the assumption that $f(x, t)$ satisfies a global uniform Lipschitz condition is shown in Kalman and Bertram [1].

The interrelations between various concepts of stability defined in Definitions 1–10 are displayed in Fig. 2-6. In this schematic representation L designates the assumption of linearity, TI/P the assumption of time invariance or periodicity and (G)ULC the assumption that $f(x, t)$ satisfies a (global) uniform Lipschitz condition. Note that each definition implies all those above it, while other conditions (L, TI/P, etc.) are required for other relationships. Thus, Definition 7 implies Definitions 6 and 3 but not 10 unless linearity is

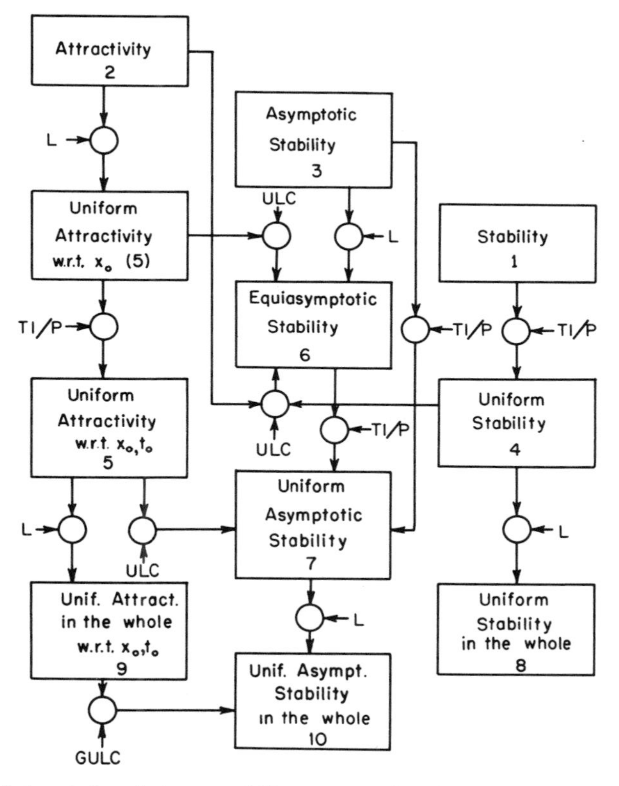

Fig. 2-6. Interrelations between stability concepts (numbers refer to Definitions).

assumed. This topic is considered in detail both in Kalman and Bertram and in Hahn; different forms of the Lipschitz condition (Chapter I, Section 1) often play an important part.

Finally it should be stressed that all of the preceding definitions refer to specific systems which can be expressed in the form (1-3) with all parameters and functional relationships specified. In the following chapters a generalization of the concept of absolute stability (Lur'e and Postnikov [1]) is used exclusively.

DEFINITION 11 ABSOLUTE $\{G_t\}$ STABILITY. The equilibrium $x \equiv 0$ of the differential equation (2-6) is absolutely $\{G_t[N, T]\}$ stable if it is uniformly asymptotically stable in the whole for all $g(\cdot, t) \in \{G_t[N, T]\}$, $g(\sigma, t)/\sigma \in [\underline{G}_N, \bar{G}_N]$. □

When $g(\sigma, t) = f(\sigma) \in \{F\}$, this definition corresponds to absolute stability which has been extensively investigated in the past. If a precise indication of the type of stability under discussion is not essential, the common usage of the term absolute stability in the extended sense is utilized.

3. Formal Problem Statement

Within the framework provided by a basic system model (Section 1) and stability definitions (Section 2), it is now possible to focus our attention on one of the many problems posed in Chapter I. The formal problem as stated below is a generalized Lur'e–Postnikov problem.

THE PROBLEM. Given a differential equation of the form (2-6), determine conditions on the transfer function $W(s)$ and on the rate of time variation of $g(\sigma_0, t)$ that are sufficient to guarantee the absolute $\{G_t\}$ stability of the system. □

In Chapter IV the frequency domain condition that is obtained to guarantee the asymptotic stability of LTI systems is shown to be equivalent to the Nyquist condition. In Chapter V similar conditions are derived that are sufficient to ensure the absolute $\{N\}$ stability of NLTI systems. The stability criteria of Chapter VI for NLTV systems contain both frequency domain restrictions on $W(i\omega)$ and (except for the circle criterion, Chapter VI, Section 1) upper bounds upon the rate of time variation of $g(\sigma_0, t)$ that are to be satisfied either at each instant or in a time-averaged sense.

III

MATHEMATICAL PRELIMINARIES

The many stability theorems that have been derived based on the philosophy of Lyapunov may in general terms be divided into three categories:

(i) *Sufficiency theorems*, which state that the existence of a Lyapunov function is sufficient to guarantee the stability of the system in question.

(ii) *Existence theorems*, which for a given class of systems [such as the LTV system $\dot{x} = A(t)x$] that are stable, assure that a specific class of Lyapunov functions [for example, $x^{\mathrm{T}}P(t)x$] exist.

(iii) *Stability criteria*, which state conditions directly in terms of system characteristics that are sufficient to ensure stability. These criteria are proved using theorems of classes (i) or (ii), but can be applied without direct recourse to Lyapunov's method.

A similar hierarchy of theorems also exists in considering the instability of dynamic systems.

The central purpose of this study is the development of stability criteria for systems of the type defined in Chapter II. First, however, it is necessary to state the sufficiency theorems that are utilized in subsequent chapters. Such a theorem consists of two parts: a statement of the properties of $v(x, t)$ [the Lyapunov function candidate] and the conditions that $\dot{v}$ [$\dot{v} \triangleq \partial v/\partial t + (\nabla v)^{\mathrm{T}}\dot{x}$, the total time derivative of v along system trajectories] must satisfy before v may be said to be a Lyapunov function for the system. The existence of a Lyapunov function is a sufficient condition for system stability in the sense defined in the theorem.

Since one of the primary unifying factors in this work is the use of Lyapunov function candidates that are special cases of one general form, this form is investigated to demonstrate that it is indeed a valid candidate. This permits the simplification of the statements of the sufficiency theorems.

The final complement to this exposition of the rigorous basis for the stability criteria to be developed in subsequent chapters is the statement of several forms of the Kalman–Yakubovich lemma. This lemma is the heart of all subsequent derivations, as it provides the link between the existence of the solutions to a certain set of matrix and vector equations and the stability conditions in the frequency domain. If the solutions exist, it is possible to guarantee stability by Lyapunov's direct method [Theorems 1 and 2, in Section 3]; the frequency domain condition that is demanded by each stability criterion devolves from the application of a form of the Kalman–Yakubovich lemma. In connection with this lemma, the properties of positive real functions are investigated.

In order to provide insight into the types of Lyapunov functions that may be used in different situations, a few noteworthy existence theorems are stated, generally without proof. Although they render assurance that the search for a Lyapunov function for problems of the type defined in Chapter II is not a vain one, these theorems often provide little or no specific guidance in this quest except in the case of LTI systems: $\dot{x} = Ax$. In treating this last problem, an effective existence theorem and several useful corollaries are given. Since the corresponding class of Lyapunov functions—the quadratic form $x^{\mathrm{T}}Px$—is of fundamental importance to the study of absolute stability, a proof of the theorem is presented, even though it is well known.

1. Sufficiency Theorems

The following theorem gives a general statement of conditions that are sufficient to guarantee the uniform asymptotic stability in the whole of the solutions of a differential equation of the form

$$\dot{x} = f(x, t), \qquad f(x, t) \in \{S\}. \tag{1-3}$$

This theorem is essentially due to Lyapunov, although he did not consider the subtle distinctions between asymptotic stability (with which he was concerned) and uniform asymptotic stability as later defined by Malkin (Definition 7).

THEOREM A (see Hahn [1, § 42]). If a function $v(x, t)$ that is defined for all x and t satisfies

(i) for all fixed t, $v(x, t)$ is continuous with respect to $\|x\|$ and bounded for all $\|x\| < \infty$, and for all fixed x, $v(x, t)$ is continuous with respect to t;

(ii) $v(x, t)$ is positive definite, that is, there exists a function $\alpha(\rho) \in \{K\}$ (Chapter I, Section 3) independent of t such that

$$0 < \alpha(\|x\|) \leqslant v(x, t)$$

for all $x \neq 0$ and t, $v(0, t) = 0$ for all t;

(iii) $v(x, t)$ is radially unbounded, that is, $\alpha(\rho)$ of part (ii) satisfies $\lim_{\rho \to \infty} \alpha(\rho) = \infty$;

(iv) $v(x, t)$ is decrescent, that is, there exists a function $\beta(\rho) \in \{K\}$ independent of t such that

$$v(x, t) \leqslant \beta(\|x\|)$$

for all t and for all $\|x\| \leqslant \rho_0 < \infty$ where ρ_0 is an arbitrary positive constant;

then a sufficient condition for the uniform asymptotic stability in the whole of the solutions to Eq. (1-3) is that

(A: v) there exists a function $\gamma(\rho) \in \{K\}$ such that the total time derivative of $v(x, t)$ along system trajectories satisfies

$$\dot{v} \triangleq \partial v/\partial t + (\nabla v)^{\mathrm{T}} f(x, t) \leqslant -\gamma(\|x\|) < 0$$

for all $x \neq 0$ and t. □

The conditions that $v(x, t)$ must be positive definite and decrescent require that $v(0, t) \equiv 0$ for all t. The time invariance of the lower and upper bounds of $v(x, t)$ with respect to $\|x\|$ required in conditions (ii) and (iv) respectively is important. The function $v(x, t) = e^{-\eta t}\|x\|^2$ is not positive definite if $\eta > 0$, even though $v(x, t) > 0$ for all t and $x \neq 0$, and it is not decrescent if $\eta < 0$, since for any fixed x, $v(x, t)$ is unbounded with respect to t.

The application of condition (iv) is simplified by the use of the following result from Hahn [1].

LEMMA A. A function $v(x, t)$ is decrescent if the elements of $\nabla v^{\mathrm{T}} \triangleq [\partial v/\partial x_1, \partial v/\partial x_2, \ldots, \partial v/\partial x_n]$ are bounded functions of t for any fixed x, $\|x\| \leqslant \rho_0 < \infty$ where ρ_0 is an arbitrary positive constant. □

In the above example $\nabla v = 2xe^{-\eta t}$, so $v(x, t) = e^{-\eta t}\|x\|^2$ is not decrescent for $\eta < 0$.

A corollary for time-invariant (autonomous) differential equations

$$\dot{x} = f(x), \qquad f(x) \in \{S\}, \tag{1-3c}$$

which notably weakens the stability requirement (A: v) is to be found in Kalman and Bertram [1].

COROLLARY A1 (LaSalle [1]). If a function $v(x)$ which is defined for all x satisfies

(i) the elements of ∇v are continuous with respect to x;

(ii) $v(x)$ is positive definite;

(iii) $v(x)$ is radially unbounded;
(iv) $v(0) = 0$;

then a sufficient condition for the uniform asymptotic stability in the whole of the solutions to Eq. (1-3c) is that

(A1 : v) $\dot{v} \leqslant 0$ for all x where $\dot{v}(x(t;\, x_0, t_0)) \equiv 0$ cannot occur along any system trajectory other than $x(t;\, x_0, t_0) \equiv 0$. □

For time-invariant systems [Eq. (1-3c)], uniform asymptotic stability and equiasymptotic stability are equivalent (see Fig. 2-6). The conditions (i) and (iv) of this corollary guarantee that $v(x)$ is decrescent by Lemma A. The relaxation of the final condition constraining dv/dt is particularly important, as it is often a simple matter to check whether $v(x(t;\, x_0, t_0)) \equiv 0$ only if $x \equiv 0$. This corollary is used exclusively in Chapters IV and V.

The terminology used by those applying the direct method of Lyapunov is somewhat diverse and contradictory. Some authors (see Hahn [1]) refer to any function $v(x, t)$ that is valid for testing the stability properties (asymptotic stability, stability, or instability) of a differential equation (1-3) as a Lyapunov function. Others, following Yoshizawa [1], have defined a Lyapunov function $v(x, t)$ entirely in terms of the properties of $\dot{v}(x, t)$. This definition does not restrict the sign of $v(x, t)$, but requires $\dot{v}(x, t)$ to be nonpositive along the trajectories of the system. We prefer a nomenclature that more clearly delineates the procedure used here in stability analysis via the direct method.

Definition A. A function $v(x, t)$ or $v(x)$ that satisfies the conditions (i)–(iv) of Theorem A or Corollary A1 is called a global Lyapunov function candidate. If for a specific system of the form (1-3) or (1-3c), the total time derivative of v satisfies condition (A : v) or (A1 : v) then the system is uniformly asymptotically stable in the whole and $v(x, t)$ is said to exist as a global Lyapunov function for the system. □

If $v(x, t)$ or $v(x)$ is not radially unbounded, then it is simply a Lyapunov function candidate which may be used to establish the local stability properties of the solutions of a differential equation. Although this distinction is seldom made, it is worthwhile if this terminology is to be used with precision. Some authors who make a distinction between Lyapunov functions and Lyapunov function candidates use other definitions [for example, that conditions (i), (ii), and (iv) must be satisfied by a candidate and that $\dot{v} \leqslant 0$ for the specific $v(x, t)$ and $f(x, t)$ for $v(x, t)$ to exist as a Lyapunov function for $\dot{x} = f(x, t)$, which guarantees only uniform stability], but Definition A is chosen to be consistent with the goals of the studies to be undertaken in subsequent chapters, as defined in Sections 2 and 3 of Chapter II.

In purely geometrical terms, $v(x, t)$ is constructed in such a way that if it can be shown that along all system trajectories v goes to zero (specifically, that $v(x, t)$ is uniformly asymptotically stable in the whole), it follows that the system trajectories display the same property. Thus Lyapunov's direct method may be considered to be a mapping of the qualitative behavior of a system with n states into that of a scalar function v whose time derivative is determined in part by the system differential equation.

From this viewpoint, it may be appreciated that condition (A: v) is a very strict differential inequality in relation to the goal of determining the uniform asymptotic stability in the whole of the candidate v; in fact, a more general condition has been established in a theorem due to Corduneanu [1], which makes the ideas implicit in Theorem A more evident.

THEOREM B. If a function $v(x, t)$ exists satisfying (i)–(iv) of Theorem A, then a sufficient condition for the uniform asymptotic stability in the whole of the solutions to Eq. (1-3) is the following.

(B: v) There exists a function $m(\eta, t)$ such that:

(a) $m(\eta, t)$ is continuous for all $\eta \geqslant 0$ and t and of such a nature that the scalar differential equation $\dot{\eta} = m(\eta, t)$ has unique solutions in this domain (the equilibrium solution being $\eta \equiv 0$), which are uniformly asymptotically stable in the whole;

(b) $\dot{v} \leqslant m(v, t)$. □

A simple special case of this theorem is particularly useful in the development of frequency domain stability criteria for NLTV systems.

COROLLARY B1. A sufficient condition for conditions (a) and (b) of Theorem B to be satisfied is as follows.

(B1: v) There exists a continuous function $p(t)$ such that

(a) $\lim_{T\to\infty} \int_{t_0}^{T} p(t)\, dt = -\infty$ uniformly with respect to t_0 if p is aperiodic, or $\int_0^T p(t)\, dt < 0$ if $p(t + T) = p(t)$ for all t;

(b) $\dot{v} \leqslant p(t)v$. □

This corollary corresponds to the special case $m(\eta, t) = p(t)\eta$; the differential equation $\dot{\eta} = p(t)\eta$ is uniformly asymptotically stable in the whole if the condition of (B1: v) is satisfied, as is evident from the solution:

$$\dot{\eta} = p(t)\eta, \qquad \eta(t_0) = \eta_0 \longrightarrow \eta(t; \eta_0, t_0) = \eta_0 \exp\left\{\int_{t_0}^{t} p(\tau)\, d\tau\right\}.$$

The latter result, used exclusively in our derivations, is essentially the theorem of Krasovskii (Theorem E, Section 6); the stability mapping concept and the generality of Theorem B are important to note, however.

In using this theorem, we modify the terminology given in Definition A.

DEFINITION B. If for a specific system of the form (1-3), the total time derivative of the global Lyapunov function candidate satisfies (B: v) or (B1: v), the system is uniformly asymptotically stable in the whole and $v(x, t)$ is said to exist as a global Corduneanu function for that system. □

In the general statements about stability analysis and the absolute stability problem made in Chapter I, we stress that traditional analysis using the direct method of Lyapunov entails a search for a specific Lyapunov function for each specific problem encountered. In most meaningful cases, many candidates must be chosen in a trial and error fashion and the corresponding derivative along system trajectories inspected until one is found that satisfies the final condition (v) of Theorem A or its corollary. If even a prolonged effort fails to unearth a Lyapunov function for the system in question, instability is not indicated, as an exhaustive search is almost never possible (refer to the existence Theorems C, D, and E, Section 6), so this procedure can be completely unproductive.

In treating the absolute stability problem, we would like to further emphasize the important departure represented by the approach instituted by Lur'e and Postnikov by using a slightly modified terminology in place of that given in Definitions A and B.

DEFINITION 1. If for a given class of systems of the form of Eq. (2-6) with $g(\sigma, t) \in \{G_i[N, T]\}$, a corresponding form of Lyapunov function candidate is used, this form is designated an absolute $\{G_i[N, T]\}$ Lyapunov function candidate. If for all members of the class of systems, $\dot{v}$ satisfies condition (A: v) or (A1: v), then $v(x, t)$ is the absolute $\{G_i[N, T]\}$ Lyapunov function, or if $\dot{v}$ satisfies Condition (B: v) or (B1: v), it is then the absolute $\{G_i[N, T]\}$ Corduneanu function. □

The existence of an absolute Lyapunov or Corduneanu function is thus a sufficient condition for absolute stability in the same extended sense (Definition 11, Chapter II, Section 2). It is never necessary to find a specific $v(x, t)$ for a specific system of any class; conditions are obtained which guarantee that some member of an appropriate form exists, and what this member might be (in terms of its parameters) is immaterial.

2. The Absolute Lyapunov Function Candidates

The most general absolute Lyapunov function candidate used in this study is

$$v = \tfrac{1}{2}x^{\mathrm{T}}Px + \sum_{i=0}^{m} \beta_i \int_0^{\sigma_i} g(\zeta, t)\, d\zeta + \tfrac{1}{2}\beta_0 \rho \tau^2, \tag{3-1}$$

where

(i) $P = P^{\mathrm{T}} > 0$ (that is, P is symmetric and positive definite),
(ii) $\beta_i \geqslant 0,\ i = 0, 1, \ldots, m;\ \rho \geqslant 0$,
(iii) $\sigma_i = r_i^{\mathrm{T}}x;\ \sigma_0 = h^{\mathrm{T}}x + \rho\tau$.

In some cases (ii) is not required; in such an instance, the validity of $v(x, t)$ is reconsidered. Only the positive definiteness of $v(x, t)$ is potentially affected by relaxing condition (ii).

The following are special cases:

(1) $v = \frac{1}{2}x^T\hat{P}x$

in general; if $\rho = 0$, then

$$v = \tfrac{1}{2}x^T\left[P + \kappa \sum_{i=0}^{m} \beta_i r_i r_i^T\right]x = \tfrac{1}{2}x^T[P + \kappa M]x$$

(used for LTI systems and in deriving the circle criterion [Chapter VI, Section 1]).

(2) $v = \frac{1}{2}\left[x^T Px + k(t)\sum_{i=0}^{m} \beta_i x^T r_i r_i^T x\right] \triangleq \frac{1}{2}[x^T Px + k(t)x^T Mx]$

(used for LTV systems with $g(\sigma, t)/\sigma = k(t) \in \{K_1\}$; again, $\rho = 0$).

(3) $$v = \tfrac{1}{2}x^T Px + \sum_{i=0}^{m} \beta_i \int_0^{\sigma_i} f(\zeta)\, d\zeta + \tfrac{1}{2}\beta_0 \rho\tau^2 \tag{3-2}$$

(used for general NLTI systems. If $f(\sigma) \in \{F\}$, then only the first integral term ($i = 0$) is taken, and we have the candidate of Lur'e and Postnikov [1] as modified by Popov [2]).

(4) $v = \frac{1}{2}x^T Px + \sum_{i=0}^{m} \beta_i k(t) \int_0^{\sigma_i} f(\zeta)\, d\zeta + \frac{1}{2}\beta_0 \rho\tau^2$

(used for NLTV systems that are separable, with $k(t) \in \{K_1\}$).

A. *The Validity of Candidates for Absolute Lyapunov Functions*

We proceed to demonstrate which forms of $v(x, t)$ may be used in treating the various classes of systems considered in later chapters. For easy reference, the results are established in the form of three lemmas.

LEMMA V1. The positive definite quadratic form $v(x) = \frac{1}{2}x^T Px$ ($P = P^T > 0$ and P real) is a valid absolute $\{G_i[N, T]\}$ Lyapunov function candidate for all classes defined in Chapter II, Section 2. □

Before proving this result, we briefly review a number of important elementary properties of quadratic forms.

(1) Given a quadratic form $x^T Px$, there is no loss in generality in assuming that P is symmetric ($P = P^T$; $p_{ij} = p_{ji}$, $i, j = 1, 2, \ldots, n$). Any matrix R may be expressed as the sum of a symmetric matrix R_s and an antisymmetric matrix R_a ($r_{aij} = -r_{aji}$, all $i \neq j$; $r_{aii} = 0$); $R_s = \frac{1}{2}(R + R^T)$, $R_a = \frac{1}{2}(R - R^T)$. For any real antisymmetric matrix R_a, $x^T R_a x \equiv 0$.

(2) A symmetric matrix P is positive definite, that is, $x^T Px > 0$ for all

$x \neq 0$ (positive semidefinite, that is, $x^{\mathrm{T}}Px \geqslant 0$ for all x) denoted by $P = P^{\mathrm{T}} > 0$ ($P = P^{\mathrm{T}} \geqslant 0$) if and only if the $2^n - 1$ principal minors are positive (nonnegative). For example, a (3×3) symmetric matrix P is positive definite if the 7 principal minors

$$p_{11}, \quad p_{22}, \quad p_{33}, \quad \begin{vmatrix} p_{11} & p_{12} \\ p_{12} & p_{22} \end{vmatrix}, \quad \begin{vmatrix} p_{11} & p_{13} \\ p_{13} & p_{33} \end{vmatrix}, \quad \begin{vmatrix} p_{22} & p_{23} \\ p_{23} & p_{33} \end{vmatrix}, \quad |P|$$

are all positive.

A special result exists for the case of positive definite matrices: $P = P^{\mathrm{T}} > 0$ if and only if the n *leading* principal minors

$$\mu_1 \triangleq p_{11}, \mu_2 \triangleq \begin{vmatrix} p_{11} & p_{12} \\ p_{12} & p_{22} \end{vmatrix}, \ldots,$$

$$\mu_{n-1} \triangleq \begin{vmatrix} p_{11} & p_{12} \cdots p_{1(n-1)} \\ p_{12} & p_{22} \cdots p_{2(n-1)} \\ \vdots & \vdots \quad \vdots \\ p_{1(n-1)} & \cdots p_{(n-1)(n-1)} \end{vmatrix}, \qquad \mu_n \triangleq |P| \tag{3-3}$$

are all positive. The conditions that $\mu_i \geqslant 0$ do *not* guarantee that $P = P^{\mathrm{T}} \geqslant 0$, as the following example shows: Choose any $\alpha > 0$, $\beta > 0$, $\gamma > 0$, and any δ; then

$$P = \begin{bmatrix} \alpha & (\alpha\beta)^{1/2} & \delta \\ (\alpha\beta)^{1/2} & \beta & \delta(\beta/\alpha)^{1/2} \\ \delta & \delta(\beta/\alpha)^{1/2} & \gamma \end{bmatrix} \longrightarrow \mu_1 = \alpha > 0, \quad \mu_2 = 0, \quad \mu_3 = 0,$$

but an investigation of the principal minors reveals that P is positive semidefinite only if $\delta^2 \leqslant \alpha\gamma$.

(3) If $P = P^{\mathrm{T}} > 0$, then the eigenvalues of P or the n roots λ_{iP} of the characteristic equation $|\lambda I - P| = 0$ are all real and positive. If they are ordered such that $0 < \lambda_{1P} \leqslant \lambda_{2P} \leqslant \cdots \leqslant \lambda_{nP}$, then

$$0 < \tfrac{1}{2}\lambda_{1P}\|x\|^2 \leqslant \tfrac{1}{2}x^{\mathrm{T}}Px \leqslant \tfrac{1}{2}\lambda_{nP}\|x\|^2$$

for all $x \neq 0$.

Proof: These observations essentially complete the proof of Lemma V1. The continuity conditions (i) of Theorem A are satisfied by inspection, and the lower and upper bounds on $\frac{1}{2}x^{\mathrm{T}}Px$ in (3) show that conditions (ii) to (iv) are satisfied with $\alpha(\rho) = \frac{1}{2}\lambda_{1P}\rho^2$ and $\beta(\rho) = \frac{1}{2}\lambda_{nP}\rho^2$, where the parameters λ_{iP} are real and satisfy $0 < \lambda_{1P} \leqslant \lambda_{nP}$. □

LEMMA V2. The form

$$v(x) = \tfrac{1}{2}x^{\mathrm{T}}Px + \sum_{i=0}^{m} \beta_i \int_0^{\sigma_i} f(\zeta)\, d\zeta + \tfrac{1}{2}\beta_0\rho[f(\sigma_0)]^2, \tag{3-2}$$

where $P = P^{\mathrm{T}} > 0$, $\beta_i \geqslant 0$, $i = 0, 1, \ldots, m$ and $\rho \geqslant 0$, and $\sigma_0 = h^{\mathrm{T}}x + \rho\tau$, $\sigma_i = r_i^{\mathrm{T}}x$, where r_i are real n vectors, $i = 1, 2, \ldots, m$, is a valid absolute $\{N\}$ Lyapunov function candidate for all classes $\{N\}$ defined in Chapter II, Section 2. □

Proof: The continuity of the elements of ∇v,

$$\partial v/\partial x_k = \sum_{j=1}^{n} p_{kj}x_j + \beta_0 f(\sigma_0)\,(\partial/\partial x_k)\,[\sigma_0 + \rho f(\sigma_0)] + \sum_{i=1}^{m} \beta_i f(\sigma_i) r_{ik}$$

follows directly from the observation that $\sigma_0 = h^{\mathrm{T}}x - \rho f(\sigma_0)$ [Eq. (2-6)] so that $(\partial/\partial x_k)\,[\sigma_0 + \rho f(\sigma_0)] = h_k$, and from the assumed continuity of $f(\sigma_0)$. Since $P = P^{\mathrm{T}} > 0$, $\beta_i \geqslant 0$, $\rho \geqslant 0$ and $f(\sigma)$ lies in the first and third quadrants of the σ, $f(\sigma)$ plane (Fig. 2-2), we have $v(x) \geqslant \frac{1}{2}x^{\mathrm{T}}Px \geqslant \frac{1}{2}\lambda_{1P}\|x\|^2 > 0$ for all $x \neq 0$. Again, choosing $\alpha(\rho) = \frac{1}{2}\lambda_{1P}\rho^2$, it follows that $v(x)$ is positive definite and radially unbounded. Also $v(0) = 0$, which according to Corollary A1 completes the proof. □

LEMMA V3. The function $v(x, t)$ defined in Eq. (3-1) is a valid absolute $\{G_i[N, T]\}$ Lyapunov function candidate for all classes $\{N\}$ and for $\{T\} = \{K_1\}$ or $\{K_2\}$ as defined in Chapter II, Section 2. □

Proof: (i) The continuity requirements of Theorem A are satisfied due to the assumed continuity of $g(\sigma_0, t)$ with respect to σ_0 and t.

(ii) and (iii) That $v(x, t)$ is positive definite and radially unbounded proceeds as in the proof of Lemma V2; all terms except $\frac{1}{2}x^{\mathrm{T}}Px$ are positive semidefinite under the assumptions made on the parameters and on $g(\sigma_0, t)$, so $v(x, t) \geqslant \frac{1}{2}x^{\mathrm{T}}Px \geqslant \frac{1}{2}\lambda_{1P}\|x\|^2$.

(iv) The elements of ∇v are $\nabla v^{\mathrm{T}} = x^{\mathrm{T}}P + \sum_{i=0}^{m} \beta_i g(\sigma_i, t) r_i^{\mathrm{T}}$, where $r_0 = h$; the decrescency of $v(x, t)$ follows from Lemma A and the assumption that $g(\sigma_0, t)$ is bounded with respect to t for all fixed σ_0. □

B. *The Total Time Derivative of* $v(x, t)$

Given the form of (∇v) found in proving Lemmas V2 and V3, it is convenient to establish the total time derivative of $v(x, t)$ [Eq. (3-1)] for systems of the form specified by the differential equation (2-6): $\dot{v} = (\nabla v)^{\mathrm{T}}\dot{x} + \partial v/\partial t$ is

$$\begin{aligned}\dot{v} = {} & \tfrac{1}{2}x^{\mathrm{T}}(A^{\mathrm{T}}P + PA)x - g(\sigma_0, t)x^{\mathrm{T}}[Pb - (\beta_0 A^{\mathrm{T}}h + \gamma_0 h)] \\ & - [\beta_0 h^{\mathrm{T}}b + \gamma_0(\rho + \bar{G}_N^{-1})]g^2(\sigma_0, t) - \gamma_0\sigma_0 g(\sigma_0, t)[1 - g(\sigma_0, t)/(\bar{G}_N\sigma_0)] \\ & + \sum_{i=1}^{m} \beta_i g(\sigma_i, t) r_i^{\mathrm{T}}[Ax + b\tau] + \sum_{i=0}^{m} \beta_i \int_0^{\sigma_i} [\partial g(\zeta, t)/\partial t]\, d\zeta \end{aligned} \tag{3-4}$$

In arriving at this expression we use

(i) $x^{\mathrm{T}}PAx = \frac{1}{2}x^{\mathrm{T}}(A^{\mathrm{T}}P + PA)x$, that is, we take only the symmetric part of the product PA, and

(ii) the identity

$$\gamma_0 g(\sigma_0, t)\{[h^{\mathrm{T}}x + \rho\tau - \sigma_0] - [g(\sigma_0, t)/\bar{G}_N - \sigma_0 g(\sigma_0, t)/\bar{G}_N\sigma_0]\} \equiv 0.$$

This last manipulation—essentially, expanding $\dot{v}(x, t)$ by adding and subtracting equal terms—is a standard practice used in obtaining $\dot{v}$ in a form convenient for subsequent analysis.

3. Restated Stability Theorems

Since various forms of $v(x, t)$ in (3-1) have been shown to be valid candidates, only $\dot{v}$ need be considered in the derivation of stability criteria. The sufficiency theorems of Section 1 reduce to:

THEOREM 1 (from Theorem A and Corollary A1).

(a) If the system (2-6) is time-invariant, then for some $v(x)$ chosen according to Lemma V2, a sufficient condition for uniform asymptotic stability in the whole is that $\dot{v} \leqslant 0$ where $\dot{v}(x(t; x_0, t_0)) \equiv 0$ only if $x(t; x_0, t_0) \equiv 0$.

(b) If the system (2-6) is not time-invariant, then for a $v(x)$ or $v(x, t)$ chosen according to Lemma V1 if $\{T\} = \{K_0\}$ or Lemma V3 if $\{T\} = \{K_1\}$ or $\{K_2\}$, a sufficient condition for uniform asymptotic stability in the whole is that $\dot{v}(x, t)$ [Eq. (3-4)] is negative definite. □

THEOREM 2 (from Corollary B1). If the system (2-6) is time-varying with $\{T\} = \{K_1\}$ or $\{K_2\}$, and if for $v(x, t)$ specified by Lemma V3 there exists a continuous real function $p(t)$ such that $\dot{v} \leqslant p(t)v$, then a sufficient condition for uniform asymptotic stability in the whole is that:

(a) $\lim_{T\to\infty} \int_{t_0}^{T} p(t)\, dt = -\infty$ uniformly with respect to t_0 if $p(t)$ is aperiodic, or

(b) $\int_0^T p(t)\, dt < 0$ if $p(t + T) = p(t)$ for all t. □

4. The Kalman–Yakubovich Lemma

Inspecting the first three terms of $\dot{v}$ [Eq. (3-4)], it is useful to note that they would form a complete square and a quadratic form if solutions to the following equations existed:

$$\left.\begin{aligned} A^{\mathrm{T}}P + PA &= -qq^{\mathrm{T}} - D; \\ Pb - k &= [2\beta_0 h^{\mathrm{T}}b + 2\gamma_0(\rho + \bar{G}_N^{-1})]^{1/2} q \triangleq \sqrt{\psi}\, q. \end{aligned}\right\} \tag{3-5}$$

The resulting terms of $\dot{v}$ would be $-\frac{1}{2}[x^{\mathrm{T}}q - \sqrt{\psi}\,\tau]^2 - \frac{1}{2}x^{\mathrm{T}}Dx$. If D were at least positive semidefinite, then much of the problem of showing that $\dot{v}$ is negative semidefinite or definite is solved. A basic lemma that establishes conditions for the existence of solutions of this form has been proven by Yakubovich [1]; this was the major step in relating the work of Popov with that of Lur'e and Postnikov. A significant contribution of Kalman [2] was a direct proof of the sufficiency part of the lemma, which has been widely used. Many refinements and specialized forms of this lemma exist; we confine our attention to those that are required in later sections. Of the five lemmas stated, only Lemma 5 is considered in detail; the proofs of Lemmas 1–4 follow along similar lines.

LEMMA 1 (Kalman). Given $A \in \{A_1\}$, (A, b) completely controllable, a real vector k and a real scalar ψ, then a real vector q and matrices $P = P^{\mathrm{T}} \geqslant 0$, $K = K^{\mathrm{T}} \geqslant 0$ satisfying

$$\left.\begin{aligned} &\text{(a)} \qquad A^{\mathrm{T}}P + PA = -qq^{\mathrm{T}} - K, \\ &\text{(b)} \qquad Pb - k = \sqrt{\psi}\, q, \end{aligned}\right\} \tag{3-6}$$

exist if and only if

$$\text{(c)} \qquad H(s) \triangleq \tfrac{1}{2}\psi + k^{\mathrm{T}}(sI - A)^{-1}b \in \{PR\}. \ \square \tag{3-7}$$

This latter condition, that $H(s)$ must be positive real, forms the core of the derivation of all the frequency domain stability criteria that follow. We should note that the statement of condition (c) was originally given as $\operatorname{Re} H(i\omega) \geqslant 0$; in the necessity proof of Lemma 5 we see that this is entirely equivalent to the constraint $H(s) \in \{PR\}$ in this context. This condition and other related constraints, that is, $H(s) \in \{SPR\}$ and $H(s) \in \{SPR_0\}$, are considered in detail in Section 5.

Kalman, and subsequently Lefschetz and Meyer (see Lemmas 2 and 3), explicitly required that $\psi \geqslant 0$. This condition, however, is seen to be implicitly ensured by the constraint (c) on $H(s)$, as $H_\infty \triangleq \lim_{\omega\to\infty} H(i\omega) = \frac{1}{2}\psi$; the term $k^{\mathrm{T}}(i\omega I - A)^{-1}b$ goes to zero at least as rapidly as $(i\omega)^{-1}$ as $\omega \longrightarrow \infty$.

A less restrictive form of this lemma may be found in Lefschetz [1].

LEMMA 2 (Lefschetz). Given $A \in \{A_1\}$, (A, b) completely controllable, a real vector k, scalars ψ and $\varepsilon > 0$ and an arbitrary real matrix $L = L^{\mathrm{T}} > 0$, then a real vector q and a real matrix $P = P^{\mathrm{T}} > 0$ satisfying

$$\left.\begin{aligned} &\text{(a)} \qquad A^{\mathrm{T}}P + PA = -qq^{\mathrm{T}} - \varepsilon L, \\ &\text{(b)} \qquad Pb - k = \sqrt{\psi}\, q, \end{aligned}\right\} \tag{3-8}$$

exist if and only if ε is sufficiently small and $H(s)$ of Eq. (3-7) satisfies

$$\text{(c)} \qquad H(s) \in \{SPR\}. \ \square$$

The relaxations vis-à-vis Lemma 1 are the conditions that P is definite and that $L = L^{\mathrm{T}} > 0$, where L is arbitrary, may appear in Eq. (3-8); it is significant, however, that requirement (c) is more strict in Lemma 2.

A second less strict form of Lemma 1 is due to Meyer [1]:

LEMMA 3 (Meyer, Lemma 1). Given $A \in \{A_1\}$, (A, b) completely controllable, a real vector k and a real scalar ψ, then a real vector q and matrices $P = P^{\mathrm{T}} > 0$, $M = M^{\mathrm{T}} \geqslant 0$ satisfying

$$\left.\begin{array}{ll} \text{(a)} & A^{\mathrm{T}}P + PA = -qq^{\mathrm{T}} - M, \\ \text{(b)} & Pb - k = \sqrt{\psi}\, q, \\ \text{(c)} & (A, q^{\mathrm{T}}) \ \text{ is completely observable,} \end{array}\right\} \tag{3-9}$$

exist if and only if $H(s)$ of Eq. (3-7) satisfies

$$\text{(d)} \qquad H(s) \in \{PR\}. \ \square$$

The condition Re $H(i\omega) \not\equiv 0$ was not stated by Meyer. If $\psi = 0$ and $k = 0$ are chosen, $H(s) \equiv 0$, which otherwise satisfies condition (d). Under this assumption, $Pb = 0$ from Eq. (3-9), so $b^{\mathrm{T}}Pb = 0$, which is incompatible with the conditions $P = P^{\mathrm{T}} > 0$ and (A, b) completely controllable. We do not consider $H(s) \equiv 0$ to be a member of $\{PR\}$.

A further extension given by Meyer [1] is the removal of the requirement that (A, b) be completely controllable; this is not required in the present work, however. The relaxation of (d) (that $H(s)$ need not be strictly positive real) with respect to Lemma 2 is a result of the *semi*definiteness of M.

It is found necessary to modify the lemma slightly in order to be able to apply Theorem 2 (Corduneanu) in the generation of absolute stability criteria (see Taylor and Narendra [2] or Chapter VI, Section 6). The following is equivalent to Lemma 3 with $\hat{A} \triangleq A + \mu I$. The relation between Lemmas 3 and 4 is the same as that between the fundamental results of Lyapunov (Theorem F) and Kalman (Corollary F1) stated in Section 6.

LEMMA 4 (Taylor and Narendra). Given $\hat{A} \triangleq A + \mu I \in \{A_1\}$, $(\hat{A}, b)$ completely controllable,† a real vector k and a real scalar ψ, then a real vector q and matrices $P = P^{\mathrm{T}} > 0$, $N = N^{\mathrm{T}} \geqslant 0$ satisfying

$$\left.\begin{array}{ll} \text{(a)} & A^{\mathrm{T}}P + PA = -qq^{\mathrm{T}} - N - 2\mu P, \\ \text{(b)} & Pb - k = \sqrt{\psi}\, q, \end{array}\right\} \tag{3-10}$$

exist if and only if

$$\text{(c)} \qquad \hat{H}(s) \triangleq \tfrac{1}{2}\psi + k^{\mathrm{T}}((s - \mu)I - A)^{-1}b \in \{PR\}. \ \square \tag{3-11}$$

Note that $\hat{H}(s) = H(s - \mu)$ of Eq. (3-7).

In all of the above formulations of the Kalman–Yakubovich lemma, it is assumed that A is a stable matrix. This evidently precludes the treatment of the particular cases (Chapter II, Section 1C), especially the case of a system with a single zero eigenvalue that arises in considering the so-called indirect control system. As this problem is of quite general interest, the following

† $(\hat{A}, b)$ is completely controllable if and only if (A, b) is completely controllable.

extension of Lemma 1 is presented as a means of unifying the stability analysis of direct and indirect control systems. The proof of this lemma is quite directly related to that given by Lefschetz [1], with suitable modifications to take into account the relaxation $A \in \{A_0\}$ and the correction given by Lefschetz, Meyer, and Wonham [1].

The zero eigenvalue presents a number of difficulties that must be resolved before such a proof may be undertaken.

LEMMA 5. Given the marginally stable matrix A, $A \in \{A_0\}$, of the form

$$A = \left[\begin{array}{c|ccc} 0 & 1 & 0 \cdots & 0 \\ \hline 0 & & & \\ \cdot & & & \\ \cdot & & \hat{A} & \\ \cdot & & & \\ 0 & & & \end{array}\right], \tag{3-12}$$

$\hat{A} \in \{A_1\}$ and (A, b) are in phase variable canonical form (Chapter II, Section 1), a symmetric matrix $L_0 = L_0^{\mathrm{T}}$ of the form

$$L_0 = \left[\begin{array}{c|ccc} 0 & 0 & \cdots & 0 \\ \hline 0 & & & \\ \cdot & & & \\ \cdot & & \hat{L} & \\ \cdot & & & \\ 0 & & & \end{array}\right], \tag{3-13}$$

where $\hat{L} = (\hat{L})^{\mathrm{T}} > 0$ is arbitrary, a real vector k, real scalars $\varepsilon > 0$ and ψ, then a real vector q and a matrix $P = P^{\mathrm{T}} > 0$ satisfying

$$\left.\begin{array}{ll} \text{(a)} & A^{\mathrm{T}}P + PA = -qq^{\mathrm{T}} - \varepsilon L_0, \\ \text{(b)} & Pb - k = \sqrt{\psi}\, q, \end{array}\right\} \tag{3-14}$$

exist if and only if ε is sufficiently small and $H(s)$ of Eq. (3-7) satisfies

$$\text{(c)} \qquad H(s) \in \{SPR_0\}. \ \square \tag{3-15}$$

It is important to note that $H(s) \in \{SPR_0\}$ requires the strict inequality $k_1 > 0$ (Section 5), which precludes the removal of the pole of $W(s)$ at $s = 0$ by pole-zero cancellation (see Chapter V, Section 2), that is, $H(s)$, which may be factored into

$$H(s) = W(s)Z(s), \tag{3-16}$$

must retain the pole at $s = 0$.

The following points are important to consider before undertaking a proof of the lemma.

(1) If $Q \triangleq A^{\mathrm{T}}P + PA$ is expanded, we have

$$Q = \begin{bmatrix} 0 & (p_{11} - a_2 p_{1n}) & (p_{12} - a_3 p_{1n}) \cdots (p_{1(n-1)} - a_n p_{1n}) \\ (p_{11} - a_2 p_{1n}) & q_{22} & \\ \cdot & & \cdot \\ \cdot & & \quad \cdot \\ \cdot & & \qquad \cdot \\ (p_{1(n-1)} - a_n p_{1n}) & & \qquad q_{nn} \end{bmatrix} \tag{3-17}$$

where only the q_{1j} terms are emphasized. Since $q_{11} = 0$, it is evident that Eq. (3-14a) becomes

$$q_1^2 + \varepsilon l_{11} = 0.$$

As $L_0 \geqslant 0$, it is clear that $l_{11} \geqslant 0$ is required, so as both terms are non-negative, $q_1 = l_{11} = 0$. This in turn results in the choice $l_{1j} = 0$ for all j; this is the form indicated in Eq. (3-13).

(2) Since A and b are in the phase variable canonical form, namely

$$A = \left[\begin{array}{c|cccc} 0 & & & & \\ 0 & & & & \\ \cdot & & \mathrm{I} & & \\ \cdot & & & & \\ \cdot & & & & \\ 0 & & & & \\ \hline 0 & -a_2 & -a_3 & \cdots & -a_n \end{array}\right], \qquad b = \begin{bmatrix} 0 \\ 0 \\ \cdot \\ \cdot \\ \cdot \\ 0 \\ 1 \end{bmatrix} \tag{3-18}$$

[which is equivalent to the assumption of complete controllability (Chapter II, Section 1)], we have that:

(a)
$$\begin{aligned} \phi(s) &\triangleq |sI - A| = s[s^{n-1} + a_n s^{n-2} + \cdots + a_3 s + a_2] \\ &\triangleq s\theta(s), \end{aligned} \tag{3-19}$$

where $\theta(s)$ is assumed to be a Hurwitz polynomial.

(b)
$$m(s) \triangleq (sI - A)^{-1} b = \begin{bmatrix} \dfrac{1}{\phi(s)} \\ \dfrac{s}{\phi(s)} \\ \cdot \\ \cdot \\ \cdot \\ \dfrac{s^{n-1}}{\phi(s)} \end{bmatrix} = \begin{bmatrix} \dfrac{1}{\phi(s)} \\ \dfrac{1}{\theta(s)} \\ \\ \\ \\ \dfrac{s^{n-2}}{\theta(s)} \end{bmatrix}, \tag{3-20}$$

(c)
$$k^{\mathrm{T}} m(s) = \frac{k_n s^{n-1} + \cdots + k_2 s + k_1}{\phi(s)}, \tag{3-21}$$

which is a real rational transfer function if k is a real vector.

(d) The scalar function

$$\pi(\omega) \triangleq m^*(i\omega)L_0m(i\omega) \triangleq m^{\mathrm{T}}(-i\omega)L_0m(i\omega) \tag{3-22}$$

is a real rational function of ω, and the first row and column of L_0, being composed of zeros [Eq. (3-13)], implies that the poles at $\omega = 0$ are removed; in particular

$$\pi(\omega) = \frac{\eta_1(\omega^2)}{\theta(i\omega)\theta(-i\omega)} = \frac{l_{nn}\omega^{2(n-2)} + \cdots}{\theta(i\omega)\theta(-i\omega)}, \tag{3-23}$$

where η_1 is of order $2(n-2)$ in ω and $\theta(i\omega)$ is of order $2(n-1)$; hence

$$\lim_{\omega\to\infty} \pi(\omega) = 0.$$

Since $\theta(s)$ has only poles with negative real parts, $\pi(\omega)$ is continuous for finite ω, and hence has finite upper and lower bounds; in fact

$$0 < \pi(\omega) \leqslant \mu_1 \tag{3-24}$$

for all finite ω since $\hat{L} > 0$.

(e) The scalar function

$$\kappa(\omega) \triangleq m^*k + k^{\mathrm{T}}m = 2\,\mathrm{Re}\{k^{\mathrm{T}}m(i\omega)\} \tag{3-25}$$

is also of the form

$$\kappa(\omega) = \frac{\eta_2(\omega^2)}{\theta(i\omega)\theta(-i\omega)}. \tag{3-26}$$

This may be seen by inspecting the expansion of the numerator of $k^{\mathrm{T}}m(i\omega)$:

$$k^{\mathrm{T}}m(i\omega) = \frac{-[k_n(i\omega)^n + \cdots + k_2(i\omega)^2 + k_1 i\omega]\theta(-i\omega)}{\omega^2\theta(i\omega)\theta(-i\omega)};$$

considering only the highest and lowest order terms of this numerator, we have

$$-\{k_n\omega^{2(n-1)}(i\omega) + [k_{n-1} - a_nk_n]\omega^{2(n-1)} + \cdots + [a_3k_1 - a_2k_2]\omega^2 + a_2k_1 i\omega\}.$$

In finding the real part of $k^{\mathrm{T}}m(i\omega)$, only the even powers of ω are taken; hence

$$\kappa(\omega) = \frac{2\{-k^{\mathrm{T}}Ab\omega^{2(n-2)} + \cdots + [a_2k_2 - a_3k_1]\}}{\theta(i\omega)\theta(-i\omega)} \triangleq \frac{\eta_2(\omega^2)}{\theta(i\omega)\theta(-i\omega)},$$

where the substitution $k^{\mathrm{T}}Ab = (k_{n-1} - a_nk_n)$ has been made. If $k^{\mathrm{T}}Ab \neq 0$, then $\eta_2(\omega^2)$ is of order $2(n-2)$, as is $\eta_1(\omega^2)$. Since $\kappa(\omega)$ is of the same form as $\pi(\omega)$, it again must have well-defined upper and lower bounds. From condition (3-15), it is seen that if $\psi > 0$, then

$$-\psi < (-\psi + \nu) \leqslant \kappa(\omega) \leqslant \mu_2 < \infty \tag{3-27a}$$

for all finite ω, where μ_2 is arbitrary but finite and $\nu > 0$ is arbitrarily small, or if $\psi = 0$, then for any $\bar{\omega} < \infty$, there exists some $\nu > 0$ such that

$$0 < \nu \leqslant \kappa(\omega) \leqslant \mu_2 < \infty, \qquad |\omega| \leqslant \bar{\omega}. \tag{3-27b}$$

Finally, inspect the identity

$$[(\sigma - i\omega)I - A]^{\mathrm{T}}P + P[(\sigma + i\omega)I - A] = -(A^{\mathrm{T}}P + PA) + 2\sigma P,$$

which may be postmultiplied by $m(s)$ and premultiplied by $m^*(s)$ to establish

$$m^*Pb + b^{\mathrm{T}}Pm = m^*qq^{\mathrm{T}}m + \varepsilon m^*L_0m + 2\sigma m^*Pm, \tag{3-28}$$

where Eq. (3-14a) is used.

Proof of Necessity: Substitute the expression for Pb [Eq. (3-14b)] into (3-28) to obtain

$$m^*[k + \sqrt{\psi}q] + [k^{\mathrm{T}} + \sqrt{\psi}q^{\mathrm{T}}]m = m^*qq^{\mathrm{T}}m + \varepsilon m^*L_0m + 2\sigma m^*Pm$$

or

$$2\,\mathrm{Re}(k^{\mathrm{T}}m) > |q^{\mathrm{T}}m|^2 - 2\sqrt{\psi}\,\mathrm{Re}(q^{\mathrm{T}}m) + 2\sigma m^*Pm, \tag{3-29}$$

where Eq. (3-24) is used and the condition $\varepsilon > 0$ ensures the strict inequality for all finite s. If the complex quantity $q^{\mathrm{T}}m(s)$ is denoted by $q^{\mathrm{T}}m(s) = \alpha + i\beta$, then adding ψ to both sides of Eq. (3-29) yields

$$2\,\mathrm{Re}\,H(s) \triangleq \psi + 2\,\mathrm{Re}\{k^{\mathrm{T}}m(s)\} > (\alpha - \sqrt{\psi})^2 + \beta^2 + 2\sigma m^*Pm \geqslant 0,$$

which essentially completes the proof of the necessity of Eq. (3-15); $\mathrm{Re}\,H(s) > 0$ is required throughout the right half of the s-plane ($\sigma \geqslant 0$) since the matrix P satisfying Eq. (3-14a) is necessarily positive definite due to $A \in \{A_1\}$ [Theorem F, Section 6]. We note below that $\mathrm{Re}\,H(i\omega)$ must not go to zero as $\omega \longrightarrow \infty$ more rapidly than ω^{-2} (refer to Section 5).

Proof of Sufficiency: Using the preliminary definitions of $\kappa(\omega)$ and $\pi(\omega)$, it is seen that

$$\psi + m^*k + k^{\mathrm{T}}m - \varepsilon m^*L_0m \geqslant \nu - \varepsilon\mu_1$$

for all finite ω where ν, ε, and μ_1 are all positive and finite. It is necessary to establish that $\varepsilon > 0$ may be found such that this is positive, that is,

$$\psi + \frac{\eta_2(\omega^2) - \varepsilon\eta_1(\omega^2)}{\theta(i\omega)\theta(-i\omega)} \triangleq \frac{\eta_3(\omega^2)}{\theta(i\omega)\theta(-i\omega)} > 0 \tag{3-30}$$

for all finite ω. It is not sufficient to use the result (3-27a) to say that $\varepsilon < \nu/\mu_1$ ensures (3–30); it is also necessary to consider the asymptotic behavior of $\psi + m^*k + km$ and εm^*L_0m as $\omega \longrightarrow \infty$. From the previous comments, the limit of the numerator of Eq. (3-30) is dominated by the terms

$$\lim_{\omega\to\infty}\{\eta_3(\omega^2)\} = \lim_{\omega\to\infty}\{\psi\omega^{2(n-1)} - 2k^{\mathrm{T}}Ab\omega^{2(n-2)} - \varepsilon l_{nn}\omega^{2(n-2)}\},$$

from which the necessity of demanding that either $\psi > 0$ (in which case $\mathrm{Re}\,H(i\omega)$ approaches $\psi > 0$ as $\omega \longrightarrow \infty$) or that $k^{\mathrm{T}}Ab < 0$ if $\psi = 0$ (in which case $\mathrm{Re}\,H(i\omega)$ goes to zero no more rapidly than ω^{-2}). If neither of these conditions is satisfied, then Re $H(i\omega)$ goes to zero more rapidly that ω^{-2} and the numerator $\eta_3(\omega^2)$ is dominated by $-\varepsilon l_{nn}\omega^{2(n-2)}$ and it is never possible to choose $\varepsilon > 0$ such that (3-30) holds. This eventuality is precluded by condition (c) of the lemma, as is shown Section 5.† In the case that $\psi > 0$, $\varepsilon < \nu/\mu_1$ guarantees the condition (3-30), and if $\psi = 0$, then $\varepsilon < \min\{\nu/\mu_1, -2k^{\mathrm{T}}Ab/l_{nn}\}$ serves this purpose.

Given that ε is sufficiently small as specified in the previous argument, the numerator $\eta_3(\omega^2)$ has no real roots, and is of order $2(n-1)$ at most (if $\psi > 0$), having as its leading term $\psi\omega^{2(n-1)}$. Since $\eta_3(\omega^2) > 0$, it may be factored into real polynomials,

$$\eta_3(\omega^2) = \zeta(i\omega)\zeta(-i\omega),$$

where the order of ζ is at most $(n-1)$ and $\zeta(i\omega)$ has as its leading term $\sqrt{\psi}(i\omega)^{n-1}$. Thus, $\zeta(s)/\theta(s)$ may be reduced to

$$\zeta(s)/\theta(s) = \sqrt{\psi} + \xi(s)/\theta(s),$$

where $\xi(s)$ is of order $(n-2)$.

We now define the vector q:

$$q^{\mathrm{T}}m(s) = \frac{q_n s^{n-1} + \cdots + q_2 s + q_1}{\phi(s)} \triangleq \frac{-\xi(s)}{\theta(s)};$$

note that by definition, $q_1 = 0$ so that the pole at $s = 0$ is removed; this is consistent with observation (i). It remains to be seen whether or not q as defined here satisfies the remaining conditions.

The definition of q substituted into Eq. (3-30) leads to

$$\begin{aligned}\psi + k^{\mathrm{T}}m + m^*k - \varepsilon m^*L_0 m &= (\sqrt{\psi} - q^{\mathrm{T}}m)(\sqrt{\psi} - m^*q) \\ &= m^*qq^{\mathrm{T}}m - \sqrt{\psi}(q^{\mathrm{T}}m + m^*q) + \psi.\end{aligned}$$

Subtracting ψ from each side and substituting for $(m^*qq^{\mathrm{T}}m + \varepsilon m^*Lm)$ using Eq. (3-28) with $s = i\omega$ $(\sigma = 0)$ yields

$$k^{\mathrm{T}}m + m^*k = m^*Pb + b^{\mathrm{T}}Pm - \sqrt{\psi}(q^{\mathrm{T}}m + m^*q);$$

† Note that this problem arises only in Lemmas 2 and 5, where L or a part of L is assumed to be positive definite ($l_{nn} > 0$ in both cases). In Lemmas 1, 3, and 4, the corresponding matrices K, M, and N, respectively, are only assumed to be semidefinite, so if $k^{\mathrm{T}}Ab = 0$ and $\psi = 0$, then k_{nn}, m_{nn}, or n_{nn} respectively is zero. For example,

$$H(s) = \frac{s + a_2}{s^2 + a_2 s + a_1} \longrightarrow \operatorname{Re} H(i\omega) = \frac{a_1 a_2}{(a_1 - \omega^2)^2 + (a_2\omega)^2}, \tag{3-31}$$

which goes to zero as ω^{-4} even though Re $H(i\omega) > 0$ for all real finite ω. By direct calculation using the standard phase variable canonical form for (k, A, b),

$$P = P^{\mathrm{T}} = \begin{bmatrix} (a_1 + a_2{}^2) & a_2 \\ a_2 & 1 \end{bmatrix} > 0 \longrightarrow A^{\mathrm{T}}P + PA = \begin{bmatrix} -2a_1a_2 & 0 \\ 0 & 0 \end{bmatrix}$$

and $Pb = k \triangleq \operatorname{col}[a_2, 1]$, so Lemmas 2 and 5 are not satisfied, whereas Lemma 3 is. Note that $k^{\mathrm{T}}Ab < 0$ is not explicitly required in Lemma 2 or 5 since the present definition of strictly positive real functions ensures this condition (Definition PR2; Taylor [1]).

hence, for all ω

$$\operatorname{Re}\{[Pb - k - \sqrt{\psi}q]^{\mathrm{T}}m\} \equiv 0.$$

Define the vector d by $d \triangleq Pb - k - \sqrt{\psi}q$. If the first element of $(Pb - k)$ is zero, then $d_1 = 0$ and

$$N(i\omega) \triangleq d^{\mathrm{T}}m(i\omega) = \frac{d_n(i\omega)^{n-2} + \cdots + d_3 i\omega + d_2}{\theta(i\omega)}.$$

$N(i\omega)$ can be expressed in terms of the matrix $\hat{A}$ [Eq. (3-12)], and reduced vectors $\hat{b} \triangleq [0, 0, \ldots, 1]^{\mathrm{T}}$ of dimension $(n - 1)$ and $\hat{d} \triangleq [d_2, d_3, \ldots, d_n]^{\mathrm{T}}$ as

$$N(i\omega) = \hat{d}^{\mathrm{T}}(i\omega I - \hat{A})^{-1}\hat{b} \triangleq \hat{d}^{\mathrm{T}}\hat{m}(i\omega).$$

This is the same form considered in Lefschetz [1]; in particular $\hat{A} \in \{A_1\}$, and the same argument that $\operatorname{Re}\{\hat{d}^{\mathrm{T}}\hat{m}(i\omega)\} \equiv 0$ for all ω guarantees that $\hat{d} = 0$ may be used to show that Eq. (3-14b) is satisfied. For completeness, we summarize this result.

Lemma. If $\hat{A} \in \{A_1\}$, $(\hat{A}, \hat{b})$ is completely controllable and $\operatorname{Re} \hat{d}^{\mathrm{T}}(i\omega I - \hat{A})^{-1}\hat{b} \equiv 0$ for all ω, then $\hat{d} = 0$. □

Proof: The function $N(s) = \hat{d}^{\mathrm{T}}\hat{m}(s)$ is real, rational and, if we assume that $d \neq 0$, then it is not identically zero. The poles of $N(s)$ lie in the open left half-plane, and at least one such pole must exist since numerator and denominator order must differ by at least unity. Under the given condition that $\operatorname{Re} N(i\omega) \equiv 0$, $N(s)$ must assume only purely imaginary values on the imaginary axis of the s-plane. Thus, the function

$$\hat{N}(s) \triangleq iN(is)$$

is a rational function of s with poles in the open upper half-plane (since $A \in \{A_1\}$) which must have real values for s real. This provides a contradiction to the assumption $d \neq 0$: $\hat{N}(s)$ can be real for s real only if its poles (which are necessarily complex) occur in conjugate pairs which is explicitly precluded; thus, $d = 0$.

It is now necessary to demonstrate that it is possible to choose P such that the first element of $(Pb - k)$ is zero. This question is closely related to the existence of $P = P^{\mathrm{T}} > 0$ satisfying Eq. (3-14a). The Lyapunov function $v = x^{\mathrm{T}}Px$ exists even for A of the form assumed in Eq. (3-18) (refer to Hahn [1]); $P = P^{\mathrm{T}} > 0$ if Q in the equation

$$A^{\mathrm{T}}P + PA = -Q$$

is a suitable positive semidefinite matrix. It may be seen that Q as defined in Eq. (3-14a) is the required form. Hahn has shown that the P obtained for such a Q is indeterminable. It is precisely this indeterminability that permits the choice of P such that $(Pb - k)_1 = 0$; from the expansion of the first row or column of $A^{\mathrm{T}}P + PA$ [Eq. (3-17)],

$$p_{11} = a_2 k_1, \qquad p_{12} = a_3 k_1, \ldots, p_{1n} = k_1.$$

This observation completes the proof of this lemma, since $P = P^{\mathrm{T}} > 0$ requires that $p_{11} > 0$, which in turn gives us the requirement that $k_1 > 0$, which is guaranteed by $H(s) \in \{SPR_0\}$, Section 5. □

The Lemmas 1–5 have several important similarities. In no case is the complete observability of the original triple (h, A, b) required; in fact, complete controllability may be dispensed with as well (see Meyer [1]), but as mentioned in Section 1 of Chapter II, no apparent gain in generality is obtained in our particular applications by dropping the use of the phase variable canonical form.

5. Positive Real Functions

Since the concept of positive real functions of a complex variable is central to the development of frequency domain stability criteria, a formal definition of this property and a discussion of some of its ramifications are in order. The points considered here may be found in any text on circuit theory. Two restrictions that are stronger than $H(s) \in \{PR\}$ are also defined and discussed.

DEFINITION PR1. A function $H(s)$ of the complex variable $s = \sigma + i\omega$ is positive real ($H(s) \in \{PR\}$) if

(i) $H(\sigma)$ is real,

(ii) $\operatorname{Re} H(s) \geqslant 0$ for all $\operatorname{Re} s > 0$. □

The adjective "real" indicates the realness of the function for real values of s and the adjective "positive" indicates the positiveness of its real part for values of s with a positive real part. This property is both necessary and sufficient to insure the realizability of $H(s)$ as the driving point impedance or admittance of a linear passive network. This definition and many of the properties of $H(s)$ that are implied by it (some of which are given below) are due to Otto Brune.

(1) The analyticity of $H(s)$ in the open right half-plane (ORHP) follows from condition (ii). If $H(s)$ is rational, this condition implies that no poles or zeros may lie in the ORHP.

(2) If the coefficients of the polynomials are all real, condition (i) is clearly satisfied.

(3) If $H(s)$ has poles on the ω-axis, they must be distinct and have positive residues.

(4) The order of the numerator of H must not differ from the order of the denominator by more than ± 1, that is,

$$k \triangleq \text{order}\,\{\text{Numerator}\} - \text{order}\,\{\text{Denominator}\} \in [-1, 1]. \tag{3-32}$$

(5) Define the path Γ_R in the s-plane:

$$\Gamma_R \triangleq \{s\colon\ \ s = i\omega, \quad -R \leqslant \omega \leqslant +R; \quad s = Re^{i\theta}, \quad -\pi/2 \leqslant \theta \leqslant +\pi/2\}$$

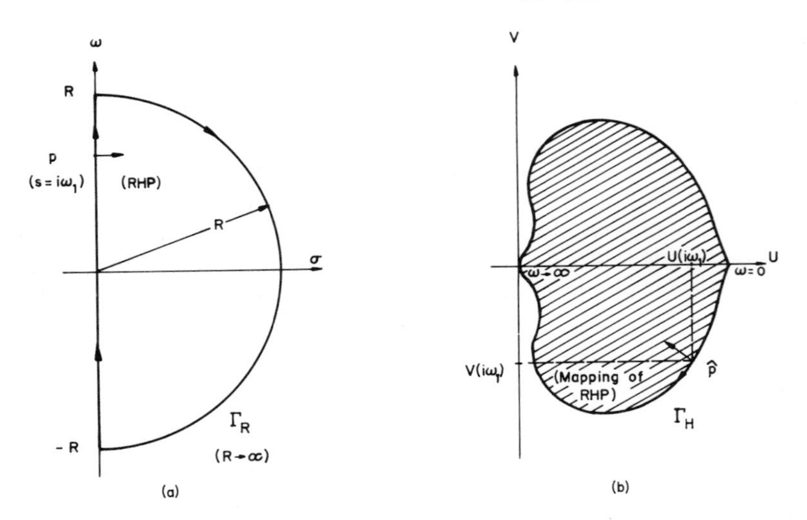

Fig. 3-1. The mapping properties of $H(s)$: (a) $s = (\sigma + i\omega)$-plane; (b) $H = (U + iV)$-plane.

(Fig. 3-1a) which is traversed in the sense shown by the arrows. The interior is always on the "right-hand side" of a point traveling along $\boldsymbol{\Gamma}_R$. In the limit as $R \longrightarrow \infty$, the curve $\boldsymbol{\Gamma}_\infty$ encloses the entire ORHP. Figure 3-1b shows the mapping of the path $\boldsymbol{\Gamma}_\infty$ into the H-plane for a typical $H(s) \in \{PR\}$; the ω-axis transforms to $\boldsymbol{\Gamma}_H$ and the "infinite semicircle" maps to the origin in this case.

Since the mapping is conformal, the interior of $\boldsymbol{\Gamma}_R$ (the ORHP, that is, the area to the right-hand side of p as the point traverses $\boldsymbol{\Gamma}_R$) maps into the area to the right of the point $\hat{p}$ (the image of point p) as $\hat{p}$ traverses $\boldsymbol{\Gamma}_H$ as shown in Fig. 3-1. Since $H(s)$ is assumed to be analytic within $\boldsymbol{\Gamma}_R$, the maximum and minimum values of $H(s)$ for s in the closed RHP must occur on the boundary $\boldsymbol{\Gamma}_H$; hence only $H(s)$ for $s \in \boldsymbol{\Gamma}_R$ and R arbitrarily large has to be inspected to determine if condition (ii) is satisfied.

The behavior of $\boldsymbol{\Gamma}_H$ for s on the infinite semicircle depends upon k, the difference between numerator and denominator order [Eq. (3-32)] as shown in Fig. 3-2. In the cases $k = 0$ and $k = -1$, the limit as $R \longrightarrow \infty$ is formally the point H_∞ indicated, while if $k = +1$, the infinite semicircle in the s-plane maps into an infinite semicircle in the H-plane (Fig. 3-2a). Since we always consider $\boldsymbol{\Gamma}_\infty$ if $k = -1$ or 0 and $\boldsymbol{\Gamma}_R$ if $k = +1$, we can use the notation $\boldsymbol{\Gamma}$ with no ambiguity.

(6) If $H(s) \in \{PR\}$, then $H^{-1}(s) \in \{PR\}$ also.

From the preceding discussion of the properties of $H(s) \in \{PR\}$, we arrive at the following alternative definition for positive realness which is found to be more useful.

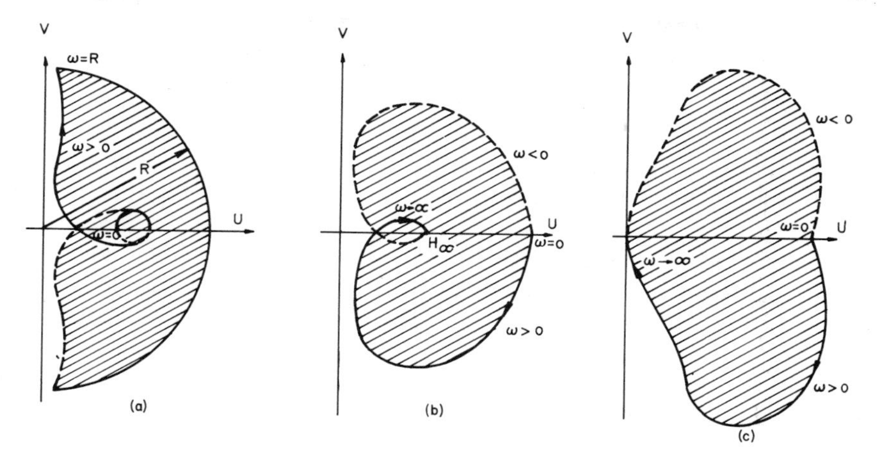

Fig. 3-2. The mapping of the infinite semicircle: (a) $k = +1$; (b) $k = 0$; (c) $k = -1$.

DEFINITION PR1′. A rational function $H(s)$ with real coefficients satisfies $H(s) \in \{PR\}$ if

(a) It is analytic in the ORHP,
(b) The mapping $\Gamma_H \triangleq H(s \in \Gamma)$ lies in the half plane $\operatorname{Re} H \geqslant 0$. □

To test a given rational function for positive realness, it is most advantageous to first determine whether or not the poles and zeros of $H(s)$ lie in the ORHP. If any coefficients of either polynomial are negative, then the function is immediately not positive real. If this simple test is passed, there are standard procedures due to Routh and Hurwitz (see Chapter IV, Section 2) which may be used to check for ω-axis and ORHP poles and zeros. Only after these points have been considered should the mapping Γ_H be investigated.

A stronger measure of passivity is required for Lemma 3; it is provided in this context by the following recently proposed definition of strictly positive real functions (Taylor [1]).

DEFINITION PR2. A function $H(s)$ is strictly positive real ($H(s) \in \{SPR\}$) if $H(s - \varepsilon) \in \{PR\}$ for some real $\varepsilon > 0$. □

An important property that a strictly positive real function exhibits in relation to the LKY Lemma is that the real part of a strictly positive real function cannot go to zero more rapidly than ω^{-2} as ω goes to infinity. To prove this, we need consider only functions which have one more pole than zero; within a constant multiplier, we have

$$H(s) = \frac{n(s)}{d(s)} = \frac{s^{n-1} + b_{n-1}s^{n-2} + \cdots + b_2 s + b_1}{s^n + a_n s^{n-1} + \cdots + a_2 s + a_1} \in \{SPR\}.$$

Consider $H(s - \varepsilon)$: the leading terms of the numerator and denominator of the real part of $H(i\omega - \varepsilon)$ are found by expansion to be

$$\operatorname{Re} H(i\omega - \varepsilon) = \operatorname{Re} \frac{n(i\omega - \varepsilon)d(-i\omega - \varepsilon)}{d(i\omega - \varepsilon)d(-i\omega - \varepsilon)} = \frac{(a_n - b_{n-1} - \varepsilon)\omega^{2(n-1)} + \cdots}{\omega^{2n} + \cdots}.$$

Thus, for $H(s - \varepsilon)$ to be positive real, we must have $(a_n - b_{n-1}) \geqslant \varepsilon$, which for $\varepsilon > 0$ proves that $\operatorname{Re} H(i\omega)$ goes to zero no more rapidly than ω^{-2} as $\omega \longrightarrow \infty$.

It is worthwhile considering Definition PR2 from the viewpoint of network theory. In terms of realizability, the class of rational functions $\{SPR\}$ may be realized as the driving point impedance or admittance of a network made up of resistance, lossy inductance [L in series with a resistance of value εL yielding the impedance $L(s + \varepsilon)$] and lossy C [C in parallel with conductance of value εC yielding the admittance $C(s + \varepsilon)$]; clearly for any $H(s) \in \{PR\}$, the replacement of every term sL by $(s + \varepsilon)L$ and every term sC by $(s + \varepsilon)C$ results in a network with a strictly positive real impedance and admittance.

The points concerning the asymptotic behavior and realizability of $H(s) \in \{SPR\}$ are best illustrated by an example taken from Guillemin [1]: consider the input impedance of a network made up of two parallel paths, the first a lossy capacitor [C in parallel with G] and the second a lossy inductor [L in series with R]. The normalized impedance is

$$H(s) = \frac{s + b_1}{s^2 + a_2 s + a_1},$$

where

$$C = 1, \qquad L = [a_1 - b_1(a_2 - b_1)]^{-1},$$
$$G = (a_2 - b_1)C, \qquad R = b_1 L.$$

When $b_1 = a_2$, we see that $G = 0$ and [referring to Eq. (3-31)] that $\operatorname{Re} H(i\omega) \longrightarrow 0$ as ω^{-4} when $\omega \longrightarrow \infty$; if $b_1 = 0$ then $R = 0$. In both cases $H(s)$ becomes only positive real and the network which has input impedance $H(s)$ cannot be realized with lossy elements. Only in the second case is $\operatorname{Re} H(i\omega)$ zero for finite ω (at $\omega = 0$); if $b_1 = a_2$ then $\operatorname{Re} H(i\omega) > 0$ for $\omega \in (-\infty, \infty)$, so this condition is not in itself a useful definition of a strictly positive real function in light of the requirements of Lemma 2.

For use in Lemma 5 we need a slightly less strict type of positive real functions, namely, those that have a single pole at $s = 0$. For this purpose, we define a third class $\{SPR_0\}$.

DEFINITION PR3. A rational function

$$H(s) = \frac{b_m s^{m-1} + \cdots + b_2 s + b_1}{s^n + a_n s^{n-1} + \cdots + a_2 s} \triangleq \frac{\chi(s)}{\phi(s)}, \qquad n \leqslant m \leqslant (n + 2)$$

is strictly positive real in the particular case of one pole at $s = 0$ ($H(s) \in$

$\{SPR_0\}$) if

$$\hat{H}(s) \triangleq \frac{\chi(s-\varepsilon)}{\phi(s-\varepsilon)-\phi(-\varepsilon)} \in \{\mathrm{PR}\}$$

for some $\varepsilon > 0$. □

By subtracting the denominator factor $[(-\varepsilon)^n + a_n(-\varepsilon)^{n-1} + \cdots + a_2(-\varepsilon)]$, we ensure that $\hat{H}(s)$ retains the pole at $s = 0$; otherwise the last or constant term in the denominator polynomial would be negative for ε arbitrarily small and $H(s)$ could never be positive real; by the same reasoning we must have $b_1 > 0$. This definition thus ensures that $H(s)$ satisfies the conditions Re $H(i\omega) > 0$ and that Re $H(i\omega)$ cannot go to zero more rapidly than ω^{-2} as $\omega \longrightarrow \infty$, and that $k_1 > 0$ in $H(s)$ [Eq. (3-7)] as required in the proof of Lemma 5. In the standard second-order example,

$$H(s) = \frac{s+b_1}{s^2+a_2 s} \in \{SPR_0\}$$

if

$$\hat{H}(s) = \frac{s+(b_1-\varepsilon)}{s^2+(a_2-2\varepsilon)s} \in \{PR\}$$

for some $\varepsilon > 0$, that is, if $0 < b_1 < a_2$.

6. Existence Theorems

A basic question in the theory of Lyapunov's direct method is the existence of a specific form of Lyapunov function as a necessary and sufficient condition for the stability of the null solution of a differential equation of a given structure. Many theorems establishing results of this sort have been derived; we are interested in reviewing those few that are relevant to the class of problems defined in Chapter II.

As a rule, it may be said that as the system under consideration becomes more general, the Lyapunov function that is guaranteed to exist becomes less well defined. This is an inevitable state of affairs that renders such theorems less useful in most specific applications. The theorems given below cover a wide range with respect to system generality and the corresponding class of Lyapunov functions.

In the first instance (Theorem C) a system formulation is assumed that is sufficiently broad that nearly all the problems of this book fall within its ambit.

THEOREM C (Massera [1]). Given a system $\dot{x} = f(x, t)$, where f satisfies the local Lipschitz condition (Chapter I, Section 1) with respect to

x and $f(0, t) \equiv 0$ for all t; a necessary and sufficient condition that the equilibrium $x \equiv 0$ is uniformly asymptotically stable in the whole is that there exists a Lyapunov function $v(x, t)$ (per Theorem A) that possesses bounded partial derivatives with respect to x and t of arbitrarily high order. Moreover, if $f(x, t)$ satisfies a uniform Lipschitz condition, the partial derivatives are bounded uniformly with respect to time, and if $f(x, t)$ is periodic or time invariant, v satisfies the same condition. □

Clearly this powerful result provides virtually no guidance in the actual choice of a Lyapunov function candidate.

Next we restrict our attention to linear time-varying situations. As might be expected, this greatly reduced generality allows us to consider a much narrower class of Lyapunov functions: the time-varying quadratic form. For LTV systems, two important theorems exist.

THEOREM D (Lyapunov–Perron–Malkin; see Kalman and Bertram [1]). Given a system $\dot{x} = A(t)x$, where $\|A(t)\|$ is bounded above for all t; a necessary and sufficient condition that the equilibrium state $x \equiv 0$ is uniformly asymptotically stable is that for any bounded positive definite matrix $Q(t)$ (that is

$$0 < \varepsilon x^{\mathrm{T}}x \leqslant x^{\mathrm{T}}Q(t)x \leqslant \delta x^{\mathrm{T}}x < \infty$$

for all t and finite nonzero x where ε and δ are both real and positive) that is continuous in t, the scalar function

$$v(x, t) \triangleq \int_t^{\infty} x_0{}^{\mathrm{T}}\Phi^{\mathrm{T}}(t, \tau)Q(\tau)\Phi(t, \tau)x_0 \, d\tau$$

exists† and is a Lyapunov function in the sense of Theorem A. □

Asymptotic stability implies asymptotic stability in the whole, due to linearity, as shown in Fig. 2-6.

THEOREM E (Krasovskii [1]). Given a system $\dot{x} = A(t)x$ where the $a_{ij}(t)$ are bounded continuous functions of time; a necessary and sufficient condition that the equilibrium $x \equiv 0$ is uniformly asymptotically stable is that $P(t)$ exists such that

(a) $v = x^{\mathrm{T}}P(t)x$ is positive definite;
(b) $\dot{v} = -x^{\mathrm{T}}Q(t)x = x^{\mathrm{T}}(A^{\mathrm{T}}P + PA + \dot{P})x \leqslant p(t)v$, where $\lim_{T\to\infty} \int_{t_0}^{T} p(t)\, dt = -\infty$ uniformly with respect to t_0. □

The first of these theorems guarantees the existence of $x^{\mathrm{T}}P(t)x$ in the sense of Theorem A (Lyapunov), while the second assures it in the sense of Theorem

† $\Phi(t, \tau)$ in v is known as the transition matrix of the system; it satisfies the matrix differential equation $d\Phi/dt = A(t)\Phi$ subject to $\Phi(\tau, \tau) = I$; thus, the solution of $\dot{x} = A(t)x$ subject to $x(t_0) = x_0$ is $x(t; x_0, t_0) = \Phi(t, t_0)x_0$.

B (Corduneanu). It is unfortunate from our point of view that neither theorem provides an effective method for generating $P(t)$. The first form involves the transition matrix $\Phi(t, \tau)$, which is unknown, and the second is specified only by the matrix differential inequality

$$dP/dt \leqslant p(t)P(t) - (A^{T}(t)P(t) + P(t)A(t)),$$

where $p(t)$ is unknown but subject to the condition that $\lim_{T\to\infty} \int_{t_0}^{T} p(t)\, dt = -\infty$ uniformly with respect to t_0.

Finally, when we consider LTI systems, a practically useful theorem is available which provides not only a specific form, but also a method of generation.

THEOREM F (Lyapunov; see LaSalle and Lefschetz [1]). Given an LTI system $\dot{x} = Ax$; the equilibrium $x \equiv 0$ is asymptotically stable if and only if for any $Q = Q^{T} > 0$ there exists a matrix $P = P^{T} > 0$ such that

$$A^{T}P + PA = -Q. \quad \square \tag{3-33}$$

Since the system is time invariant, this also implies uniform asymptotic stability in the whole.

If one chooses a simple form for Q (usually, $Q = I$), it is a relatively straightforward matter to solve Eq. (3-33) to obtain the $\frac{1}{2}n(n + 1)$ parameters of the symmetric matrix P. The test for asymptotic stability then entails showing that $P > 0$, that is, that $\mu_1 \triangleq p_{11} > 0$, $\mu_2 \triangleq (p_{11}p_{22} - p_{12}^2) > 0$, $\ldots$, $\mu_n \triangleq |P| > 0$, as mentioned following Lemma V1 (Section 2).

An interesting corollary to this theorem due to Kalman provides an example of a situation where the rate of decrease of v expressed in terms of $\dot{v}/v$ can be related to the eigenvalues of the system.

COROLLARY F1 (Kalman; see Kalman and Bertram [1]). Given an LTI system $\dot{x} = Ax$; the real parts of the roots of the characteristic equation $|\lambda I - A| = 0$ are all strictly less than $(-\mu)$ if and only if for any $Q = Q^{T} > 0$ there exists a matrix $P = P^{T} > 0$ such that

$$A^{T}P + PA = -Q - 2\mu P;$$

μ may be positive (asymptotically stable systems) or negative (exponentially unstable systems). $\square$

Note that the above relation demonstrates that for $v = x^{T}Px$, the corresponding time derivative is

$$\dot{v} = -x^{T}Qx - 2\mu x^{T}Px < -2\mu v;$$

since v satisfies the differential inequality $\dot{v} < -2\mu v$, the preliminary comment regarding the rate of variation of v and the eigenvalues of A is verified. It should also be noted that the matrix equation of this corollary is intimately related to Eq. (3-9a) of Lemma 4, and that the above comment is prompted by Theorem 2.

A second corollary to Theorem F is useful when, for computational simplicity, a negative semidefinite form $-x^{\mathrm{T}}Qx$, $Q = Q^{\mathrm{T}} \geqslant 0$ is used as $\dot{v}$ in accordance with Corollary A1.

COROLLARY F2. Given an LTI system $\dot{x} = Ax$; the equilibrium $x \equiv 0$ is asymptotically stable if and only if for any $Q = Q^{\mathrm{T}} \geqslant 0$ such that $x^{\mathrm{T}}Qx \equiv 0$ can occur along no trajectory other than $x \equiv 0$, there exists a matrix $P = P^{\mathrm{T}} > 0$ such that

$$A^{\mathrm{T}}P + PA = -Q. \square \tag{3-33}$$

Since Theorem F and its corollaries are of quite general significance in the study of absolute stability, proofs are in order. Theorem F and Corollary F1 both follow directly from the proof of Corollary F2, so this result is considered first.

Sufficiency: Assume that some positive definite symmetric matrix $P = P^{\mathrm{T}} > 0$ exists so that the matrix Q of Eq. (3-33) is symmetric and positive semidefinite ($Q = Q^{\mathrm{T}} \geqslant 0$; the symmetricity follows directly from $P = P^{\mathrm{T}}$). Use as a Lyapunov function candidate $v(x) = x^{\mathrm{T}}Px$; by Lemma V1, it is valid. For the differential equation $\dot{x} = Ax$, the total time derivative of $v(x)$ is

$$\dot{v} = x^{\mathrm{T}}P(Ax) + (Ax)^{\mathrm{T}}Px = x^{\mathrm{T}}(A^{\mathrm{T}}P + PA)x \triangleq -x^{\mathrm{T}}Qx,$$

which is postulated to satisfy condition (A1: v) of Corollary A1, which is a sufficient condition for uniform asymptotic stability in the whole, that is, $A \in \{A_1\}$.

Necessity: Assume that $A \in \{A_1\}$. The solutions to $\dot{x} = Ax$ are

$$x(t; x_0, t_0) = \Phi(t - t_0)x_0 \triangleq \exp[A(t - t_0)]x_0,$$

where Φ is the transition matrix which for LTI systems is a function of $(t - t_0)$ and is expressed in terms of the matrix exponential function

$$\exp(At) \triangleq L^{-1}[(sI - A)^{-1}] \triangleq \sum_{k=0}^{\infty} (k!)^{-1}A^k t^k.$$

This solution satisfies $\lim_{t\to\infty} x(t; x_0, t_0) = 0$ uniformly with respect to x_0 and t_0, by definition; also, $x(t; x_0, t_0) \equiv 0$ if and only if $x_0 = 0$.

Consider a matrix $Q = Q^{\mathrm{T}} \geqslant 0$, which for solutions to $\dot{x} = Ax$ satisfies the condition that $x^{\mathrm{T}}Qx \equiv 0$ only if $x \equiv 0$. Consider the symmetric matrix

$$\begin{aligned} P(t) &\triangleq \int_t^{\infty} \exp[A^{\mathrm{T}}(\tau - t)]Q \exp[A(\tau - t)]\, d\tau, \\ &= \exp(-A^{\mathrm{T}}t) \left\{ \int_t^{\infty} \exp(A^{\mathrm{T}}\tau)Q \exp(A\tau)\, d\tau \right\} \exp(-At), \end{aligned} \tag{3-34}$$

which must exist since $A \in \{A_1\}$. Since $\Psi \triangleq \exp(Bt)$ satisfies the matrix differential equation $\dot{\Psi} = B\Psi$ for any B, and the matrices B and $\exp(Bt)$

commute, that is,

$$B\exp(Bt) \triangleq (B)\sum_{k=0}^{\infty} k!^{-1}B^k t^k = \left[\sum_{k=0}^{\infty} k!^{-1}B^k t^k\right](B) \triangleq [\exp(Bt)]B,$$

direct differentiation of Eq. (3-34) yields

$$dP/dt = -Q - A^{\mathrm{T}}P - PA.$$

Making the change of variable $\zeta \triangleq \tau - t$ in Eq. (3-34) yields

$$P = \int_0^{\infty} \exp(A^{\mathrm{T}}\zeta)Q\exp(A\zeta)\,d\zeta, \tag{3-35}$$

which shows that P is a constant matrix or $dP/dt \equiv 0$. Hence, the matrix defined by Eq. (3-34) satisfies the relation

$$A^{\mathrm{T}}P + PA = -Q.$$

Consider the quadratic form $x^{\mathrm{T}}Px$ for $x = \exp[A(t - t_0)]x_0$, that is,

$$x^{\mathrm{T}}Px = \int_t^{\infty} \{\exp[A(\tau - t_0)]x_0\}^{\mathrm{T}}Q\exp[A(\tau - t_0)]x_0\,d\tau$$

from Eq. (3-34). The integrand is $x^{\mathrm{T}}(\tau;\, x_0, t_0)Qx(\tau;\, x_0, t_0) \geqslant 0$, which is not identically zero unless $x \equiv 0$, so the integral must be strictly positive unless $x \equiv 0$, which proves that the matrix P is positive definite. □

In proving Theorem F, the integrand $x^{\mathrm{T}}(\tau;\, x_0, t_0)Qx(\tau;\, x_0, t_0)$ is strictly positive, which does not affect the proof. Corollary F1 follows directly by making the substitution $\hat{A} \triangleq A + \mu I$ in Theorem F.

IV

LINEAR TIME-INVARIANT SYSTEMS AND ABSOLUTE STABILITY

The most general LTI system that may be obtained from the specified system model [Eq. (2-6)] is

$$\dot{x} = Ax + b\tau, \qquad \sigma_0 = h^{\mathrm{T}}x + \rho\tau, \qquad \tau = -\kappa\sigma_0, \tag{4-1}$$

or, equivalently,

$$L(\sigma_0)/L(\tau) \triangleq W(s) = \rho + h^{\mathrm{T}}(sI - A)^{-1}b \triangleq \rho + n(s)/d(s),$$
$$\tau = -\kappa\sigma_0$$

in our standard feedback control system configuration where $n(s)$ is of lower order than $d(s)$. In differential equation form the system can be reduced to

$$\dot{x} = Ax - b[\kappa/(1 + \rho\kappa)]h^{\mathrm{T}}x \triangleq A_\kappa x \tag{4-2}$$

by eliminating the variables σ_0 and τ. If $\kappa = -1/\rho$, then we clearly have a degenerate situation, since the matrix A_κ has elements that are unbounded. If we consider the total closed loop transfer function (Fig. 2-1)

$$\frac{L(\sigma_0)}{L(v)} \triangleq W_\kappa(s) \triangleq \frac{W(s)}{1 + \kappa W(s)} = \frac{n(s) + \rho d(s)}{(1 + \rho\kappa)d(s) + \kappa n(s)},$$

we also note that W_κ for $\kappa = -1/\rho$ has more zeros than poles, that is, $W_\kappa(s)$ is not proper. We specifically preclude this situation by defining Eq. (4-1) to be asymptotically stable if $A_\kappa \in \{A_1\}$ for some value or values of

κ (Chapter II, Section 1). Since we can always express Eq. (4-2) in the form

$$\dot{x} = [A - \kappa bh^{\mathrm{T}}]x \tag{4-3}$$

under these circumstances, that is, we can always take $\rho = 0$, we only consider this simpler case in this chapter.

Determining the stability properties of the equilibrium $x \equiv 0$ of LTI systems of this form is the only general problem of the sort considered in this study for which useful necessary and sufficient conditions are known. Not only is this case far more tractable than that involving nonlinear and/or time-varying parameters, but many applied mathematicians, most notably Hurwitz, Routh and Nyquist, have dealt extensively with this question. The techniques of stability analysis that were developed involve the generation of conditions that are necessary and sufficient to guarantee that all roots λ_i of the characteristic equation

$$p(\lambda) \triangleq |\lambda I - A + \kappa bh^{\mathrm{T}}| = 0$$

(eigenvalues of $A - \kappa bh^{\mathrm{T}}$) have negative real parts, that is, that the characteristic polynomial $p(\lambda)$ is Hurwitz. This in turn is necessary and sufficient for the asymptotic stability of the equilibrium $x \equiv 0$.

1. Relations between Linear Time-Invariant and Nonlinear Time-Varying Systems

Despite the existence of classical solutions to this question, there is a compelling reason for the inclusion of this problem in the present work: the stability analysis of an NLTV system is fundamentally and inextricably related to the analogous LTI case. Virtually every technique for treating an NLTV system is an extension of a method for LTI system analysis, and the resultant stability criterion for the NLTV system is always measured against the equivalent result for the LTI case. The following points should serve to elucidate this relationship further.

A. *Principle of Stability in the First Approximation*

Let us consider the uniform asymptotic stability of the equilibrium $x \equiv 0$ of the nonlinear time-varying differential equation

$$\dot{x} = f(x, t), \qquad f \in \{S\}, \tag{4-4}$$

where we assume that $f(x, t)$ possesses a power series expansion that is valid in some neighborhood of $x = 0$;

$$\dot{x} = A(t)x + f^{(2)}(x, t),$$

where $f^{(2)}$ contains terms that are of at least second order in x. This implies that $f(x, t)$ in Eq. (4-4) satisfies a local Lipschitz condition (Section 1 of Chapter I). Following the terminology of Hahn [1],

$$\dot{x} = A(t)x \tag{4-5}$$

is called the reduced differential equation and the terms $f^{(2)}(x, t)$ are perturbation terms. We then can use the following result.

THEOREM (Hahn [1]). If the equilibrium of the reduced system (4-5) is uniformly asymptotically stable, then so is the equilibrium $x \equiv 0$ of Eq. (4-4). □

The special case of determining the stability of an NLTI system using linearization by Taylor series expansion is due to Lyapunov.

The proof of this theorem is based upon the existence of a quadratic Lyapunov function

$$v(x, t) = x^{\mathrm{T}} P_0(t) x \geqslant \eta x^{\mathrm{T}} x, \qquad \eta > 0 \tag{4-6}$$

for the reduced system. If $A(t)$ is bounded and $Q(t)$ is chosen so that positive constants ε and δ exist such that

$$0 < \varepsilon x^{\mathrm{T}} x \leqslant x^{\mathrm{T}} Q_0(t) x \leqslant \delta x^{\mathrm{T}} x$$

for all $x \neq 0$ and t, then by the existence theorem D (Chapter III, Section 6), the function $v(x, t)$ of Eq. (4-6) exists and $P_0(t)$ satisfies

$$\dot{v} = x^{\mathrm{T}}[A^{\mathrm{T}}(t)P_0(t) + P_0(t)A(t) + (d/dt)P_0(t)]x \triangleq -x^{\mathrm{T}} Q_0 x \tag{4-7}$$

if and only if $\dot{x} = A(t)x$ is uniformly asymptotically stable. If the same function (4-6) is used as a candidate for the perturbed differential equation (4-4), then

$$\dot{v} = -x^{\mathrm{T}} Q_0(t) x + w(x, t), \tag{4-8}$$

where $w(x, t)$ is at least of order three with respect to x. For sufficiently small x, the quadratic negative definite terms of $\dot{v}$ dominate the higher order terms represented by $w(x, t)$, and hence guarantee the negative definiteness of $\dot{v}$ in some neighborhood of $x \equiv 0$.

This outline of the proof of the theorem allows us to see the importance of the linearized system and the existence of the corresponding quadratic Lyapunov function. However, from the viewpoint of our goals, this method of stability analysis is of no avail: The region of uniform asymptotic stability is not global with respect to x. It is generally very difficult to even estimate how small $\|x_0\|$ must be so that $x(t; x_0, t_0)$ is a uniformly asymptotically stable solution. The type of stability ensured by this technique is often called infinitesimal stability for this reason.

B. *The Conjectures of Aizerman and Kalman*

The possibility of the validity of the Aizerman conjecture (refer to Section 6 of Chapter I) provided great impetus to the study of absolute stability. In its most general form, due to Krasovskii [1], this surmise may be stated in the following manner.

Consider the system defined by

$$\dot{x} = Ax + BF(x), \qquad F(x) \triangleq [f_{ij}(x_j)]; \tag{4-9}$$

the corresponding linear system is

$$\dot{x} = Ax + BHx. \tag{4-10}$$

Assume that Eq. (4-10) is asymptotically stable for $\underline{h}_{ij} < h_{ij} < \bar{h}_{ij}$. These parameter ranges may be determined by the application of the Hurwitz conditions, Section 2, for example. The generalized Aizerman conjecture, then, is that the system of nonlinear differential equations (4-9) is equiasymptotically stable in the whole if

$$\underline{h}_{ij} < f_{ij}(x_j)/x_j < \bar{h}_{ij}. \tag{4-11}$$

This conjecture does not present a linearization procedure as in *A*; rather than replacing $f_{ij}(x_j)$ with a single linear term $h_{ij}x_j$, it considers *all* linear gains having the same range as $f_{ij}(x_j)/x_j$.

Even in the case of second-order systems, this conjecture has proven to be false in the context of the stability analysis to be undertaken in this book; the Hurwitz inequalities alone cannot guarantee equiasymptotic stability in the whole. Extensive study of the second order system

$$\dot{x}_1 = a_{11}(x_1)x_1 + a_{12}(x_2)x_2, \qquad \dot{x}_2 = a_{21}(x_1)x_1 + a_{22}(x_2)x_2 \tag{4-12}$$

was undertaken by Malkin [1], Erugin [1], Pliss [1], and Krasovskii [1] in the early 1950's. The characteristic equation and resulting Hurwitz inequalities for the corresponding LTI system are

$$\lambda^2 - (a_{11} + a_{22})\lambda + (a_{11}a_{22} - a_{12}a_{21}) = 0 \tag{4-13}$$

and

$$\text{(i)} \quad a_{11} + a_{22} < 0, \qquad \text{(ii)} \quad a_{11}a_{22} - a_{12}a_{21} > 0.$$

The latter conditions are necessary and sufficient for asymptotic stability in the LTI system.

The first complete investigation of Eq. (4-12) treated the special case where only $a_{11}(x_1)$ is nonlinear and a_{12}, a_{21}, and a_{22} are constants. Malkin and Erugin were able to show that the Hurwitz inequalities ensure the equiasymptotic stability of solutions near the origin for this case; however, it is further necessary to require either that the integral

$$I(x) \triangleq \int_0^x [a_{11}(\zeta)a_{22} - a_{12}a_{21}]\zeta \, d\zeta$$

diverge ($\lim_{x\to\infty} I(x) = \infty$), or that $a_{22}^2 + a_{12}a_{21} \neq 0$, in order to be certain that the system is equiasymptotically stable in the whole. Krasovskii generated a counterexample to the Aizerman conjecture by violating both of these conditions: consider

$$A(x) = \begin{bmatrix} g(x_1) & +1 \\ -1 & -1 \end{bmatrix},$$

where

$$g(x_1) \triangleq \begin{cases} 1 - e^{-2}/(1 + e^{-1}) = 0.9011, & x_1 < 1, \\ 1 - e^{-2x_1}/[x_1(1 + e^{-x_1})], & x_1 \geqslant 1. \end{cases}$$

The range of $g(x_1)$ is $[0.9011, 1)$, which is within the Hurwitz range $(-\infty, 1)$. The solution corresponding to the initial conditions $x_1(0) = 1$, $x_2(0) = (e^{-1} - 1)$ is found from the transcendental equations

$$x_1 + e^{x_1} = t + 1 + e, \qquad x_2 = e^{-x_1} - x_1,$$

which demonstrate that both $|x_1|$ and $|x_2|$ grow without bound.

This solution satisfies the condition $x_1 > 1$ for $t > 0$, so it is also a solution to the same differential equation with $g(x_1)$ replaced by $\hat{g}(x_1)$,

$$\hat{g}(x_1) \triangleq \begin{cases} 1 - \dfrac{e^{-2}}{1 + e^{-1}} = 0.9011, & |x_1| < 1, \\ 1 - \dfrac{e^{-2|x_1|}}{|x_1|(1 + e^{-|x_1|})}, & |x_1| \geqslant 1, \end{cases}$$

where the nonlinearity $\hat{g}(x_1)$ has been made symmetric $[\hat{g}(x_1) = \hat{g}(-x_1)]$. To place this system in the context of the system description adopted in this book, make the nonsingular transformation

$$y = \begin{bmatrix} 0 & +1 \\ -1 & -1 \end{bmatrix} x$$

to obtain

$$\dot{y} = \begin{bmatrix} 0 & 1 \\ -1 & -1 \end{bmatrix} y - \begin{bmatrix} 0 \\ 1 \end{bmatrix} \hat{f}(\sigma_0), \qquad \sigma_0 = [-1 \quad -1]y,$$

$$\hat{f}(\sigma_0) = \begin{cases} \left(1 - \dfrac{e^{-2}}{1 + e^{-1}}\right)\sigma_0 = 0.9011\sigma_0, & |\sigma_0| < 1, \\ \left(1 - \dfrac{e^{-2|\sigma_0|}}{|\sigma_0|(1 + e^{-|\sigma_0|})}\right)\sigma_0, & |\sigma_0| \geqslant 1. \end{cases} \tag{4-14}$$

The forward path or plant is represented by $W(s) = -(s + 1)/(s^2 + s + 1)$;

this system is shown in Fig. 4-1. Although the Nyquist (Hurwitz) range ensuring asymptotic stability for the linearized system is $(-\infty, +1)$, the fact that $\hat{f}(\sigma_0)/\sigma_0 \in [0.9011, 1.0)$ for all finite σ_0 does not imply equiasymptotic stability in the whole for this system.

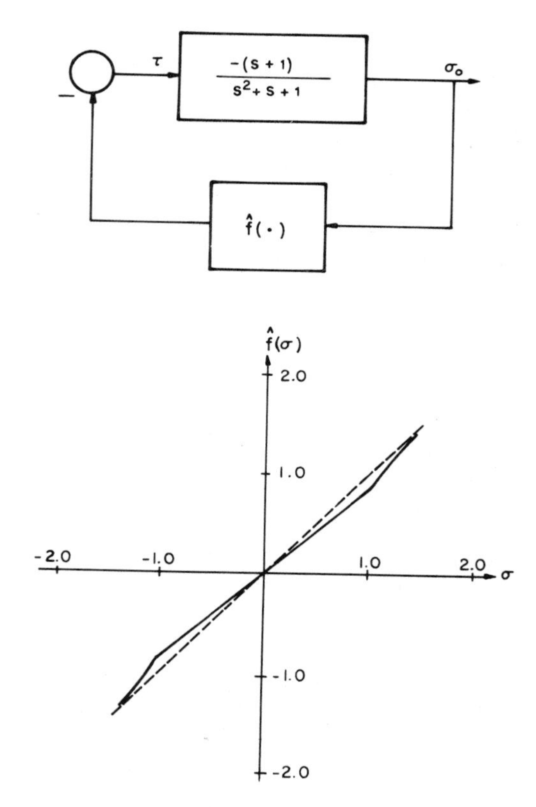

Fig. 4-1. A counterexample to Aizerman's conjecture.

For $n \geqslant 3$ the Aizerman conjecture was first studied by Pliss [1] and more recently by Dewey and Jury [1]. Although many special cases have been found where the conjecture is valid (see Bergen and Willems [1], for example), it is not true even locally in general. Pliss studied the system corresponding to

$$W(s) = (s^2 - \beta)/((s^2 + 1)(s + \alpha))$$

in our notation and developed conditions on $f(\sigma_0)$ lying in the Hurwitz range $(0, \alpha/\beta)$ that are sufficient for the existence of periodic solutions. Dewey and Jury simulated a specific system for which $\beta = \frac{1}{2}\alpha^2$, so the Hurwitz range is $(0, 2/\alpha)$. Choosing a nonlinearity that was within the Hurwitz range, they obtained a system that exhibited sustained oscillations, thus providing a concrete example of a system that violated the Aizerman conjecture as had been predicted by Pliss†.

A similar conjecture of Kalman [1] replaces the conditions (4-11) with the stronger restrictions

$$\underline{h}_{ij} < df_{ij}(x_j)/dx_j < \bar{h}_{ij}.$$

By constraining the slope of each nonlinear function to lie in the Hurwitz range of the LTI system (4-10), a broad class of functions f_{ij} is excluded. Neither the nonlinearities of Krasovskii nor of Dewey and Jury satisfy this condition; hence, those examples are not counterexamples to the Kalman conjecture.

The Kalman conjecture has, however, been disproven experimentally by the generation of counterexamples. Fitts [1] simulated the fourth-order nonlinear system described by

$$W(s) = \frac{s(s+\alpha)}{[(s+\beta)^2 + (0.9)^2][(s+\beta)^2 + (1.1)^2]} \qquad (4\text{-}15)$$
$$f(\sigma_0) = \kappa\sigma_0^{\ 3}$$

on an analogue computer where α was chosen to be sufficiently small so that the Hurwitz range was $[0, \infty)$. Both the range and slope of the nonlinearity satisfy the Hurwitz inequalities, yet by the proper choice of the parameter κ it was found to be possible to generate sustained oscillations (limit cycles) in the system, invalidating the Kalman conjecture.

If either of these conjectures were correct, then the absolute stability problem would be of little significance. We would have a method of analysis which is quite simply implemented [replacing a nonlinearity $f(\sigma_0)$ by linear functions $\kappa\sigma_0$ where the linear system that results must be asymptotically stable for $\kappa \in [\underline{F}, \bar{F}]$ if $\underline{F} \leqslant f(\sigma_0)/\sigma_0 \leqslant \bar{F}$ in the case of the Aizerman conjecture or for $\kappa \in [\underline{M}, \bar{M}]$ if $\underline{M} \leqslant df(\sigma_0)/d\sigma_0 \leqslant \bar{M}$ for the Kalman conjecture]. That neither conjecture is valid does very little to detract from the point that the stability analysis of LTI systems has evidently exerted a strong influence on the treatment of NLTI systems.

C. *The Describing Function Method*

The describing function method is well established as an approximate method of considerable utility in explaining quite general questions about the behavior of nonlinear systems. In this approach to nonlinear systems analysis, nonlinearities are replaced with quasilinear approximating functions,

† The stability range predicted by the Popov criterion (GSC 1, Chapter V) for this problem is $[\varepsilon, \alpha^{-1}]$. In the example chosen by Dewey and Jury, the nonlinearity lies outside the Popov range but within the Hurwitz range for small values of σ_0.

called describing functions, which approximately represent their transfer characteristics. In view of the power and generality of this linearization technique as a valuable aid to the design and analysis of nonlinear systems, a comparison with rigorous methods of stability analysis is in order.

The technique suffers from several limitations, the most fundamental of which is that the form of the signal at the input of the nonlinearity must be guessed in advance. For example, in the sinusoidal-input describing function method applied to a feedback system with a single nonlinear element, only the fundamental component of the output of the nonlinearity is taken into account, since it is assumed that harmonic components are small and that the linear elements of the system further attenuate them to such an extent that the input signal to the nonlinearity is very nearly sinusoidal. This assumption obviously implies that the results obtained are only approximate and that additional means must be sought to determine the accuracy of the results.

As far as our study of absolute stability is concerned, we are principally interested in pointing out that the describing function method is not a rigorous method of stability analysis, so we confine our attention to two specific examples in which the describing function method would seem to predict asymptotic stability in the whole, while the systems are actually unstable, with one of them permitting the existence of limit cycle oscillations.

A closed loop system of the standard form may consist of an LTI plant $W(s)$ in the forward path and an odd (symmetric) single-valued nonlinearity $f(\sigma_0)$ in the feedback path (Fig. 2-1). If the input to the nonlinearity is of the form

$$\sigma_0 = \eta \sin \omega t,$$

the output is of the form $f(\eta \sin \omega t)$, and can be expressed by a Fourier series expansion.

$$f(\eta \sin \omega t) = \sum_{k=1}^{\infty} a_k \sin(k\omega t).$$

The fact that $f(\sigma_0)$ is single valued precludes the generation of a cosine component, as the output is always in phase with the input, and since $f(\sigma_0)$ is an odd function, the constant term in the output of $f(\cdot)$ is zero. Assume that the fundamental component of the output has an amplitude $a_1 \triangleq \eta\kappa(\eta)$, that is,

$$\kappa(\eta) \triangleq (\pi\eta)^{-1} \int_0^{2\pi} f(\eta \sin \zeta) \sin \zeta \, d\zeta.$$

If the plant $W(s)$ is such that higher order harmonics at the output of the nonlinearity are all attenuated ($W(s)$ is said to be low pass) and only the fundamental component of the signal is transmitted, it is possible to replace the nonlinear gain with an amplitude sensitive linear gain for purposes of analysis. For the simple example that we are considering, $\kappa(\eta)$ is called the describing function of the nonlinearity. In any analysis of the behavior of the feedback system, all the relevant information is assumed to be conveyed by the describing function $\kappa(\eta)$ and the frequency response $W(i\omega)$ of the linear part of the system.

If a closed-loop linear system with gain $\hat{\kappa}$ permits sustained oscillations, then the frequency response diagram in polar form (the Nyquist diagram; see Section 3 for a more complete discussion) must pass through the point $-1/\hat{\kappa}$ (Fig. 4-2b); oscillations of any amplitude may exist in this system at the frequency ω_0 defined by the intersection of $\Gamma_W(\omega)$ with the point $-1/\hat{\kappa}$. In the describing function approach for a nonlinear feedback system, the locus of points $-1/\kappa(\eta)$ on the real axis is located as in Fig. 4-2c; where this locus intersects $\Gamma_W(\omega)$, oscillations are possible with both amplitude and frequency specified; amplitude is specified by the equivalent gain curve (Fig. 4-2a) and frequency by $\Gamma_W(\omega)$. This represents limit cycle operation, since oscillations of amplitudes other than η_1 and η_2 cannot be sustained in this example.

The stability of each limit cycle is determined by considering small perturbations in amplitude around η_1 and η_2.

(1) *Stability:* If a small positive increment δ moves the operating point outside Γ_W (the inside of Γ_W being defined as the region in the W-plane

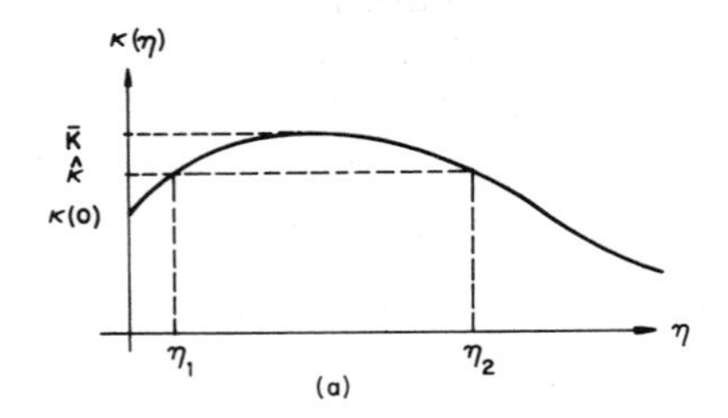

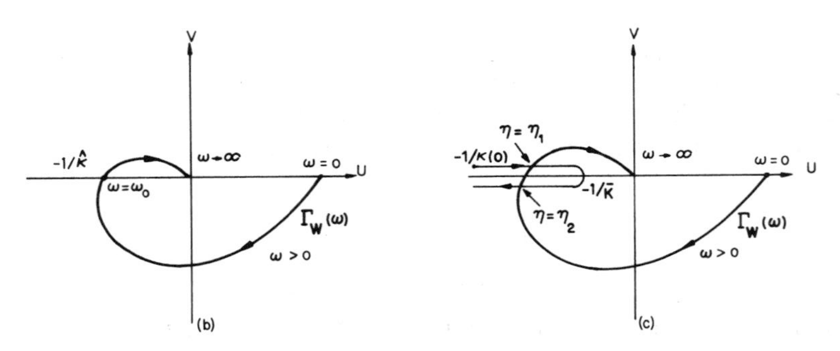

Fig. 4-2. The describing function method: (a) the equivalent gain; (b) an oscillatory linear system; (c) limit cycle operation.

corresponding to the RHP of the s-plane; see Chapter III, Section 5, and Section 3 of this chapter) then by appealing to the Nyquist criterion, the oscillation at $(\eta_i + \delta)$ must decrease in amplitude, moving back toward η_i, and if a negative increment $(-\delta)$ moves the operating point inside Γ_W, the resulting instability causes the oscillation to grow and thus approach η_i from below.

(2) *Instability:* By reversing the above, that is, by having a positive increment in η correspond to a movement along the locus $-1/\kappa(\eta)$ into the inside of Γ_W and a negative increment taking the point $-1/\kappa(\eta)$ outside Γ_W, the limit cycle must be unstable.

In Fig. 4-2 the limit cycles at η_1 and η_2 are unstable and stable respectively.

By continuing the analogy with linear systems analysis further, the following criteria may suggest themselves as informal methods of stability analysis:

(i) If the locus of $-1/\kappa(\eta)$ always lies outside the curve Γ_W, then the system is asymptotically stable in the whole;

(ii) Only if $-1/\kappa(\eta)$ intersects Γ_W in such a way that the locus is outside Γ_W for all large η (say for $\eta > \eta_n$ where η_n is the largest value of η that leads to an intersection with Γ_W) is the system marginally stable or oscillatory;

(iii) If $-1/\kappa(\eta)$ passes inside Γ_W for large η but falls outside Γ_W for small η, local stability is guaranteed;

(iv) If $-1/\kappa(\eta)$ lies inside Γ_W for all $\eta \in [0, \infty)$, the system is unstable for all initial conditions.

In the discussion of the conjectures of Aizerman and Kalman, we have counterexamples which invalidate both (i) and (ii). The system defined by Eq. (4-14), which is not asymptotically stable in the whole, is a counterexample to (*i*); $W(s) = -(s + 1)/(s^2 + s + 1)$ is definitely low-pass, and the nonlinearity is quite regular (Fig. 4-1), having a describing function $\kappa(\eta) \in [0.9011, 1.0)$ by inspection. The plant of system (4-15) is also low-pass, although the nonlinearity is not as well behaved. The equivalent gain is $\kappa(\eta) = \frac{3}{4}\kappa\eta^2 \in [0, \infty)$, which is well defined, however. This system permits the existence of limit cycle operation, in contradiction to (ii).

From the existence of several systems that behave in manners contrary to the expectations based on the describing function approach, it is clear that while this technique is very useful for explaining many observed phenomena, considerable care must be exercised prior to using it as a predictive tool.

D. *The Common Lyapunov Function*

Closely related conceptually to the Aizerman conjecture is the question of the existence of common Lyapunov functions for nonlinear time-varying systems [Eq. (2-6)]. The idea of such functions arose from the following line of reasoning.

For any LTI system

$$\dot{x} = A_0 x - b_0 \kappa_0 h_0^{\mathrm{T}} x \triangleq B_0 x, \tag{4-16}$$

where A_0, b_0, κ_0, and h_0 are specified so that B_0 is an asymptotically stable matrix, it is known that an infinite number of Lyapunov functions exists. For convenience, this infinity of functions may first be subdivided into forms, for example, the quadratic form or the quartic form; then functions of each form may be further categorized by specific parameter values.

The particular form that has been most fully investigated as a common Lyapunov function is the quadratic form, $v(x) = x^{\mathrm{T}}Px$. Each such function may be specified by the set of $m = \frac{1}{2}n(n+1)$ parameters p_{ij} of the symmetric matrix P; hence each matrix may be represented by a point p in an m-dimensional parameter space $\mathcal{P}$. We denote this as

$$P \triangleq [p \in \mathcal{P}].$$

A subset of the space $\mathcal{P}$ is $\mathcal{R}$ ($\mathcal{R} \subset \mathcal{P}$), which defines points in $\mathcal{P}$ that correspond to positive definite matrices, that is points $p \triangleq \{p_{ij}\}$ where the parameters $\mu_k(p_{ij})$ defined in Eq. (3-3) satisfy the given standard conditions for $P = P^{\mathrm{T}} > 0$. A necessary condition that $x^{\mathrm{T}}Px$ be a Lyapunov function for $\dot{x} = B_0 x$ is that $P \triangleq [p \in \mathcal{R}]$.

The reason that this form has been used so often lies in Theorem F of Chapter III, Section 6: The system $\dot{x} = B_0 x$ is asymptotically stable [$B_0 \in \{A_1\}$] if and only if for any $Q = Q^{\mathrm{T}} \triangleq [q \in \mathcal{R}]$ there exists a $P = P^{\mathrm{T}} \triangleq [p \in \mathcal{R}]$ such that

$$B_0^{\mathrm{T}}P + PB_0 = -Q. \tag{4-17}$$

Thus, if B_0 is an asymptotically stable matrix, each point $q \in \mathcal{R}$ is mapped into a point $p \in \mathcal{R}$ by relation (4-17). The set of all points $p \in \mathcal{R}$ corresponding to every $q \in \mathcal{R}$ for a given B_0 is denoted $\mathcal{R}_0(B_0)$; $\mathcal{R}_0 \subset \mathcal{R}$. This is the only situation where we can readily determine all the members of a family of Lyapunov functions for a given system defined by B_0, which is the great utility of the quadratic form.

This development and notation is best illustrated by a concrete example: Assume

$$B_0 = \begin{bmatrix} 0 & 1 \\ -4 & -4 \end{bmatrix}. \tag{4-18}$$

The characteristic equation is $p(\lambda) = (\lambda + 2)^2 = 0$; so B_0 is asymptotically stable. First let us define $\mathcal{R}$:

$$v = x^{\mathrm{T}}Px = p_{11}x_1^2 + 2p_{12}x_1x_2 + p_{22}x_2^2$$

is positive definite if (i) $p_{11} > 0$, and (ii) $p_{11}p_{22} - p_{12}^2 > 0$. Without any loss of generality, we can set $p_{nn} = 1$ (if $p \in \mathcal{R}$, then $\alpha p \in \mathcal{R}$ when α is a positive constant). If this is done for the case $n = 2$, $\mathcal{R}$ is defined by the single requirement

$$p_{11} > p_{12}^2, \tag{4-19}$$

where p_{12} is any real number. This gives the boundary of $\mathcal{R}$ in the reduced (2-dimensional) parameter space (p_{12}, p_{11}) as the parabola $p_{11} = p_{12}^2$. If $Q \triangleq [q \in \mathcal{R}]$, then since B_0 is an asymptotically stable matrix, the matrix P determined by Eq. (4-17) must correspond to some $p \in \mathcal{R}$; hence, we can ascertain $\mathcal{R}_0(B_0)$ by solving for the parameters q_{ij} in terms of the p_{ij}, then determining the constraints on p_{ij} such that Q is positive definite:

$$Q = \begin{bmatrix} 8p_{12} & (4 + 4p_{12} - p_{11}) \\ (4 + 4p_{12} - p_{11}) & 2(4 - p_{12}) \end{bmatrix}. \tag{4-20}$$

The boundary of $\mathcal{R}_0$ can be found by choosing p_{11} and p_{12} such that Q is positive semidefinite. The conditions $q_{11} \geqslant 0$ and $q_{22} \geqslant 0$ are satisfied if $0 \leqslant p_{12} \leqslant 4$, and the determinant of Q is

$$|Q| = 16p_{12}(4 - p_{12}) - (4 + 4p_{12} - p_{11})^2.$$

Thus, the boundary of $\mathcal{R}_0$ is defined by the ellipse

$$p_{11}^2 - 2p_{11}(4 + 4p_{12}) + 16(2p_{12}^2 - 2p_{12} + 1) = 0, \tag{4-21}$$

which is just tangent to the lines $p_{12} = 0$ and $p_{12} = 4$, as is shown in Fig. 4-3. This demonstrates the manner in which we can determine $\mathcal{R}_0$, and hence for any B_0 find the entire family of quadratic Lyapunov functions

$$\{v_q{}^0\} \triangleq \{v = x^{\mathrm{T}}Px; \quad P \triangleq [p \in \mathcal{R}_0(B_0)]\}.$$

From the continuity of Eq. (4-17) [each q_{ij} is linearly related to the parameters of B_0], it is clear that a perturbation of the matrix B_0 for a given P would result in a perturbation of the matrix Q. If $\hat{v} = x^{\mathrm{T}}\hat{P}x$ is a member of $\{v_q{}^0\}$ chosen as a Lyapunov function for a given matrix B_0 (where $\hat{p}$ lies in the interior of $\mathcal{R}_0(B_0)$), then $\hat{v}(x)$ is also a member of $\{v_q{}^1\}$, where $\{v_q{}^1\}$ is the family of quadratic Lyapunov functions for a stable matrix B_1 obtained by perturbing B_0. In particular, we observe that there must exist some range $(\underline{K}, \bar{K})$ such that $\hat{v} \in \{\hat{v}_q\}$ for $\hat{B}$ defined by

$$\hat{B} \triangleq A_0 - b_0\kappa h_0{}^{\mathrm{T}}, \qquad \kappa \in (\underline{K}, \bar{K}); \tag{4-22}$$

only the gain κ is allowed to take on different values. Hence the specific function $\hat{v}(x)$ is common to all LTI systems defined by $\hat{B}$ in Eq. (4-22).

To return to our example, assume that B_0 in Eq. (4-18) corresponds to $\kappa = 0$, that is, $B_0 = A_0$. The perturbed system

$$\dot{x} = \hat{B}x = [A_0 - b_0\kappa h_0{}^{\mathrm{T}}]x \tag{4-23}$$

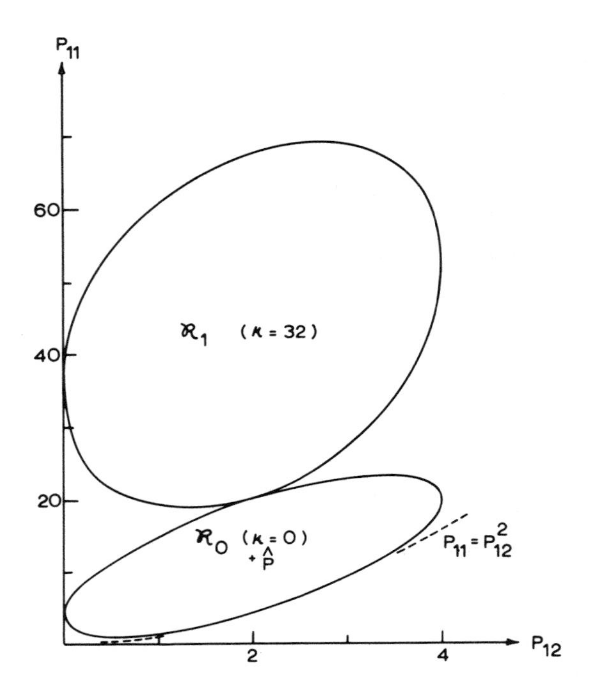

Fig. 4-3. Common Lyapunov functions.

is completed by choosing $h_0^{\mathrm{T}} = [1, 0]$, $b_0^{\mathrm{T}} = [0, 1]$ in the usual phase variable form; hence the forward path and closed loop transfer functions are

$$W(s) = (s^2 + 4s + 4)^{-1}, \quad W_\kappa(s) = (s^2 + 4s + (4 + \kappa))^{-1}. \qquad \text{(4-24)}$$

From $W_\kappa(s)$ it is seen that $\dot{x} = \hat{B}x$ of Eq. (4-23) is asymptotically stable for $\kappa \in (-4, \infty)$. The gains $\underline{K}$ and $\bar{K}$ in the range $(-4, \infty)$ can now be found such that

$$\hat{v}(x) = x^{\mathrm{T}}\hat{P}x = x^{\mathrm{T}}\begin{bmatrix} 12 & 2 \\ 2 & 1 \end{bmatrix}x \qquad \text{(4-25)}$$

is a Lyapunov function for (4-23) with $\kappa \in (\underline{K}, \bar{K})$; this matrix $\hat{P}$ was chosen to be interior to $\mathcal{R}_0$ determined by (4-21); see Fig. 4-3. By inspection,

$$-\dot{\hat{v}} = x^{\mathrm{T}}\begin{bmatrix} 4(4+\kappa) & \kappa \\ \kappa & 4 \end{bmatrix}x \triangleq x^{\mathrm{T}}Qx.$$

Since $q_{22} > 0$ we need only find the range of κ for which $|Q| > 0$; inspect

$$|Q| = 16(4 + \kappa) - \kappa^2 = 0.$$

The bounds $\underline{K}$ and $\bar{K}$ are roots of this equation, that is, $(\underline{K}, \bar{K}) = (-3.312, 19.312)$.

The importance of the common Lyapunov function [for example, $\hat{v}$ of Eq. (4-25)] lies not only in its existence for a range of $\kappa \in (\underline{K}, \bar{K})$ for the LTI system, but in the fact that it is independent of κ. We thus may replace $\kappa\sigma_0$ in Eq. (4-23), where $\sigma_0 = h^{\mathrm{T}}x$, with any nonlinear time-varying function $g(\sigma_0, t)$ without affecting the sign of $\dot{\hat{v}}$ as long as

$$\underline{K} < g(\sigma_0, t)/\sigma_0 < \bar{K}. \tag{4-26}$$

The search for common Lyapunov functions for the LTI system corresponding to the most general NLTV system [Eq. (2-6)] has resulted in the circle criterion described in Chapters VI and VII. By this criterion we can guarantee that a quadratic common Lyapunov function $v(x) = x^{\mathrm{T}}Px$ exists for the system (2-6) provided that a simple frequency domain condition is satisfied by $W(s)$. For the example of Eq. (4-24), the maximum range of the form $[0, \bar{K}]$ for asymptotic stability is found to be $[0, 32 - \varepsilon]$ and a common Lyapunov function is

$$v = x^{\mathrm{T}}Px = x^{\mathrm{T}}\begin{bmatrix} 20 & 2 \\ 2 & 1 \end{bmatrix}x. \tag{4-27}$$

The boundary of $\mathcal{R}_1$ corresponding to

$$B_1 = \begin{bmatrix} 0 & 1 \\ -36 & -4 \end{bmatrix}$$

(or to $\kappa = 32$) also is shown in Fig. 4-3, and the matrix P of Eq. (4-27) is the only common point of $\mathcal{R}_0$ $(\kappa = 0)$ and $\mathcal{R}_1$ $(\kappa = 32)$.

E. The κ-Dependent Lyapunov Function

In this approach to generating Lyapunov functions for the linear system

$$\dot{x} = Ax - b\kappa h^{\mathrm{T}}x, \tag{4-1}$$

a candidate that depends linearly on κ is used to span a range $(\underline{K}', \bar{K}')$ of gains κ. As is usually the case, the quadratic form has proven to be most useful, namely

$$v(x) = \tfrac{1}{2}[x^{\mathrm{T}}Px + \kappa x^{\mathrm{T}}Mx];$$

$$P = P^{\mathrm{T}} > 0; \qquad M = M^{\mathrm{T}} = \sum_{i=0}^{m} \beta_i r_i r_i^{\mathrm{T}} \geqslant 0, \qquad r_0 \triangleq h. \tag{4-28}$$

Although this function does not allow us to substitute a nonlinear time-varying gain $g(\sigma_0, t)$ for $\kappa h^{\mathrm{T}}x \triangleq \kappa\sigma_0$ in the system (4-1) as simply as in the case of the common Lyapunov function, the form $v(x)$ of (4-28) is the linear special case of the generalized Lur'e–Postnikov form for $\rho = 0$, namely

$$v(x) = \tfrac{1}{2}x^{\mathrm{T}}Px + \sum_{i=0}^{m} \beta_i \int_0^{r_i^{\mathrm{T}}x} f(\zeta)\, d\zeta, \tag{4-29}$$

as is considered in Section 2 of Chapter III (Lemma V2) and utilized in Chapter V. The time derivative of $v(x)$ in Eq. (4-29) is exactly the same as that for Eq. (4-28), with the nonlinear function $f(r_i^{\mathrm{T}}x)$ substituted for the linear term $\kappa r_i^{\mathrm{T}}x$; hence the use of (4-28) for the LTI system (4-1) has an important bearing on the use of the Lyapunov function candidate (4-29) for NLTI systems.

One reason that the κ-dependent Lyapunov function is of such utility and appeal is that in the LTI case it may be shown to exist as a necessary and sufficient condition for asymptotic stability over any entire Hurwitz range of κ (but not over all ranges simultaneously, if the LTI system is stable in more than one range). To illustrate this point, we return to the example treated under the common Lyapunov function:

$$W(s) = (s^2 + 4s + 4)^{-1}, \tag{4-24}$$

for which the Hurwitz range is $\kappa \in (-4, \infty)$. We may take as a κ-dependent Lyapunov function candidate

$$v(x) = \tfrac{1}{2}x^{\mathrm{T}}\left\{\begin{bmatrix} \delta(3 - (8-\delta)^{1/2}) & \delta/2 \\ \delta/2 & 1 \end{bmatrix} + (4 + \kappa - \delta)\begin{bmatrix} 1 & 0 \\ 0 & 0 \end{bmatrix}\right\}x. \tag{4-30}$$

The first matrix is real and positive definite for $\delta \in (0, 4)$ and $\delta \in (4, 8]$, and the second term is positive semidefinite if $\kappa \geqslant -(4 - \delta) > -4$. Hence, since δ may be chosen to be arbitrarily small, we have a Lyapunov function candidate (4-30) that is valid for $\kappa \in (-4, \infty)$. The time derivative is

$$dv/dt = -\tfrac{1}{2}[\delta x_1 + (8 - \delta)^{1/2}x_2]^2, \tag{4-31}$$

which for $\delta \in (0, 8)$ is nonpositive and is equal to zero only if $x_1 = -((8-\delta)^{1/2}/\delta)x_2 \triangleq -\mu x_2$. It is easily shown that dv/dt cannot be identically equal to zero on any trajectory of

$$\dot{x} = \begin{bmatrix} 0 & 1 \\ (-4-\kappa) & -4 \end{bmatrix}x,$$

as $x_1 \equiv -\mu x_2$ implies $\dot{x}_1 = x_2 \equiv -\mu\dot{x}_2$. This is only possible if $\kappa = -4$ and $\mu = \frac{1}{4}$, that is, if

$$\dot{x} = \begin{bmatrix} 0 & 1 \\ 0 & -\dfrac{1}{\mu} \end{bmatrix}x.$$

But $\kappa = -4$ is outside the Hurwitz range we are concerned with, so $\dot{v} \equiv 0$

only if $x \equiv 0$. Hence the Lyapunov function (4-31) exists by Corollary F2 (Chapter III, Section 6) for the system for the entire Hurwitz range $\kappa \in (-4, \infty)$.

2. The Existence of the Quadratic Lyapunov Function $x^T Px$ and the Hurwitz Condition

Consider the LTI ordinary vector differential equation

$$\dot{x} = A_0 x, \tag{1-3a}$$

where A_0 is assumed to be in the phase variable canonical form (2-2). The corresponding nth order ordinary scalar differential equation (2-5) is $[D^n + a_n D^{n-1} + \cdots + a_2 D + a_1]\xi = 0$, having the characteristic equation

$$p(\lambda) = |\lambda I - A_0| = \lambda^n + a_n \lambda^{n-1} + \cdots + a_2 \lambda + a_1 = 0. \tag{4-32}$$

If its roots $\lambda_i(A_0)$ all have negative real parts, then asymptotic stability is guaranteed; if some roots have positive real parts, then exponential instability results; roots on the imaginary axis are said to be critical. If all critical roots are distinct, then marginal stability results [$\lambda_i = \pm i\omega_0$ yields solutions of the form $\cos(\omega_0 t)$ or $\sin(\omega_0 t)$; $\lambda_i = 0$ yields constant solutions], whereas critical roots of multiplicity $m \geqslant 2$ yield unstable solutions [$t^{m-1}\cos(\omega_0 t)$, $t^{m-1}\sin(\omega_0 t)$, or t^{m-1}]. Multiple roots in the LHP do not result in instability, as $t^{m-1}e^{-\alpha t}$ tends to zero as t tends to infinity for finite m.

A characteristic polynomial (4-32) is defined to be Hurwitz if all of its roots lie in the open LHP. The problem of stability analysis for $\dot{x} = A_0 x$ thus reduces to determining constraints on the parameters a_i of (4-32) that ensure that $p(\lambda)$ is Hurwitz.

A simple necessary condition is that all a_i must be positive; to see that this is so, assume that $(-\lambda_1, -\lambda_2, \ldots, -\lambda_{m_1})$ are real negative roots and $(-\lambda_{m_1+1} \pm i\mu_{m_1+1}) \cdots (-\lambda_{m_2} \pm i\mu_{m_2})$ are complex roots in the LHP. Then

$$p(\lambda) = \prod_{i=1}^{m_1} (\lambda + \lambda_i) \prod_{i=m_1+1}^{m_2} [\lambda^2 + 2\lambda_i \lambda + (\lambda_i^2 + \mu_i^2)];$$

since all parameters are positive, it is evident that the indicated product would yield only positive values of a_i.

A. *The Hurwitz Conditions*

One well-known technique that yields both necessary and sufficient conditions that $p(\lambda)$ has roots only in the LHP is due to Hurwitz. Each Hurwitz determinant Δ_i, $i = 1, 2, \ldots, n$ must be positive where Δ_i is the ith leading principal minor of the $n \times n$ matrix

$$\begin{bmatrix} a_n & 1 & 0 & 0 & 0 & \cdots & 0 & 0 \\ a_{n-2} & a_{n-1} & a_n & 1 & 0 & \cdots & 0 & 0 \\ a_{n-4} & a_{n-3} & \cdot & \cdot & \cdot & \cdots & 0 & 0 \\ \cdot & \cdot & & & & & \cdot & \cdot \\ \cdot & \cdot & & & & & \cdot & \cdot \\ \cdot & \cdot & & & & & \cdot & \cdot \\ 0 & 0 & \cdot & \cdot & \cdot & \cdots & a_2 & a_3 \\ 0 & 0 & \cdot & \cdot & \cdot & \cdots & 0 & a_1 \end{bmatrix}, \tag{4-33}$$

with the parameters a_i defined as in Eq. (4-32). Thus asymptotic stability of $\dot{x} = A_0 x$ is guaranteed if and only if

$$\begin{array}{lll} \text{(i)} & a_n > 0, & [\Delta_1 > 0], \\ \text{(ii)} & a_n a_{n-1} > a_{n-2}, & [\Delta_2 > 0], \\ & \vdots & \vdots \\ \text{(n)} & a_1 > 0, & [\Delta_n > 0]. \end{array}$$

B. The Existence of the Quadratic Lyapunov Function $x^T P x$

A second criterion that yields necessary and sufficient conditions for the stability of the system (1-3a) is based on Lyapunov's direct method. Associated with Eq. (1-3a) is the Lyapunov function candidate

$$v(x) = x^T P x, \tag{4-34}$$

whose derivative along the trajectories of $\dot{x} = A_0 x$ is

$$dv/dt = x^T(A_0{}^T P + P A_0)x \triangleq -x^T Q x. \tag{4-35}$$

A necessary and sufficient stability condition for asymptotic stability has been stated in Corollary F2 (Chapter III, Section 6). If we choose any $Q = Q^T \geq 0$ such that $x^T Q x \equiv 0$ can occur along no trajectory other than $x \equiv 0$, then Eq. (1-3a) is asymptotically stable if and only if there exists a matrix $P = P^T > 0$ satisfying Eq. (4-35).

C. The Equivalence

Since the two preceding conditions are necessary and sufficient, it is natural to assume that they are equivalent. This was established formally by Parks

[1], as follows: Consider

$$\dot{y} = B_0 y, \tag{4-36}$$

where B_0 is in the Schwarz canonical form, namely,

$$B_0 = \begin{bmatrix} 0 & 1 & 0 & 0 & \cdots & 0 & 0 \\ -b_n & 0 & 1 & 0 & \cdots & 0 & 0 \\ 0 & -b_{n-1} & 0 & 1 & \cdots & 0 & 0 \\ \cdot & \cdot & & & & & \\ \cdot & \cdot & & & & & \\ \cdot & \cdot & & & & & \\ 0 & 0 & \cdot & \cdot & \cdots & 0 & 1 \\ 0 & 0 & \cdot & \cdot & \cdots & -b_2 & -b_1 \end{bmatrix}. \tag{4-37}$$

Parks demonstrated that the system (4-36) has the same stability properties as (1-3a) [specifically, that the characteristic equation $|\lambda I - B_0| = 0$ has the form

$$|\lambda I - B_0| = (-1)^{n-1} p(\lambda) = 0,$$

and hence the same roots as $p(\lambda)$ of Eq. (4-32)] if

$$b_1 = \Delta_1, b_2 = \frac{\Delta_2}{\Delta_1}, b_3 = \frac{\Delta_3}{\Delta_1 \Delta_2}, \ldots, b_r = \frac{\Delta_r \Delta_{r-3}}{\Delta_{r-1} \Delta_{r-2}},$$

where Δ_i are the Hurwitz determinants of the matrix A_0 previously defined. Thus a necessary and sufficient condition for asymptotic stability of $\dot{y} = B_0 y$ is that $b_i > 0$, $i = 1, 2, \ldots, n$, as was shown by Schwarz [1]; in fact he proved that for any matrix B_0 such that $b_i \neq 0$, $i = 1, 2, \ldots, n$, the number of eigenvalues with negative real parts is equal to the number of positive terms in the sequence $\{b_n, b_n b_{n-1}, \ldots, \prod_{i=1}^{n} b_i\}$.

If we consider

$$Q_0 = Q_0^{T} \triangleq \operatorname{diag}[0, 0, \ldots, 0, 2b_1^{2}] \geqslant 0,$$

then the unique P_0 satisfying $Q_0 = -(B_0^{T} P_0 + P_0 B_0)$ is

$$P_0 = P_0^{T} = \operatorname{diag}\left[\left(\prod_{i=1}^{n} b_i\right), \left(\prod_{i=1}^{n-1} b_i\right), \ldots, (b_1 b_2), b_1\right],$$

which is clearly positive definite if and only if $b_i > 0$, $i = 1, 2, \ldots, n$. To complete the proof of the equivalence of the Hurwitz conditions and the existence of $x^{T}Px$, the semidefiniteness of Q necessitates showing that it is not possible for $y_n(t)$ [the nth scalar component of $y(t)$ satisfying (4-36)] to be zero along any trajectory other than $y \equiv 0$. From the form of B_0 [Eq. (4-37)], this result follows by inspection: if $y_n \equiv 0$, then $y_{n-1} \equiv 0$, and so on.

3. The Existence of the Quadratic Lyapunov Function $x^T Px + \kappa x^T Mx$ and the Nyquist Criterion

The alternative system description corresponding to $\dot{x} = A_0 x$ in transfer function formulation is

$$\frac{L(\sigma_0)}{L(\tau)} \triangleq W(s) = h^T(sI - A)^{-1}b$$
$$= \frac{h_n s^{n-1} + \cdots + h_2 s + h_1}{s^n + a_n s^{n-1} + \cdots + a_2 s + a_1}, \qquad A \in \{A_1\} \tag{4-38}$$
$$\tau(t) = -\kappa\sigma_0(t) \triangleq -\kappa h^T x.$$

In the latter context, the usual stability problem that is treated is somewhat different from that considered in Section 2; rather than finding the ranges (often interdependent) of all the parameters of A_0 that are necessary and sufficient for the asymptotic stability of $\dot{x} = A_0 x$, it is assumed that $W(s)$ defines a given stable plant and the stability range of the single parameter κ is to be determined.

A. The Nyquist Criterion

The application of the Nyquist criterion is a standard technique for determining the range of κ such that the system (4-38) is asymptotically stable. For reference we note that a recent and quite general statement of this criterion is to be found in Desoer and Wu [1]; since the systems under consideration have plants $W(s)$ that are simply real rational functions of s, we do not require this degree of sophistication.

For systems described by Eq. (4-38), where $W(s)$ has at least one more pole than zero (since $\rho = 0$) and $W(s)$ is asymptotically stable, the frequency response diagram Γ_W [$V \triangleq \text{Im}\, W(i\omega)$ plotted versus $U \triangleq \text{Re}\, W(i\omega)$ for all real $\omega \in (-\infty, +\infty)$] allows a direct determination of stability for any κ. First locate that portion of the U, V plane representing the mapping of the RHP of the s-plane [refer to Section 5 of Chapter III, especially Fig. 3-1, for a more detailed discussion of the mapping properties of $W(s)$]. This is accomplished by determining the direction of increasing ω on $\Gamma_W(\omega)$; the region to the right of a point traversing Γ_W in the direction of increasing ω corresponds to the RHP in the s-plane (Fig. 4-4).

THEOREM (Nyquist, special case). The closed loop system described by Eq. (4-38) is asymptotically stable if the point $U = -1/\kappa$, $V = 0$ lies neither on Γ_W nor in the region that corresponds to the mapping of the RHP of the s-plane. □

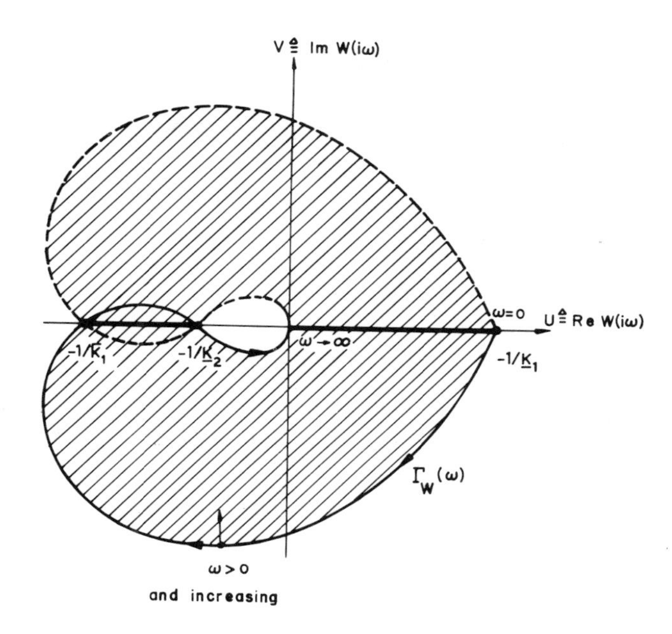

Fig. 4-4. Application of Nyquist's criterion.

The specification of some $\rho \neq 0$ does not substantially alter this criterion; as shown in Fig. 3-2b, only the location of $W_\infty \triangleq \lim_{\omega \to \infty} W(i\omega) = \rho$ is changed. Then we must not allow $(-1/\kappa, 0)$ to lie on Γ_W even in the limit as $\omega \longrightarrow \infty$ to preclude $\kappa = -1/\rho$.

This criterion may be used with equal facility to determine the complete ranges of κ that lead to asymptotically stable closed loop systems. The values of κ are determined such that the point $(-1/\kappa, 0)$ lies on Γ_W ($\underline{K}_1$, $\bar{K}_1$ and $\underline{K}_2$ in Fig. 4-4); these form the endpoints of stability ranges. In this example, $\kappa \in (\underline{K}_1, \bar{K}_1)$ where $\underline{K}_1 < 0 < \bar{K}_1$ and $\kappa \in (\underline{K}_2, \infty)$ specify all values of κ that yield asymptotically stable closed loop systems. The segments of the real axis that lie on or inside Γ_W (in the sense defined in the theorem) correspond to all values of κ that lead to the instability or marginal stability of the closed loop system.

Since $W(s)$ has only real coefficients the functions $U(\omega)$ and $V(\omega)$ are even and odd functions of ω respectively; $U(-\omega) = U(\omega)$ and $V(-\omega) = -V(\omega)$. Thus the plot for $\omega < 0$ is a reflection of the plot for positive frequency about the real axis, and since no new information is conveyed by $\Gamma_W(\omega)$ for $\omega < 0$, this part of Γ_W is often omitted.

B. The Existence of a Shifted LC Multiplier

Consider a transfer function $W(s)$ that has as one of its Nyquist ranges $[0, \bar{K})$. A conjecture due to Narendra and Neuman [2] is that for any system of this form (where $A \in \{A_1\}$ and (A, b) is completely controllable) a κ-dependent Lyapunov function (4-28) must exist as a necessary and sufficient condition for the asymptotic stability of the system (4-1) with $\kappa \in [0, \bar{K})$.

In proving the validity of this conjecture, use is made of a constructional technique similar to that of Brockett and Willems [1]. The equilibrium $\xi = 0$ of

$$[p(D) + \kappa q(D)]\xi = 0$$

(where p and q are polynomials in the operator $D \triangleq d/dt$) is asymptotically stable for $\kappa \in [0, \bar{K})$ if and only if a rational strictly positive real† function $Z(s)$ exists such that

$$H(s) \triangleq W_{\bar{K}}(s)Z(s) \triangleq [q(s)/p(s) + \bar{K}^{-1}]Z(s) \in \{\text{PR}\}. \tag{4-39}$$

In proving this result, it is noted that the Nyquist criterion under these circumstances guarantees that $\phi_{\bar{K}}(\omega) \triangleq \arg[W_{\bar{K}}(i\omega)]$ satisfies

$$-\pi < \phi_{\bar{K}}(\omega) < +\pi,$$

and hence a preliminary transfer function $Z(s)$ satisfying Eq. (4-39) can be determined by inspecting $\phi_{\bar{K}}(\omega)$, Fig. 4-5. Define two sets of frequencies λ_i and μ_i such that:

$$\begin{aligned}
&\text{(i)} \quad \phi_{\bar{K}}(\lambda_i) = 0 \quad \text{and} \quad \left[\frac{d}{d\omega}\phi_{\bar{K}}(\omega)\right]_{\omega=\lambda_i} < 0, \\
&\qquad\qquad\qquad\qquad\qquad\qquad i = 1, 2, \ldots, l, \\
&\text{(ii)} \quad \phi_{\bar{K}}(\mu_i) = 0 \quad \text{and} \quad \left[\frac{d}{d\omega}\phi_{\bar{K}}(\omega)\right]_{\omega=\mu_i} > 0, \\
&\qquad\qquad\qquad\qquad\qquad\qquad i = 1, 2, \ldots, k
\end{aligned} \tag{4-40}$$

(frequencies ω such that $\phi_{\bar{K}}(\omega) = (d/d\omega)\phi_{\bar{K}} = 0$ may be ignored). Considering a function $Z(s) \in \{Z_{LC}\}$,

$$Z(s) = s^{\pm 1}\frac{\prod_{i=1}^{l}(s^2 + \lambda_i^2)}{\prod_{i=1}^{k}(s^2 + \mu_i^2)}; \qquad \begin{aligned}&\text{if } s^{+1}\text{: } l = k-1,\ k;\ 0 < \mu_1 < \lambda_1 \cdots \\ &\text{if } s^{-1}\text{: } l = k,\ k+1;\ 0 < \lambda_1 < \mu_1 \cdots\end{aligned} \tag{4-41}$$

where s^{+1} is taken if $\phi_{\bar{K}}(0^+) < 0$ and s^{-1} if $\phi_{\bar{K}}(0^+) > 0$, we see in the example of Fig. 4-5 that $Z(s)\, W_{\bar{K}}(s)$ is positive real.

† If $Z(s)$ and $H(s)$ are both only constrained to be positive real, then marginal stability could result for some value or values of $\kappa \in [0, \bar{K})$.

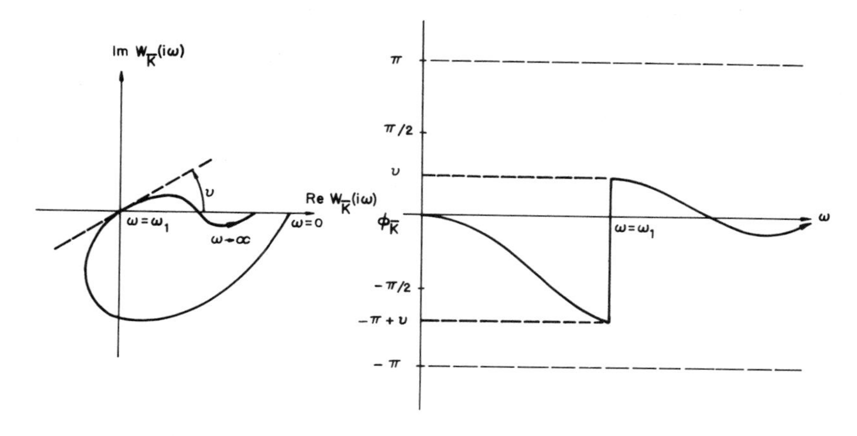

Fig. 4-5. The LC multiplier for $[0, \bar{K})$.

Since this function $Z(s)$ is only positive real, it is not satisfactory for our purposes. Thus, a slight modification of this argument must be used: For any $\kappa \in [0, \bar{K})$ define

$$W_\kappa(s) \triangleq W(s) + \kappa^{-1}$$

and

$$\phi_\kappa(\omega) \triangleq \arg\{W_\kappa(i\omega)\}.$$

Since there are no imaginary poles or zeros of $W_\kappa(s)$ [no poles because $W_\kappa(s)$ is asymptotically stable; no zeros, as seen in Fig. 4-6], $\phi_\kappa(\omega)$ is continuous as shown in Fig. 4-6b.

Since this is the case, a frequency domain multiplier $Z(s)$ of the form (4-41) that has a discontinuous argument is not required to satisfy Eq. (4-39).

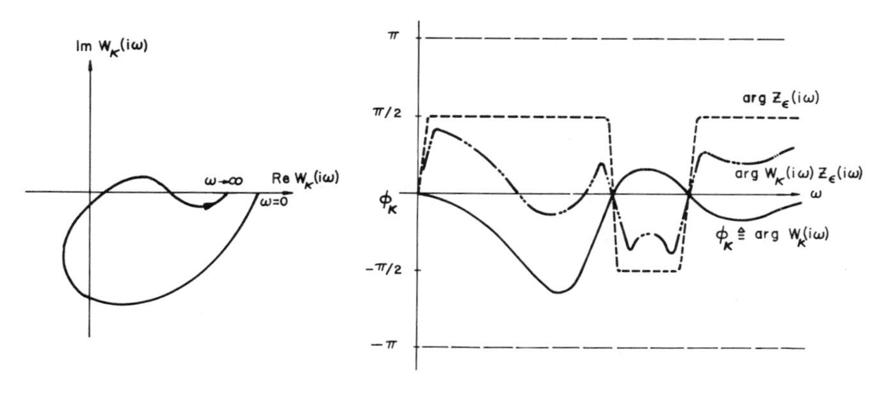

Fig. 4-6. The shifted LC multiplier.

Instead, the shifted function $Z_\varepsilon(s) \triangleq Z(s+\varepsilon)$ can be chosen where ε is an arbitrary positive constant:

$$Z_\varepsilon(s) \triangleq (s+\varepsilon)^{\pm 1} \frac{\prod_{i=1}^{l} [(s+\varepsilon)^2 + \lambda_i^2]}{\prod_{i=1}^{k} [(s+\varepsilon)^2 + \mu_i^2]}. \tag{4-42}$$

By choosing ε sufficiently small, it can always be guaranteed that

$$W_\kappa(s) Z_\varepsilon(s) \in \{\mathrm{PR}\} \tag{4-43}$$

since as $\varepsilon \longrightarrow 0$ the argument of $Z_\varepsilon(s)$ approaches that of $Z(s)$ arbitrarily closely.

An important observation is that given a value of κ, only the ε of the shifted multiplier needs to be determined; the frequencies λ_i and μ_i as determined by Eq. (4-40) are independent of κ. If $\phi_\kappa(\lambda_i) = 0$, $d\phi_\kappa/d\omega|_{\omega=\lambda_i} < 0$ for one value of κ, then the same is true for any $\kappa \in [0, \bar{K})$, and similarly for $\phi_\kappa(\mu_i)$. As $\kappa \longrightarrow \bar{K}$, then it is necessary that $\varepsilon \longrightarrow 0$ if the argument of $W_{\bar{K}}(i\omega)$ is discontinuous; if, however, $\kappa \in [0, \bar{K})$, then we take ε arbitrarily small, and $Z_\varepsilon(s)$ exists satisfying (4-43).

C. *The Equivalence*

The fact that one multiplier $Z(s)$ exists for the whole range $[0, \bar{K})$ allows us to guarantee that a single κ-dependent Lyapunov function

$$v(x) = \tfrac{1}{2} x^{\mathrm{T}} [P + \kappa M] x \tag{4-29}$$

exists over the entire Hurwitz range of κ, as the matrices $P (P = P^{\mathrm{T}} > 0)$ and $M (M = M^{\mathrm{T}} \geqslant 0)$ are determined solely by the parameters of $W(s)$ [given in Eq. (4-38)] and of $Z(s)$.

The actual proof that this Lyapunov function exists for (4-1) whenever (4-43) is satisfied and hence that it exists for the entire Hurwitz range $[0, \bar{K})$ as argued above is undertaken in Chapter V as a special case of the stability of NLTI systems. It is shown that as the nonlinearity $f(\sigma_0)$ is constrained the class of frequency domain multipliers $Z(s)$ that may be used to guarantee absolute stability becomes more general until for linear systems the shifted LC multiplier of Eq. (4-42) is permitted with ε arbitrarily small. For this reason the proof in its entirety is not repeated here.

The validity of this conjecture was essentially proved by Thathachar and Srinath [1], using an approach very similar to that presented here. The κ-dependent Lyapunov function was shown to exist whenever

$$Z(s) = \delta + s \frac{\prod_{i=1}^{l} (s^2 + \lambda_i^2)}{\prod_{i=1}^{k} (s^2 + \mu_i^2)}$$

exists satisfying $Z(s)W_K(s) \in \{PR\}$. This proof is applicable only for systems such that $\phi(0^+) < 0$.

This result has important implications for our analysis of NLTI systems in Chapter V and of NLTV systems in Chapter VI, as the κ-dependent Lyapunov function is only a special case of the most general Lyapunov function used in these situations. The fact that in this special case the resulting stability conditions are necessary and sufficient reinforces the idea that the generalized Lur'e–Postnikov form is of fundamental significance and that the resulting stability criteria are thus not adventitious.

V

STABILITY OF NONLINEAR SYSTEMS

The approach used in this chapter to generate conditions that are sufficient to guarantee the equiasymptotic stability in the whole of various classes of nonlinear time-invariant systems may be viewed as a logical extension of the developments of the previous chapter, although the original derivations were not necessarily prompted by such considerations. In actuality, many of the results pertaining to the stability of LTI systems obtained via the direct method of Lyapunov were formally proved more recently than some of those presented in this chapter for nonlinear systems; however, the fundamental concepts of the common and κ-dependent Lyapunov functions were recognized and unquestionably exerted an influence on these developments, particularly in the choice of absolute $\{N\}$ Lyapunov function candidates.

We start with a derivation of the Popov frequency domain solution to the problem of Lur'e and Postnikov. This result—which is obtained using the Lur'e–Postnikov Lyapunov function (as modified by Popov) and the Meyer form of the Kalman–Yakubovich lemma (the MKY lemma, Chapter III, Section 4)—is very general in that the continuous nonlinearity is constrained only with respect to its range; $f(\sigma_0) \in \{F\}$ implies only that

$$0 \leqslant f(\sigma_0)/\sigma_0 \leqslant \bar{F} \quad (\text{or} < \infty).$$

The philosophy that is developed in considering this problem and the practical utility of the resulting stability criterion lead us to a treatment of NLTI systems with nonlinear gains belonging to more restricted classes of functions.

All systems dealt with in this chapter are described by the state vector differential equation

$$\dot{x} = Ax + b\tau, \qquad \sigma_0 = h^{\mathrm{T}}x + \rho\tau, \qquad \tau = -f(\sigma_0), \tag{5-1}$$

where the triple (h, A, b) is in phase variable canonical form. There are generally two types of problems considered in terms of the specification of the nonlinearity $f(\sigma_0)$ and the A matrix: the principal case,

$$\begin{aligned} f(\sigma_0) &\in \{N\}, & f(\sigma_0)/\sigma_0 &\in [\underline{F}_N, \bar{F}_N], \\ A_{\underline{F}_N} &\in \{A_1\}, & A_{\bar{F}_N} &\in \{A_1\}, \end{aligned} \tag{5-2a}$$

and the particular case of one zero eigenvalue,

$$\begin{aligned} f(\sigma_0) &\in \{N\}, & f(\sigma_0)/\sigma_0 &\in (\underline{F}_N, \bar{F}_N], \\ A_{\underline{F}_N} &\in \{A_0\}, & A_{\bar{F}_N} &\in \{A_1\}, \end{aligned} \tag{5-2b}$$

where the lower bound $\underline{F}_N$ is zero in all sections except Section 5. We formally treat the case $f(\sigma_0)/\sigma_0 \in [0, \bar{F})$, $A \in \{A_1\}$ for the problem of Popov in Section 1, but not for other classes of NLTI gains.

As a preliminary to all developments, we must assume that A_κ satisfies

$$A_\kappa \triangleq [A - (\kappa/(1 + \kappa\rho))bh^{\mathrm{T}}] \in \{A_1\} \tag{4-2}$$

for all values of κ in the range of $f(\sigma_0)/\sigma_0$, in order to ensure that $x \equiv 0$ is the only equilibrium of Eq. (5-1), as shown in Chapter II, Section 1D. The frequency domain condition that ensures absolute stability is a sufficient condition to guarantee that $A_\kappa \in \{A_1\}$ for $\kappa \in (\underline{F}_N, \bar{F}_N)$, so this constraint need only be retained for those extreme values of $f(\sigma_0)/\sigma_0$ that may be achieved, that is, for $\kappa = \underline{F}_N$ and $\bar{F}_N$ in Eq. (5-2a) or only for $\kappa = \bar{F}_N$ in Eq. (5-2b).

As a final general point, we note that in the particular case with $\underline{F}_N = 0$, the condition $h_1 > 0$ is required for $x \equiv 0$ to be the only equilibrium of Eq. (5-1) and for stability in the limit (see Sections 1D in Chapter II and 1 in this chapter, respectively). The frequency domain condition for absolute stability in the particular case always guarantees that $h_1 > 0$ as indicated in the derivation of Criterion 1c, so we need not be explicitly concerned with this condition. The same is true for the case when $\underline{F}_N \neq 0$ in Eq. (5-2b).

1. The Popov Stability Criterion

The problem considered by V. M. Popov is identical to the absolute stability problem posed by Lur'e and Postnikov, except for a minor difference in system model. He considered the vector differential equation

$$\dot{x} = Ax - bf(\sigma), \qquad A \in \{A_1\}, \qquad \dot{\xi} = f(\sigma), \qquad \sigma = m^{\mathrm{T}}x - \gamma\xi$$

which alternatively may be described in block diagram form (Fig. 2-1) with $\hat{W}(s)$ in the forward path and $f(\cdot)$ in the return path in the standard negative feedback configuration where

$$\hat{W}(s) = \gamma/s + m^{\mathrm{T}}(sI - A)^{-1}b.$$

If $\gamma = 0$, then this system is in the form of a direct control system with $\rho = 0$ [Eq. (2-6)]. If $\gamma \neq 0$, then $\hat{W}(s)$ may be realized without transformation by a direct control system with $A \in \{A_0\}$. As the Popov model is thus formally equivalent to the standard form of Eq. (2-6), it is not treated here.

The system whose stability properties are to be analyzed in the first instance is represented by the vector differential equation (5-1) with the constraints

$$f(\sigma_0) \in \{F\}, \qquad f(\sigma_0)/\sigma_0 \in [0, \bar{F}), \qquad A \in \{A_1\}. \tag{5-3}$$

The absolute Lyapunov function candidate considered is the Lur'e–Postnikov form with the additional term $\frac{1}{2}\hat{\beta}_0\rho\tau^2$ introduced by Popov,

$$v(x) = \tfrac{1}{2}x^{\mathrm{T}}Px + \hat{\beta}_0\left\{\int_0^{\sigma_0} f(\zeta)\, d\zeta + \tfrac{1}{2}\rho\tau^2\right\}, \tag{5-4}$$

which, as has been demonstrated in Lemma V2 (Chapter III, Section 2), is a valid absolute Lyapunov function candidate for NLTI systems, provided that $P = P^{\mathrm{T}} > 0$, $\hat{\beta}_0 \geqslant 0$, and $\rho \geqslant 0$. From the general expression for $\dot{v} = (\nabla v)^{\mathrm{T}}\dot{x}$ [Eq. (3-4)], we have

$$\begin{aligned}\dot{v} = {} & \tfrac{1}{2}x^{\mathrm{T}}(A^{\mathrm{T}}P + PA)x - f(\sigma_0)x^{\mathrm{T}}[Pb - \hat{\beta}_0A^{\mathrm{T}}h - \gamma_0 h] \\ & - [\hat{\beta}_0 h^{\mathrm{T}}b + \gamma_0(\rho + \bar{F}^{-1})]f^2(\sigma_0) - \gamma_0\sigma_0 f(\sigma_0)[1 - f(\sigma_0)/\bar{F}\sigma_0].\end{aligned}$$

The only additional algebraic manipulation performed on $\dot{v}$ is the inclusion of $\gamma_0[\sigma_0 f(\sigma_0)(f(\sigma_0)/\bar{F}\sigma_0) - f^2(\sigma_0)/\bar{F}]$, which is identically equal to zero. The first three terms of $\dot{v}$ are in a form that may be guaranteed to be negative semidefinite by requiring that the frequency domain restriction on $W(s)$ that devolves from the application of the MKY lemma is satisfied. We identify the parameters

$$\begin{aligned}\tfrac{1}{2}\psi &\triangleq \hat{\beta}_0 h^{\mathrm{T}}b + \gamma_0(\rho + \bar{F}^{-1}), \\ k &\triangleq \hat{\beta}_0 A^{\mathrm{T}}h + \gamma_0 h.\end{aligned} \tag{5-5}$$

The lemma then states that there is some matrix P, $P = P^{\mathrm{T}} > 0$; a matrix M, $M = M^{\mathrm{T}} \geqslant 0$; and a real vector q satisfying

(a) $A^{\mathrm{T}}P + PA = -qq^{\mathrm{T}} - M$,
(b) $Pb - k = \sqrt{\psi}q$,
(c) (q^{T}, A) is completely observable,

if and only if

(d) $H(s) = \frac{1}{2}\psi + k^{\mathrm{T}}(sI - A)^{-1}b \in \{PR\}$.

Since P in Eq. (5-4) has not been previously specified, we can thus state that some valid candidate of the form given in this equation exists such that

$$\dot{v} = -\tfrac{1}{2}[x^T q + \sqrt{\psi}\, f(\sigma_0)]^2 \\ -\tfrac{1}{2}x^T M x - \gamma_0 \sigma_0 f(\sigma_0)[1 - f(\sigma_0)/\bar{F}\sigma_0], \tag{5-6}$$

if and only if

$$\hat{H}(s) \triangleq \hat{\beta}_0(h^T b + h^T A(sI - A)^{-1} b) \\ + \gamma_0(\rho + \bar{F}^{-1} + h^T(sI - A)^{-1} b) \in \{\mathrm{PR}\}.$$

The condition $\hat{H}(s) \in \{PR\}$ precludes $H(s) \equiv 0$, so $\hat{\beta}_0$ and γ_0 cannot simultaneously be zero.

The constraint on the transfer function $\hat{H}(s)$ may be reduced to

$$\hat{H}(s) = \hat{\beta}_0 s h^T(sI - A)^{-1} b + \gamma_0[W(s) + \bar{F}^{-1}] \in \{\mathrm{PR}\}, \tag{5-7}$$

where we recall that $W(s) = \rho + h^T(sI - A)^{-1} b$ represents the transfer function of the LTI plant [Eq. (2-1)], and we make use of the relation

$$h^T b + h^T A(sI - A)^{-1} b = h^T[(sI - A) + A](sI - A)^{-1} b \\ = s h^T(sI - A)^{-1} b.$$

Thus far we have shown that $v(x)$ is a valid absolute Lyapunov function candidate for Eqs. (5-1) and (5-3), and $\dot{v} \leqslant 0$ if

(i) $\hat{\beta}_0 \geqslant 0$, $\rho \geqslant 0$ (for $v(x)$ to be positive definite),
(ii) $\gamma_0 \geqslant 0$ (for $\dot{v}$ to be negative semidefinite),
(iii) either $\hat{\beta}_0 \neq 0$ or $\gamma_0 \neq 0$ ($\hat{H}(s) \not\equiv 0$; the MKY lemma),
(iv) $\hat{H}(s) \in \{PR\}$ (Eq. (5-7); the MKY lemma),

and hence these conditions, in conjunction with

(v) $A_\kappa \in \{A_1\}$, $\kappa \in [0, \bar{F})$,
(vi) $\dot{v} \not\equiv 0$ unless $x \equiv 0$ (Theorem 1, Chapter III, Section 3),

suffice to guarantee the absolute stability of the system.

The only constraints that need to be kept as sufficient conditions for absolute stability for the system determined by Eq. (5-3) are

(a) $\gamma_0 > 0$,
(b) $H(s) \triangleq [W(s) + \bar{F}^{-1}](\beta_0 s + \gamma_0)^{\pm 1} \in \{PR\}$ where $\beta_0 \geqslant 0$, $\gamma_0 > 0$.

Conditions (i)–(vi) may be eliminated because they are unnecessarily strict or because they are subsumed by (a) and (b), as we proceed to demonstrate.

Conditions (ii) and (iii) are satisfied if $\gamma_0 > 0$, and thus they may be discarded. We see shortly that there is no loss in generality in this restriction.

The requirement that $\gamma_0 > 0$ also implies that condition (vi) is satisfied. Since no term of $\dot{v}$ [Eq. (5-6)] can ever be positive, clearly $\dot{v} \equiv 0$ requires that each term is identically zero. The last term contains the factors $\gamma_0 > 0$ and $1 - f(\sigma_0)/(\bar{F}\sigma_0) > 0$, since $f(\sigma_0)/\sigma_0 \in [0, \bar{F})$, so it is necessary that $f(\sigma_0) \equiv 0$ if the last term is to be identically zero. This reduces Eqs. (5-1) and (5-6) to

$$\dot{x} = Ax, \qquad A \in \{A_1\},$$
$$\dot{v} = -\tfrac{1}{2}(x^{\mathrm{T}}q)^2 - \tfrac{1}{2}x^{\mathrm{T}}Mx$$

along any trajectory where $\dot{v} \equiv 0$. The MKY lemma guarantees that (q^{T}, A) is completely observable, which means that the only trajectory of $\dot{x} = Ax$ for which $x^{\mathrm{T}}q \equiv 0$ is $x \equiv 0$.

It next may be demonstrated that (i), that is, $\hat{\beta}_0 \geqslant 0$ and $\rho \geqslant 0$, is not necessary. These two conditions are clearly not required to guarantee that $\dot{v} \leqslant 0$ and $\dot{v} \not\equiv 0$ unless $x \equiv 0$, so $\hat{H}(s) \in \{PR\}$ guarantees that

$$v(x) = \tfrac{1}{2}x^{\mathrm{T}}Px + \hat{\beta}_0\left\{\int_0^{\sigma_0} f(\zeta)\, d\zeta + \tfrac{1}{2}\rho\tau^2\right\} \tag{5-4}$$

corresponds to the form

$$\dot{v} = -\tfrac{1}{2}[x^{\mathrm{T}}q + \sqrt{\psi}\, f(\sigma_0)]^2$$
$$-\tfrac{1}{2}x^{\mathrm{T}}Mx - \gamma_0\sigma_0 f(\sigma_0)[1 - f(\sigma_0)/\bar{F}\sigma_0], \tag{5-6}$$

in that q, $P = P^{\mathrm{T}} > 0$ and $M = M^{\mathrm{T}} \geqslant 0$ exist by the MKY lemma. It is thus only necessary to show that conditions (a) and (b) imply that $v(x)$ is positive definite irrespective of the signs of $\hat{\beta}_0$ and ρ.

It has been assumed that the linear system corresponding to $f(\sigma_0) = \kappa\sigma_0$, $\kappa \in [0, \bar{F})$ is asymptotically stable [condition (v)]. The candidate $v(x)$ and its derivative are, by inspection, quadratic forms under the assumption of linearity, and $\dot{v}(x) \triangleq -x^{\mathrm{T}}\hat{Q}x$ satisfies $\hat{Q} = \hat{Q}^{\mathrm{T}} \geqslant 0$ and $x^{\mathrm{T}}\hat{Q}x \not\equiv 0$ unless $x \equiv 0$ when $\gamma_0 > 0$. Thus we may appeal to the theorem of Kalman (Corollary F2, Chapter III, Section 6), which states that the corresponding quadratic form of $v(x)$ for $\kappa \in [0, \bar{F})$ must be positive definite; $v(x) = x^{\mathrm{T}}\hat{P}x$, $\hat{P} = \hat{P}^{\mathrm{T}} > 0$, whatever the signs of $\hat{\beta}_0$ and ρ might be.

For the nonlinear system, it is seen that if $f(\sigma_0) = \kappa(\sigma_0)\sigma_0$, the form of the time derivative $\dot{v}$ is the same function of κ as when κ is a constant, that is, the expression for $\dot{v}$ in Eq. (5-6) is the same function of κ for both linear and nonlinear systems. This is due to the fact that

$$d/dt \int_0^{\sigma} \zeta\kappa(\zeta)\, d\zeta = \sigma\kappa(\sigma)\, d\sigma/dt$$

whether κ is a constant or a function of σ. For the linear system with $f(\sigma_0) = \kappa\sigma_0$, it is observed that $v(x)$ in Eq. (5-4) is a Lyapunov function for all values of $\kappa \in [0, \bar{F})$ whatever the signs of $\hat{\beta}_0$ and ρ might be, so for $f(\sigma_0) = \kappa(\sigma_0)\sigma_0$, $\kappa(\sigma_0) \in [0, \bar{F})$, $v(x)$ in Eq. (5-4) is of the same form and hence is positive definite.

Returning to the frequency domain condition (b), we note that if

$$H_1(s) \triangleq [W(s) + \bar{F}^{-1}](\hat{\beta}_0 s + \gamma_0)$$

or

$$H_2(s) \triangleq [W(s) + \bar{F}^{-1}](-\hat{\beta}_0 s + \gamma_0)^{-1}$$

satisfy $\operatorname{Re} H_i(i\omega) \geqslant 0$, then the same condition is satisfied by $\hat{H}(i\omega)$ [Eq. (5-7)] and vice versa, since

$$\operatorname{Re} H_1(i\omega) = \operatorname{Re} \hat{H}(i\omega),$$

$$\operatorname{Re} H_2(i\omega) = \frac{\operatorname{Re} \hat{H}(i\omega)}{(\hat{\beta}_0\omega)^2 + \gamma_0{}^2}; \quad [(\hat{\beta}_0\omega)^2 + \gamma_0{}^2] > 0.$$

Furthermore, if $\hat{\beta}_0 \geqslant 0$, then $H_1(s) \in \{\text{PR}\}$ guarantees that $\hat{H}(s) \in \{\text{PR}\}$ and vice versa, whereas if $\hat{\beta}_0 < 0$, then $H_2(s) \in \{\text{PR}\}$ accomplishes the same end. These results follow from the condition that the nonlinearity range $[0, \bar{F})$ lies in a Nyquist range of the corresponding LTI system, so both the zeros and poles of $W(s) + \bar{F}^{-1}$ must lie in the closed left half of the s-plane.† Thus we obtain the condition (b) that $H(s) \in \{\text{PR}\}$, where

$$H(s) \triangleq [W(s) + \bar{F}^{-1}](\beta_0 s + \gamma_0)^{\pm 1}, \qquad \beta_0 \geqslant 0, \quad \gamma_0 > 0. \tag{5-8}$$

Finally, condition (v) may be eliminated in light of this last constraint. The argument of the term $(\beta_0 i\omega + \gamma_0)^{\pm 1}$ (denoted $\measuredangle(\beta_0 i\omega + \gamma_0)^{\pm 1}$), given $\gamma_0 > 0$, is in the range $(-\pi/2, 0]$ for all real ω if -1 is taken, or in the range $[0, +\pi/2)$ if $+1$ is taken. Thus, since $\measuredangle H(i\omega) \in [-\pi/2, +\pi/2]$ by the condition restricting Eq. (5-8), it is necessary from the additive property of arguments,

$$\measuredangle H(i\omega) = \measuredangle[W(i\omega) + \bar{F}^{-1}] + \measuredangle(\beta_0 i\omega + \gamma_0)^{\pm 1},$$

that

$$\phi_F(\omega) \triangleq \measuredangle[W(i\omega) + \bar{F}^{-1}] \in (-\pi, +\pi),$$

which is an alternative statement of the Nyquist criterion for $\kappa \in [0, \bar{F})$. If γ_0 were permitted to be zero, condition (b) would not prevent the polar plot of $W(i\omega)$ from touching the negative real axis to the left of $(-1/\bar{F}, 0)$ in the U, V plane (Chapter IV, Section 3). Since this must not happen for absolute stability to be guaranteed (in fact, if this did occur some linear gain $\kappa_1 \in [0, \bar{F})$ would exist such that oscillations occur), $\gamma_0 > 0$ must be assumed

† If $\bar{F}$ is not in a Nyquist range, this equivalence is not valid. Consider $W(s) = -\frac{1}{2} + (s + 1)^{-1}$, which has a Nyquist range $(-2, +2)$. Taking $\bar{F} = \infty$ and $\gamma_0 = 2$, $\hat{\beta}_0 = 1$, we have $\hat{H}(s) = (s + 1)^{-1} \in \{\text{PR}\}$ from Eq. (5-7), whereas $H(s) = -\frac{1}{2}(s - 1)(s + 2)/(s + 1)$ in Eq. (5-8), which is not positive real. In this case, $\dot{v}$ is negative definite since the scalar form of the MKY lemma is satisfied, but v is not positive definite since the argument given above does not hold. This problem does not exist if $\rho \geqslant 0$, as shown in Chapter VII, Section 3.

or the additional condition that $|\phi_F(i\omega)| < \pi$ must be introduced. Since $A \in \{A_1\}$ (or later, $A \in \{A_0\}$) is assumed, the frequency at which $|\phi_F(i\omega)|$ approaches π must be either (i) finite, $\omega_0 \in (0, \infty)$, or (ii) infinite (that is, in the limit as $\omega \longrightarrow \infty$); $A \in \{A_i\}$ precludes $|\phi_F| \longrightarrow \pi$ as $\omega \longrightarrow 0$. In the first case, the above constraint is equivalent to $|\phi_F(i\omega)| \leqslant (\pi - \varepsilon)$ with $\varepsilon > 0$, which permits the use of $(\beta_0 s + \gamma_0)^{\pm 1}$ in Eq. (5-8) with $\gamma_0 > 0$. If $\phi_F(i\omega) \longrightarrow \pi$ as $\omega \longrightarrow \infty$, then

$$W(i\omega) + \bar{F}^{-1} = h^{\mathrm{T}}(i\omega I - A)^{-1}b,$$

that is, $\rho = -1/\bar{F}$, and furthermore, $h_n = 0$ so that

$$\lim_{s\to\infty} W(s) = \lim_{s\to\infty} \frac{h_{n-1}}{s^2 + a_n s + a_{n-1}},$$

where $a_n > 0$, $a_{n-1} > 0$, and $h_{n-1} > 0$. In this case, $\beta_0 s + \gamma_0$, where $\gamma_0 > 0$ and $\gamma_0/\beta_0 \leqslant a_n$ gives an overall function $H(s)$ which is positive real for large values of s. Since the use of $(\beta_0 s)^{\pm 1}$ is thus never required for $A \in \{A_1\}$ (or even for $A \in \{A_0\}$, by the same argument), we do not consider it; for those instances when $(\beta_0 s)^{\pm 1}$ can profitably be utilized, the interested reader may refer to Kalman [2] or to Aizerman and Gantmacher [1].

These observations have reduced conditions (i)–(vi) to the very simple frequency domain stability theorem of Popov and successors.

CRITERION 1a (Popov, principal case). The system defined by Eqs. (5-1) and (5-3) is absolutely $\{F\}$ stable if $H(s)$ [Eq. (5-8)] satisfies $H(s) \in \{\mathrm{PR}\}$. □

If one additional condition is added to the system constraints, a result for closed ranges $[0, \bar{F}]$ ensues. Consider

$$\begin{aligned} f(\sigma_0) &\in \{F\}, & f(\sigma_0)/\sigma_0 &\in [0, \bar{F}], \\ A &\in \{A_1\}, & A_{\bar{F}} &\in \{A_1\}. \end{aligned} \tag{5-9}$$

The constraint $A_{\bar{F}} \in \{A_1\}$, in conjunction with the condition $H(s) \in \{\mathrm{PR}\}$, guarantees that $A_\kappa \in \{A_1\}$ for $\kappa \in [0, \bar{F}]$ and that $(1 + \bar{F}W)$ has as many poles as zeros. The only remaining step that must be altered in the proof of Criterion la is the demonstration that $\dot{v} \equiv 0$ only if $x \equiv 0$. In the present case, the last term of $\dot{v}$ [Eq. (5-6)] is identically zero if $f(\sigma_0) \equiv 0$ or if $f(\sigma_0) \equiv \bar{F}\sigma_0$. In either case, the proof proceeds as before, since such trajectories are described by the asymptotically stable LTI differential equations $\dot{x} = Ax$ or $\dot{x} = A_{\bar{F}}x$.

CRITERION 1b (Popov, principal case, alternative conditions). The system described by Eqs. (5-1) and (5-9) is absolutely $\{F\}$ stable if $H(s)$ of Eq. (5-8) satisfies $H(s) \in \{\mathrm{PR}\}$. □

In treating the particular case of the A matrix having a single zero eigenvalue, we consider

$$\begin{aligned} f(\sigma_0) &\in \{F\}, & f(\sigma_0)/\sigma_0 &\in (0, \bar{F}], \\ A &\in \{A_0\}, & A_F &\in \{A_1\}. \end{aligned} \tag{5-10}$$

The condition $h_1 > 0$, which we see below is guaranteed by the absolute stability criterion, ensures stability in the limit, that is, the differential equation (5-1) as constrained by Eq. (5-10) is asymptotically stable for $\tau = -\delta\sigma_0$, $\delta > 0$ but arbitrarily small. This is evident from the characteristic polynomial $|\lambda I - A_\delta|$ or the numerator of $1 + \delta W(s)$, namely,

$$\begin{aligned} s^n(1 + \delta p) + s^{n-1}[a_n + \delta(pa_n + h_n)] + \cdots \\ + s[a_2 + \delta(pa_2 + h_2)] + \delta h_1, \end{aligned}$$

which must have positive coefficients; thus $h_1 > 0$ is a necessary condition for absolute stability.

The proof of an absolute stability criterion for such a system proceeds as in the principal case, except that the replacement of the MKY lemma (which is no longer applicable) by Lemma 5 (Chapter III, Section 4) requires that $H(s) \in \{\mathrm{SPR}_0\}$ rather than the weaker condition obtained in the principal case. The form of $\dot{v}$ is not essentially altered by the use of this different lemma,

$$\begin{aligned} \dot{v} = -\tfrac{1}{2}\{[x^{\mathrm{T}}q + \sqrt{\psi}\, f(\sigma_0)]^2 + \varepsilon x^{\mathrm{T}} L_0 x\} \\ -\gamma_0\sigma_0 f(\sigma_0)[1 - f(\sigma_0)/\bar{F}\sigma_0], \qquad \varepsilon > 0, \end{aligned}$$

except that $L_0 = L_0^{\mathrm{T}} \geqslant 0$ and q are of the specific forms

$$L_0 = \left[\begin{array}{c|ccc} 0 & 0 & \cdots & 0 \\ \hline 0 & & & \\ \cdot & & & \\ \cdot & & \hat{L} & \\ \cdot & & & \\ 0 & & & \end{array}\right], \qquad q = \begin{bmatrix} 0 \\ q_2 \\ \cdot \\ \cdot \\ \cdot \\ q_n \end{bmatrix},$$

where $\hat{L} = \hat{L}^{\mathrm{T}} > 0$ is arbitrary. Thus $\dot{v} \equiv 0$ requires that $x^{\mathrm{T}} L_0 x \equiv 0$ or $x^{\mathrm{T}} \equiv [x_1, 0, 0, \ldots, 0]$. From the form of A and b in Eq. (5-1) (the phase variable canonical form), this implies that $\dot{x}_1 = x_2 \equiv 0$ or $x_1 = x_{1e}$. The existence of such an equilibrium state for $x_{1e} \neq 0$ is not possible, so $\dot{v} \equiv 0$ only if $x \equiv 0$.

Finally, we note that $H(s) \in \{\mathrm{SPR}_0\}$ guarantees that $h_1 > 0$. We see in Chapter III, Section 5, Definition PR3, that the numerator polynomial of $H(s)$, which is of the form

$$b_m s^{m-1} + \cdots + b_2 s + b_1,$$

must satisfy $b_1 > 0$; by referring to Eq. (5-8) we observe that this is only possible if $h_1 > 0$.

The remainder of the steps taken in proving Criterion 1c remain unchanged. Thus we can directly state the stability condition applicable to the system specified by Eq. (5-10).

CRITERION 1c (Popov; particular case). The system defined by Eqs. (5-1) and (5-10) is absolutely $\{F\}$ stable if $H(s)$ of Eq. (5-8) satisfies $H(s) \in \{\text{SPR}_0\}$. □

If one is willing to tighten the restriction of the bounds on $f(\sigma_0)/\sigma_0$ to $[\varepsilon, \bar{F}]$, where ε is arbitrarily small in the particular case just treated, the frequency domain condition becomes that of Criterion 1b. With $f(\sigma_0)/\sigma_0 \in [\varepsilon, \bar{F}]$ and $h_1 > 0$, we may use the standard general finite sector transformation (Section 5) to obtain the condition

$$\hat{H}(s) \triangleq \frac{1 + \bar{F}W(s)}{1 + \varepsilon W(s)} \cdot (\beta_0 s + \gamma_0)^{\pm 1} \in \{\text{PR}\}, \qquad \beta_0 \geqslant 0, \quad \gamma_0 > 0,$$

which is satisfied if $H(s) = [W(s) + \bar{F}^{-1}](\beta_0 s + \gamma_0)^{\pm 1} \in \{\text{PR}\}$. Since in the particular case we are only interested in the possibility of $f(\sigma_0)/\sigma_0$ approaching zero (the problem $f(\sigma_0)/\sigma_0 \in [\varepsilon, \bar{F}]$ with $A \in \{A_0\}$ is actually a principal case problem since $A_\varepsilon \in \{A_1\}$), this type of result is not considered further.

In applying these criteria, it is actually only necessary to consider the factor $(1 + \alpha_0 s)^{\pm 1}$ in Eq. (5-8), since $\gamma_0 > 0$ is always assumed with no loss in generality. It is then necessary to determine analytically whether some $\alpha_0 \geqslant 0$ exists such that $[W(s) + \bar{F}^{-1}](1 + \alpha_0 s)^{\pm 1}$ satisfies the requisite frequency domain condition. Having the freedom to choose two parameters (β_0, γ_0) as in Eq. (5-8), while of no utility in the present case, is maintained throughout this development because it is required for future derivations (Sections 2 and 3).

It is also useful to observe that if $W_\infty \triangleq \lim_{s\to\infty} W(s) = \rho$ satisfies the condition $W_\infty = -\bar{F}^{-1}$, then only Criterion 1a can be applied and $(\beta_0 s + \gamma_0)^{+1}$ must be used. This is the case because $\rho = -\bar{F}^{-1}$ yields $[W(s) + \bar{F}^{-1}] = h^{\mathrm{T}}(sI - A)^{-1}b$, which has one or two more poles than zeros, so $(\beta_0 s + \gamma_0)^{-1}$ cannot be used to obtain a positive real $H(s)$.

Finally, we would like to emphasize that as in the case of the criterion of Nyquist for LTI system stability analysis, the great merit of the various forms of the Popov criterion lie in the simple and direct geometric interpretation (Popov [3]) of the frequency domain conditions for absolute $\{F\}$ stability. This is considered in detail in Chapter VII.

The original result of Popov [1, 2] appeared in 1960 and 1961. Popov also investigated the relation between his frequency domain criterion and the existence of a Lyapunov function for the system. Yakubovich [1] and Kalman [2] treated this problem extensively, and for the principle case both showed that Popov's criterion is necessary and sufficient for the existence of a Lyapunov function that is a quadratic form in the state variables and τ plus an integral of the nonlinearity. Yakubovich [2, Part I] treated discontinuous nonlinearities and considered the problem of stability of motion with an input present; in this latter case, slope restrictions were found to be necessary. The historical credits for all the developments related to Popov's criterion are well documented in Aizerman and Gantmacher [1], Lefschetz [1], and Hahn [1].

It is noted in Chapter I that following the pioneering work of Popov, several new approaches have been suggested for the stability analysis of dynamic systems. Among these, the functional analysis approach has proved to be very successful. The stability definitions using this approach are not the same as those of Lyapunov; instead, they are based on the input–output analysis of feedback control systems. Many authors, notably Zames [1], Sandberg [3, 5], and Desoer [1], have contributed significantly in this field and have derived stability conditions for systems with different classes of inputs.

2. Stability Criteria for Monotonic Nonlinearities

The nonlinear function $f(\sigma_0)$ considered in the above section is so unrestricted that the frequency domain condition that must be satisfied by the LTI plant $W(s)$ in order to guarantee absolute stability is sometimes very conservative, that is, the sector of absolute stability may be small compared to the Hurwitz sector. It is thus to be expected that as $f(\sigma_0)$ is more strictly constrained (or more closely specified), the frequency domain conditions on $W(s)$ may be relaxed. This idea provides the basis for the problems investigated in Section 2 to Section 4. In the subsequent developments, criteria of the basic forms given in Criteria 1b and 1c are derived.

The conditions obtained in Section 1 may be viewed as requiring the determination of a frequency domain multiplier $Z(s)$ such that

$$Z^{\pm 1}(s) \in \{Z_F(s)\} \triangleq \{Z(s) = (\beta_0 s + \gamma_0), \quad \beta_0 \geqslant 0, \ \gamma_0 > 0\}, \qquad (5\text{-}11)$$

and one of the following conditions is met:

$$[W(s) + \bar{F}^{-1}]Z(s) \in \{\mathrm{PR}\}, \qquad A \in \{A_1\},$$

or

$$[W(s) + \bar{F}^{-1}]Z(s) \in \{\mathrm{SPR}_0\}, \qquad A \in \{A_0\}. \qquad (5\text{-}12)$$

It may thus be said that $\{Z_F\}$ defines a set of frequency domain multipliers for the problem considered by Popov, where $f(\sigma_0) \in \{F\}$.

The generalization of Criteria 1b and 1c that is sought for restricted classes of nonlinear gains, $f(\sigma_0) \in \{N\}$, is that absolute $\{N\}$ stability is guaranteed if some $Z(s)$ exists such that $Z^{\pm 1}(s) \in \{Z_N(s)\}$ and condition (5-12) is satisfied.

This implies that as the nonlinearity $f(\sigma_0)$ is restricted by assuming that it belongs to some class $\{N\}$, a frequency domain multiplier of some class must be chosen in an attempt to satisfy condition (5-12); the main problem of the remainder of this chapter is essentially that of ascertaining the correspondence between nonlinearity classes and multiplier classes.

In Section 5 of Chapter III, the properties of general rational positive real functions are investigated. The class $\{Z_F(s)\}$, defined in Eq. (5-11), is a specific subset of strictly positive real functions. The general multiplier classes $\{Z_N(s)\}$ generated in the following sections also consist of strictly positive real functions and include $\{Z_F(s)\}$ as a subclass, since $\{N\} \subset \{F\}$. Since rational positive real functions may be realized in network theoretic terms as driving point impedances of electrical networks consisting of passive elements [resistance (R), inductance (L), and capacitance (C)], we sometimes designate a class by the elements required for its realization. For example, the multiplier $Z(s)$ used in the Popov criterion is realizable either as the impedance of a series combination of resistance and inductance $(R + s\mathrm{L})$ or as a parallel combination of resistance and capacitance $(\mathrm{R}^{-1} + s\mathrm{C})^{-1}$, and hence can be considered to be a member of a subclass of $\{Z_{\mathrm{RL}}\} \cup \{Z_{\mathrm{RC}}\}$. By convention, we consider the RL function $(\beta_0 s + \gamma_0)$ to define the fundamental class of frequency domain multipliers for the problem of Lur'e and Postnikov and the RC multiplier to be subsidiary; this practice leads to a more convenient formulation of later results (see General Stability Criterion 2). The realization of a class of multipliers is particularly important in considering monotonic gains.

The general system considered next is described by the state vector differential equation (5-1), constrained by

$$\begin{aligned} &f(\sigma_0) \in \{F_m\}, && \Delta f(\sigma_0)/\Delta\sigma_0 \in [0, \bar{M}], \\ &A \in \{A_1\}, && A_{\bar{M}} \in \{A_1\}. \end{aligned} \tag{5-13}$$

The analysis of such a system is necessarily somewhat more complex than that just completed, so the presentation is subdivided into several steps.

A. System augmentation is introduced to allow for the generation of frequency domain multipliers $Z(s)$ with arbitrary poles for the infinite sector case.

B. The infinite sector problem, that is $0 \leqslant \Delta f(\sigma_0)/\Delta\sigma_0 < \infty$ for all finite σ_0, is treated, and the class $\{Z_{F_m}\}$ is obtained.

C. A transformation is presented which converts a finite sector problem, namely, $\Delta f(\sigma_0)/\Delta\sigma_0 \in [0, \bar{M})$, into an infinite sector problem that may be analyzed as in B; the case $\Delta f(\sigma_0)/\Delta\sigma_0 \in [0, M]$ is then considered.

D. An inversion–transformation procedure is given that allows an inverse multiplier to be used in the stability analysis of some systems, that is, if $Z^{-1}(s) \in \{Z_{F_m}\}$, then $Z(s)$ may be used in condition (5-12).

A. System Augmentation

We are ultimately interested in obtaining frequency domain multipliers that are rational; as a simple example, say $Z = (s + \lambda)/(s + \eta)$. The frequency domain condition for the infinite sector problem that $W(s)Z(s) \in \{\text{PR}\}$ arises from the MKY lemma requirement that $H(s) \triangleq \{\frac{1}{2}\psi + k^{\text{T}}(sI - A)^{-1}b\} \in \{\text{PR}\}$. The denominator of $H(s)$ is determined solely by the A matrix, so as matters stand, $W(s)$ and $W(s)Z(s)$ must have the same poles. This problem has been circumvented in two ways.

(1) *Pole-Zero Cancellation*

Choose $Z(s)$ such that poles of $Z(s)$ [$s = -\eta$ in our example] are zeros of $W(s)$; then $Z(s)W(s)$ differs from $W(s)$ only in its numerator terms (zeros), which are determined by ψ and k:

$$W(s) = \rho + h^{\text{T}}(sI - A)^{-1}b \triangleq \frac{n(s)}{d(s)} \triangleq \frac{n'(s) \cdot (s + \eta)}{d(s)}$$

$$Z(s) = (s + \lambda)/(s + \eta)$$

$$\therefore \; Z(s)W(s) \triangleq \tfrac{1}{2}\psi + k^{\text{T}}(sI - A)^{-1}b = \frac{n'(s) \cdot (s + \lambda)}{d(s)}.$$

In a simplified system model with $\rho = 0$, a lemma due to Narendra and Neuman [1] may be used to accomplish this objective; here it is modified to become applicable to situations when $\rho \neq 0$.

LEMMA 5-1a. If $[W(s)]_{s=-\eta} = \rho + h^{\text{T}}(-\eta I - A)^{-1}b = 0$, then define

$$c^{\text{T}} \triangleq h^{\text{T}}(\eta I + A)^{-1}; \tag{5-14}$$

this vector† satisfies

(a) $$c^{\text{T}}b = \rho \tag{5-15}$$

(b) $$c^{\text{T}}(sI - A)^{-1}b = W(s)/(s + \eta) \tag{5-16}$$

(c) $$c^{\text{T}}Ax = h^{\text{T}}x - \eta c^{\text{T}}x. \;\square \tag{5-17}$$

Proof:

(a) $c^{\text{T}}b = -[h^{\text{T}}(sI - A)^{-1}b]_{s=-\eta} \triangleq \rho$;

(b) $h^{\text{T}}(\eta I + A)^{-1}(sI - A)^{-1}b$

$$= h^{\text{T}}(\eta I + A)^{-1}\left[\frac{(\eta I + A) + (sI - A)}{s + \eta}\right](sI - A)^{-1}b$$
$$= (s + \eta)^{-1}[h^{\text{T}}(sI - A)^{-1}b + c^{\text{T}}b],$$

which by substitution of $c^{\text{T}}b = \rho$ establishes the result;

(c) $$c^{\text{T}}Ax = h^{\text{T}}(\eta I + A)^{-1}[(\eta I + A) - \eta I]x$$
$$= h^{\text{T}}x - \eta h^{\text{T}}(\eta I + A)^{-1}x \triangleq h^{\text{T}}x - \eta c^{\text{T}}x. \;\square$$

† The vector c exists only if $(\eta I + A)^{-1}$ exists, that is, only if $(-\eta)$ is not an eigenvalue of A.

This approach is often quite restrictive. If $W(s)$ has only complex zeros, then a multiplier with real poles could not be used, although there is no reason to believe that a multiplier $Z(s)$ with real poles cannot exist such that $W(s)Z(s) \in \{PR\}$. For this reason, a different technique is used.

(2) *Increased System Dimensionality*

The problem that is encountered when pole-zero cancellation may not be used is in achieving the correct denominator for $Z(s)W(s)$; since we note that poles of ZW must be eigenvalues of the A matrix, a second solution is to modify the A matrix. This is accomplished by creating an augmented system defined by $\{\rho_a, h_a, A_a, b_a\}$; for our example the parameters of this set must satisfy

$$|sI - A_a| = |sI - A| \cdot (s + \eta),$$

and

$$W_a(s) \triangleq \rho_a + h_a^{\mathrm{T}}(sI - A_a)^{-1}b_a = W(s) \cdot (s + \eta)/(s + \eta);$$

clearly

$$\rho_a = \rho, \qquad h_a = \begin{bmatrix} 0 \\ \hline h \end{bmatrix} + \eta \begin{bmatrix} h \\ \hline 0 \end{bmatrix}, \qquad b_a = \begin{bmatrix} 0 \\ \hline b \end{bmatrix},$$

and

$$A_a = \left[\begin{array}{c|c} 0 & \\ 0 & \\ \cdot & I \\ \cdot & \\ \cdot & \\ 0 & \\ \hline -\eta a_1 & -(a_1 + \eta a_2), \ldots, -(a_n + \eta) \end{array}\right]$$

completely specify the desired augmented system indicated below [Eq. (5-1a)] in phase variable canonical form. Once this system is established, it is possible to find a vector c as in Lemma 5-la that satisfies $c^{\mathrm{T}}(sI - A_a)^{-1}b_a = W_a(s)/(s + \eta)$; by inspection,

$$c = \begin{bmatrix} \\ h \\ \\ \hline 0 \end{bmatrix} + \rho \begin{bmatrix} a_1 \\ a_2 \\ \cdot \\ \cdot \\ \cdot \\ a_n \\ 1 \end{bmatrix}.$$

It is not possible to express c in the form (5-14), as $(-\eta)$ is an eigenvalue of

A_a and thus $[\eta I + A_a]^{-1}$ does not exist. The vector c defined above may be seen to satisfy Eqs. (5-15) and (5-17), however.

The use of this second approach does not preclude the use of pole-zero cancellation in considering $H(s) \triangleq W_a(s)Z(s)$. It can be appreciated that pole-zero cancellation is of great practical utility where it is possible to use it, since the difficulty of showing that $H(s) \in \{PR\}$ increases rapidly as $(n + m)$ increases where m is the number of poles added in $W_a(s)$.

With the preceding motivation, the procedure of augmentation and its implications can be considered precisely. From Eqs. (2-3) and (2-4),

$$W(s) = \rho + \frac{h_n s^{n-1} + \cdots h_2 s + h_1}{s^n + a_n s^{n-1} + \cdots + a_2 s + a_1} = \frac{\rho s^n + p_n s^{n-1} + \cdots + p_1}{s^n + a_n s^{n-1} + \cdots + a_2 s + a_1}. \tag{5-18}$$

The transfer function of the augmented system may be expressed as

$$W_a(s) = W(s) \cdot \theta(s)/(\theta(s)), \tag{5-19}$$

where

$$\theta(s) \triangleq s^m + g_m s^{m-1} + \cdots + g_2 s + g_1 \tag{5-20}$$

is the polynomial with roots (real and negative in this instance) that are the desired poles of $Z(s)$ which are not zeros of $W(s)$. The state vector representation of the augmented system is given by

$$\dot{z} = A_a z + b_a \tau_a, \qquad \sigma_a = h_a^{\mathrm{T}} z + \rho_a \tau_a, \quad \tau_a = -f(\sigma_a), \tag{5-1a}$$

where ρ_a, h_a, A_a and b_a are defined as

$$\rho_a = \rho, \qquad h_a = \begin{bmatrix} 0 \\ 0 \\ \cdot \\ \cdot \\ \cdot \\ 0 \\ 0 \\ \hline h \end{bmatrix} + g_m \begin{bmatrix} 0 \\ 0 \\ \cdot \\ \cdot \\ \cdot \\ 0 \\ \hline h \\ \hline 0 \end{bmatrix} + \cdots + g_2 \begin{bmatrix} 0 \\ \hline h \\ \hline 0 \\ \cdot \\ \cdot \\ \cdot \\ 0 \\ 0 \end{bmatrix} + g_1 \begin{bmatrix} h \\ \hline 0 \\ 0 \\ \cdot \\ \cdot \\ \cdot \\ 0 \\ 0 \end{bmatrix},$$

$$b_a = \begin{bmatrix} 0 \\ 0 \\ \cdot \\ \cdot \\ \cdot \\ 0 \\ 0 \\ \hline b \end{bmatrix}, \qquad A_a = \left[\begin{array}{c|c} \begin{matrix} 0 \\ 0 \\ \cdot \\ \cdot \\ \cdot \\ 0 \\ 0 \end{matrix} & I \\ \hline -a_1 g_1 & -(a_1 g_2 + a_2 g_1), \ldots, -(g_m + a_n) \end{array}\right]. \tag{5-21}$$

As mentioned earlier, a difficulty encountered in using the augmented system arises in the application of Lemma 5-1a. Since the vector c as defined in Eq. (5-14) cannot be explicitly determined in this manner, the following constructive procedure is adopted. If $(s + \eta_i)$, $i = 1, 2, \ldots, m$, are factors of $\theta(s)$, then

$$\theta(s) = (s + \eta_i)\theta^i(s) \triangleq (s + \eta_i)[s^{m-1} + g^i_{m-1}s^{m-2} + \cdots + g_1{}^i].$$

Appropriate vectors c_i may be defined as follows.

LEMMA 5-1b. If

$$c_i \triangleq \begin{bmatrix} 0 \\ 0 \\ \cdot \\ \cdot \\ \cdot \\ 0 \\ 0 \\ p_1 \\ p_2 \\ \cdot \\ \cdot \\ \cdot \\ p_n \\ \rho \end{bmatrix} + g^i_{m-1} \begin{bmatrix} 0 \\ 0 \\ \cdot \\ \cdot \\ \cdot \\ 0 \\ p_1 \\ p_2 \\ \cdot \\ \cdot \\ \cdot \\ p_n \\ \rho \\ 0 \end{bmatrix} + \cdots + g_1{}^i \begin{bmatrix} p_1 \\ p_2 \\ \cdot \\ \cdot \\ \cdot \\ p_n \\ \rho \\ 0 \\ 0 \\ \cdot \\ \cdot \\ \cdot \\ 0 \\ 0 \end{bmatrix}; \tag{5-22}$$

then this vector satisfies

(i) $$c_i{}^{\mathrm{T}}b_a = \rho \tag{5-23}$$

(ii) $$c_i{}^{\mathrm{T}}(sI - A_a)^{-1}b_a = W_a(s)/(s + \eta_i) = W(s)/(s + \eta_i) \tag{5-24}$$

(iii) $$c_i{}^{\mathrm{T}}A_a z = h_a{}^{\mathrm{T}}z - \eta_i c_i{}^{\mathrm{T}}z. \quad \square \tag{5-25}$$

Relations (5-23)–(5-25), which may be proved by direct expansion, correspond to the relations (5-15)–(5-17), so that identical procedures may be used for both the original and the augmented systems.

Before considering the absolute stability of the augmented system, we must prove the equivalence of the stability properties of Eqs. (5-1) and (5-1a). Since this same procedure is used in considering time-varying systems, a lemma regarding its validity is stated in general terms. The time-invariant case used in this chapter is due to Brockett and Willems [1], who utilized it for the same purpose.

First we express the system equations in the form of ordinary scalar differential equations: If $W(s) \triangleq n(s)/d(s)$, then as in Eq. (2-5), the most general system modeled by Eq. (2-6) is equivalent to

$$d(D)\xi + g[n(D)\xi, t] = 0 \tag{5-26}$$

in differential operator notation, where $D^m \triangleq d^m/dt^m$. The augmented system, similarly, is of the form

$$d(D)\theta(D)\zeta + g[n(D)\theta(D)\zeta, t] = 0, \tag{5-26a}$$

where $\theta(D)$ is defined in Eq. (5-20).

LEMMA 5-2. If $\theta(D)\zeta = 0$ is an asymptotically stable differential equation, then the original system (5-26) is uniformly asymptotically stable in the whole if and only if the augmented system (5-26a) has the same property. □

Proof: If $\xi(t;\, \xi_0, t_0)$ is a solution to Eq. (5-26), then $\zeta(t)$, defined by the differential equation

$$\theta(D)\zeta(t) = \xi(t), \tag{5-27}$$

is a solution to Eq. (5-26a). Since $\xi(t)$ and $\zeta(t)$ are related by this linear time-invariant differential equation, it is well known that all solutions $\xi(t)$ correspond to all solutions $\zeta(t)$, that is, the set of solutions to Eq. (5-26) corresponds in a one-to-one manner to that of (5-26a).

(a) If $\zeta(t)$ is an asymptotically stable solution to Eq. (5-26a), then $\theta(D)\zeta(t)$ is also bounded and goes to zero as $t \longrightarrow \infty$; this property is not dependent upon the stability of $\theta(D)\zeta = 0$ as $\xi(t)$ is obtained as a linear combination of $\zeta(t), D\zeta, \ldots, D^m\zeta$.

(b) If $\xi(t)$ is an asymptotically stable solution to Eq. (5-26) and $\theta(D)\zeta = 0$ is asymptotically stable, then $\zeta(t)$, the solution to the nonhomogeneous linear time-invariant differential equation (5-27), is likewise asymptotically stable. □

B. *The Infinite Sector Problem*

Now consider the system defined by Eq. (5-1a) with

$$f(\sigma_a) \in \{F_m\}, \qquad \Delta f(\sigma_a)/\Delta\sigma_a \in [0, \infty), \qquad A \in \{A_1\},$$

and the absolute Lyapunov function candidate

$$v(z) = \tfrac{1}{2}z^{\mathrm{T}}Pz + \beta_0 \int_0^{\sigma_a} f(\zeta)\, d\zeta + \tfrac{1}{2}\beta_0\rho[f(\sigma_a)]^2 + \sum_{i=1}^{k} \beta_i \int_0^{\sigma_i} f(\zeta)\, d\zeta, \tag{5-28}$$

where $P = P^{\mathrm{T}} > 0$, $\beta_0 \geqslant 0$, $\rho \geqslant 0$, $\beta_i \geqslant 0$, $i = 1, 2, \ldots, k$, and $f(\sigma_a) \in \{F_m\} \subset \{F\}$ ensure that it is valid. Choose the upper limits of the newly introduced integral terms to be

$$\sigma_i \triangleq r_i^{\mathrm{T}}z = (\gamma_i/\beta_i)c_i^{\mathrm{T}}z, \qquad i = 1, 2, \ldots, k, \tag{5-29}$$

where c_i are the vectors defined in Eq. (5-14) or (5-22). From Eq. (3-4), $\dot{v}$ is

$$\dot{v} = \tfrac{1}{2}z^{\mathrm{T}}(A_a{}^{\mathrm{T}}P + PA_a)z - f(\sigma_a)z^{\mathrm{T}}[Pb_a - \beta_0 A_a{}^{\mathrm{T}}h_a - \gamma_0 h_a] - (\beta_0 h_a{}^{\mathrm{T}}b_a + \gamma_0\rho)[f(\sigma_a)]^2 - \gamma_0\sigma_a f(\sigma_a) + \sum_{i=1}^{k} \gamma_i f(\sigma_i)c_i{}^{\mathrm{T}}[A_a z - b_a f(\sigma_a)].$$

The last term of $\dot{v}$ may be simplified by using Lemma 5-1a or Lemma 5-1b and Eq. (5-29) to become

$$\sum_{i=1}^{k} \gamma_i f(\sigma_i)[\sigma_a - \eta_i(\beta_i/\gamma_i)\sigma_i],$$

which may in turn be expanded to obtain

$$\begin{aligned}\sum_{i=1}^{k} \gamma_i f(\sigma_i)[\sigma_a - \eta_i(\beta_i/\gamma_i)\sigma_i] &= -f(\sigma_a)z^{\mathrm{T}}\Big[-\sum_{i=1}^{k}\gamma_i(h_a - r_i)\Big] \\ &\quad - \sum_{i=1}^{k}\gamma_i\rho[f(\sigma_a)]^2 - \sum_{i=1}^{k}(\beta_i\eta_i - \gamma_i)\sigma_i f(\sigma_i) \\ &\quad - \sum_{i=1}^{k}\gamma_i(\sigma_a - \sigma_i)[f(\sigma_a) - f(\sigma_i)].\end{aligned}$$

The reasons behind this more complex formulation are seen by inspecting the right hand side term-by-term: when substituted into $\dot{v}$, the first term becomes part of the expression $-f(\sigma_a)z^{\mathrm{T}}[Pb - k]$, the second term becomes part of $-\tfrac{1}{2}\psi[f(\sigma_a)]^2$, and the third and fourth terms are negative semidefinite if $\beta_i\eta_i \geqslant \gamma_i \geqslant 0$. That the last term is never negative is a direct consequence of the definition of the class $\{F_m\}$. By substitution, then,

$$\begin{aligned}\dot{v} &= \tfrac{1}{2}z^{\mathrm{T}}(A_a{}^{\mathrm{T}}P + PA_a)z - f(\sigma_a)z^{\mathrm{T}}[Pb_a - \beta_0 A_a{}^{\mathrm{T}}h_a - \gamma_0 h_a - \sum_{i=1}^{k}\gamma_i(h_a - r_i)] \\ &\quad - (\beta_0 h_a{}^{\mathrm{T}}b_a + \sum_{i=0}^{k}\gamma_i\rho)[f(\sigma_a)]^2 - \gamma_0\sigma_a f(\sigma_a) \\ &\quad - \sum_{i=1}^{k}(\eta_i\beta_i - \gamma_i)\sigma_i f(\sigma_i) - \sum_{i=1}^{k}\gamma_i(\sigma_a - \sigma_i)[f(\sigma_a) - f(\sigma_i)],\end{aligned} \tag{5-30}$$

which is now in a form appropriate for the application of the MKY lemma. Identify

$$\begin{aligned} k &\triangleq \beta_0 A_a{}^{\mathrm{T}}h_a + \gamma_0 h_a + \sum_{i=1}^{k}\gamma_i(h_a - r_i), \\ \tfrac{1}{2}\psi &\triangleq \beta_0 h_a{}^{\mathrm{T}}b_a + \sum_{i=0}^{k}\gamma_i\rho;\end{aligned} \tag{5-31}$$

then corresponding to some $P = P^{\mathrm{T}} > 0$ [Eq. (5-28)], a real vector q and a matrix $M = M^{\mathrm{T}} \geqslant 0$ exist such that

$$\begin{aligned}\dot{v} &= -\tfrac{1}{2}[z^{\mathrm{T}}q + \sqrt{\psi}\, f(\sigma_a)]^2 - \tfrac{1}{2}z^{\mathrm{T}}Mz - \gamma_0\sigma_a f(\sigma_a) \\ &\quad - \sum_{i=1}^{k}(\beta_i\eta_i - \gamma_i)\sigma_i f(\sigma_i) - \sum_{i=1}^{k}\gamma_i(\sigma_a - \sigma_i)[f(\sigma_a) - f(\sigma_i)],\end{aligned} \tag{5-32}$$

if and only if

$$\begin{aligned}\hat{H}(s) \triangleq\ & \beta_0[h_a{}^{\mathrm{T}}b_a + h_a{}^{\mathrm{T}}A_a(sI - A_a)^{-1}b_a] \\ & + \gamma_0[\rho + h_a{}^{\mathrm{T}}(sI - A_a)^{-1}b_a] \\ & + \sum_{i=1}^{k} \gamma_i[\rho + (h_a - r_i)^{\mathrm{T}}(sI - A_a)^{-1}b_a]\end{aligned}$$

satisfies $\hat{H}(s) \in \{\mathrm{PR}\}$. The first two terms yield the Popov multiplier term as in Section 1. The last summation may be simply evaluated by using Lemma 5-1a or 5-1b to be

$$\sum_{i=1}^{k} \gamma_i(1 - (\gamma_i/\beta_i)(s + \eta_i)^{-1})W_a(s).$$

The zero at $s = -\eta_i + \gamma_i/\beta_i$ lies in the range $(-\eta_i, 0]$, as we must choose β_i under the constraint $\beta_i\eta_i - \gamma_i \geqslant 0$ to keep the fourth term of $\dot{v}$ negative semidefinite. The frequency domain condition thus may be expressed in the standard form:

$$\begin{aligned}H(s) &\triangleq W_a(s)Z(s) \\ &= W_a(s)\left[(\gamma_0 + \beta_0 s) + \sum_{i=1}^{k} \gamma_i \frac{s + (\eta_i - \gamma_i/\beta_i)}{s + \eta_i}\right] \in \{\mathrm{PR}\}. \qquad (5\text{-}33)\end{aligned}$$

The simplicity of the statement of this frequency domain condition for absolute stability stems from the observation that the class of RL functions is defined by

$$\{Z_{\mathrm{RL}}(s)\} \triangleq \left\{Z(s) = \prod_{i=1}^{l} (s + \lambda_i)\Big/\prod_{i=1}^{k} (s + \eta_i); \quad l = k \ \text{ or } \ k + 1, \ \ 0 < \lambda_1 < \eta_1 < \lambda_2 < \eta_2 \ldots\right\}$$

(within a constant factor) if $\{Z_{\mathrm{RL}}\} \subset \{\mathrm{SPR}\}$ is assumed; otherwise $\lambda_1 = 0$ is possible, but that case is not considered for the same reason $\gamma_0 = 0$ is not treated in Section 1. The most important points are that the first singularity is a (real) zero, and that poles and zeros alternate. This form may readily be expanded into

$$\{Z_{\mathrm{RL}}(s)\} \triangleq \left\{Z = (s + \lambda_1)\left[\beta_0 + \sum_{i=1}^{k} \gamma_i/(s + \eta_i)\right]; \quad 0 < \lambda_1 < \min_i\{\eta_i\}, \quad \beta_0 = 0 \ \text{ if } \ l = k, \ \ \beta_0 = 1 \text{ if } l = k + 1; \quad \gamma_i > 0\right\}. \qquad (5\text{-}34)$$

Returning to $H(s)$ [Eq. (5-33)] and defining the parameters γ_0 and β_i appropriately,

$$\gamma_0 = \beta_0\lambda_1 \geqslant 0, \qquad \beta_i = \gamma_i/(\eta_i - \lambda_1) > 0, \qquad (5\text{-}35)$$

where $\beta_i > 0$ is guaranteed by the condition that λ_1 is smaller than any pole η_i, it is demonstrated that $\{Z_{F_m}(s)\} = \{Z_{\mathrm{RL}}(s)\}$ as it is defined in Eq. (5-34). Hence for the absolute stability of the system under consideration, a sufficient condition is that some $Z(s) \in \{Z_{RL}\}$ exists such that $W_a(s)Z(s)$ is positive real. If this condition is satisfied, then

$$\dot{v} = -\tfrac{1}{2}[z^T q + \sqrt{\psi}\, f(\sigma_a)]^2 - \tfrac{1}{2} z^T M z - \beta_0 \lambda_1 \sigma_a f(\sigma_a)$$
$$-\sum_{i=1}^{k} \beta_i \lambda_1 \sigma_i f(\sigma_i) - \sum_{i=1}^{k} \gamma_i (\sigma_a - \sigma_i)[f(\sigma_a) - f(\sigma_i)],$$

all terms of which are nonpositive given the parameter constraints of $\{Z_{RL}\}$, that is, $\gamma_i > 0$, $\beta_i > 0$, $\beta_0 \geqslant 0$, $\lambda_1 > 0$. For absolute stability to be ensured by the frequency domain condition, it remains only to show that $\dot{v} \not\equiv 0$ unless $z \equiv 0$. This can be done following the same procedure used in the derivation of the Popov criterion.

C. *The Finite Sector Problem*

A finite sector problem, $f(\sigma_a) \in \{F_m\}$, $\Delta f(\sigma_a)/\Delta\sigma_a \in [0, \bar{M})$, may be analyzed using the result of Section B directly. The property that

$$0 \leqslant \frac{f(\sigma_a) - f(\hat{\sigma}_a)}{\sigma_a - \hat{\sigma}_a} < \bar{M}$$

for all σ_a and $\hat{\sigma}_a$ (Chapter II, Section 1) makes the following elementary transformation due to Rekasius and Gibson [1] (see also Thathachar, Srinath, and Krishna [1]) valid.

Define $W_1(s)$, σ_1 and $f_1(\sigma_1)$ by

$$\begin{aligned} W_1(s) &\triangleq L(\sigma_1)/L(\tau) \triangleq W(s) + \bar{M}^{-1}, \\ \sigma_1 &\triangleq \sigma_a - f(\sigma_a)/\bar{M}, \\ f_1(\sigma_1) &= f(\sigma_a). \end{aligned} \tag{5-36}$$

Figure 5-1 demonstrates that the feedback system having $W_1(s)$ in the forward

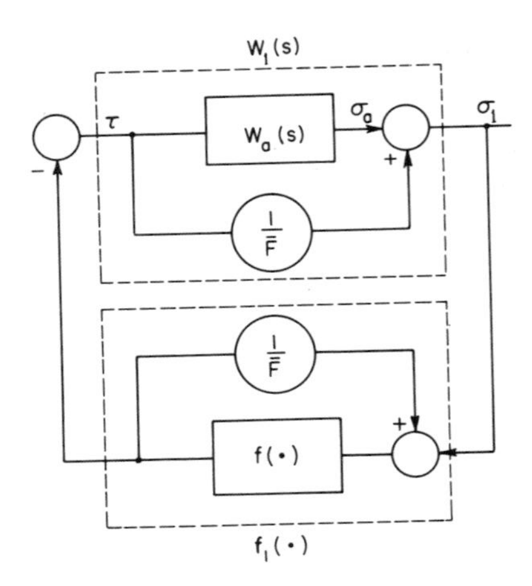

Fig. 5-1. The finite sector/infinite sector transformation.

path and $f_1(\cdot)$ in the reverse path is equivalent to the original system (5-1a) in both its behavior and its mathematical form:

$$\begin{aligned} \dot{z} &= Az + b\tau, \quad A \in \{A_1\}, \\ \sigma_1 &= h^{\mathrm{T}}z + (\rho + \bar{M}^{-1})\tau \\ \tau &= -f_1(\sigma_1). \end{aligned} \tag{5-37}$$

It only remains to be shown that $f_1(\sigma_1)$ is a monotonic gain lying in the infinite sector: define

$$m \triangleq \frac{f(\sigma_a) - f(\hat{\sigma}_a)}{\sigma_a - \hat{\sigma}_a}, \qquad m_1 \triangleq \frac{f_1(\sigma_1) - f_1(\hat{\sigma}_1)}{\sigma_1 - \hat{\sigma}_1},$$

where

$$\sigma_1 = \sigma_a - f(\sigma_a)/\bar{M}, \qquad \hat{\sigma}_1 = \hat{\sigma}_a - f(\hat{\sigma}_a)/\bar{M}.$$

By direct substitution, $m_1 = m/(1 - m/\bar{M})$; if $m \in [0, \bar{M})$, then $m_1 \in [0, \infty)$.

The result for the finite sector problem $\Delta f(\sigma_a)/\Delta\sigma_a \in [0, \bar{M})$ follows directly using the transformation (5-36); absolute stability is guaranteed if $Z(s) \in \{Z_{\mathrm{RL}}(s)\}$ exists such that

$$W_1(s)Z(s) \triangleq [W(s) + \bar{M}^{-1}]Z(s) \in \{\mathrm{PR}\}. \tag{5-38}$$

This condition is of the desired form (5-12), and is the direct extension of Criterion 1a.

As in the alternative case of the Popov criterion (Criterion 1b), the sector within which the nonlinearity is constrained to lie may be less restricted if one further frequency domain condition is applied: $\Delta f(\sigma_a)/\Delta\sigma_a \in [0, \bar{M}]$ is permitted if we assume that $[1 + \bar{M}W(i\omega)] \neq 0$, $\omega \in [-\infty, \infty]$, that is, that $A_{\bar{M}} \in \{A_1\}$. In the developments below, which allow $Z^{-1}(s)$ to be used as a frequency domain multiplier, we see that a certain degree of symmetry is revealed if this type of criterion is considered.

D. *Multiplier Inversion*

We note in the Popov criterion that $(\beta_0 s + \gamma_0)^{\pm 1}$ can be used as a frequency domain multiplier. In the case that the Nyquist diagram of $[W(i\omega) + \bar{F}^{-1}]$ lies in quadrants 1, 3, and 4 [that is, $\measuredangle[W(i\omega) + \bar{F}^{-1}] \triangleq \phi_F(\omega) \in (-\pi, +\pi/2)$], the special RL multiplier $(\beta_0 s + \gamma_0)$ is used, while if $\phi_F(\omega)$ lies in the range $(-\pi/2, +\pi)$, the RC multiplier $(\beta_0 s + \gamma_0)^{-1}$ must be utilized. Similarly, for $f(\sigma_0) \in \{F_m\}$, it would be expected that $Z(s) \in \{Z_{\mathrm{RC}}\}$ could be used to guarantee absolute stability, as $\{Z_{\mathrm{RC}}(s)\} \triangleq \{Z_{\mathrm{RL}}^{-1}(s)\}$.

The standard approach used to obtain this last result is inversion, which necessitates three transformations. For monotonic gains in the finite sector, we consider

(1)
$$W_1(s) \triangleq \frac{W(s)}{1 - \varepsilon W(s)};$$
$$f_1(\sigma_a) \triangleq f(\sigma_a) + \varepsilon\sigma_a;$$
$$f_1(\sigma_a) \in \{F_m\}, \qquad \Delta f_1(\sigma_a)/\Delta\sigma_a \in [\varepsilon, \bar{M} + \varepsilon].$$

This rearrangement of the system elements does not affect the stability properties of the system (see Fig. 5-3). Hence, the original system $\{W, f\}$ is absolutely stable if $\{W_1, f_1\}$ is guaranteed to be absolutely stable.

(2)
$$\begin{aligned} W_2(s) &\triangleq (W_1)^{-1} = (1 - \varepsilon W(s))/W(s) \triangleq L(\sigma_b)/L(\tau_b), \\ \tau_b &\triangleq -f_2(\sigma_b) \triangleq -f_2(f_1(\sigma_a)) \triangleq -\sigma_a; \\ f_2(\sigma_b) &\in \{F_m\}, \quad \Delta f_2(\sigma_b)/\Delta\sigma_b \in [(\bar{M} + \varepsilon)^{-1}, \varepsilon^{-1}]. \end{aligned} \tag{5-39}$$

This transformation represents the actual inversion procedure. If the system $\{W_1(s), f_1(\sigma_a)\}$ is represented by the nth-order scalar differential equation

$$[D^n + \hat{a}_n D^{n-1} + \cdots + \hat{a}_2 D + \hat{a}_1]\xi + f[(\hat{p}D^n + \hat{p}_n D^{n-1} + \cdots + \hat{p}_2 D + \hat{p}_1)\xi] = 0,$$

which is the NLTI form of Eq. (2-5), then the inverted system is

$$[\hat{p}D^n + \hat{p}_n D^{n-1} + \cdots + \hat{p}_2 D + \hat{p}_1]\xi + f^{-1}[(D^n + \hat{a}_n D^{n-1} + \cdots + \hat{a}_2 D + \hat{a}_1)\xi] = 0,$$

which is alternatively achieved by inverting $W(s)$ and interchanging the roles of σ and τ. The reason for the preliminary transformation is apparent; if the original gain range is not shifted by ε where $\varepsilon > 0$, then the inversion is not valid for $f(\sigma_0)/\sigma_0 \in [0, \bar{M}]$. The justification for treating the inverted system $\{W_2, f_2\}$ in order to determine the stability properties of the preceding system $\{W_1, f_1\}$ was first given by Brockett and Willems [1], using the latter scalar differential equation formulation.

(3)
$$\begin{aligned} W_3(s) &\triangleq \frac{W_2(s)}{1 + W_2(s)/(\bar{M} + \varepsilon)} = (\bar{M} + \varepsilon)\frac{[1 - \varepsilon W(s)]}{[1 + \bar{M}W(s)]} \\ f_3(\sigma_b) &\triangleq f_2(\sigma_b) - \sigma_b/(\bar{M} + \varepsilon), \\ f_3(\sigma_b) &\in \{F_m\}, \qquad \Delta f_3(\sigma_b)/\Delta\sigma_b \in [0, \bar{M}/(\varepsilon(\bar{M} + \varepsilon))]. \end{aligned} \tag{5-40}$$

This final alteration is used to shift the range of $f_3(\sigma_b)$ to have a lower bound of zero; it is the same type of transformation made in (1).

Before applying the result of Section C to the final system $\{W_3(s), f_3(\sigma_b)\}$, it is necessary to guarantee that this system is of the required form. The assumption that $A_{\bar{M}} \in \{A_1\}$ ensures that $W_3(s)$ is asymptotically stable and that it has no more zeros than poles. We should also be certain that $f_3(\sigma_b)$ is indeed a monotonic function lying in the indicated sector, that is, that

$$m_3 \triangleq \frac{f_3(\sigma_b) - f_3(\hat{\sigma}_b)}{\sigma_b - \hat{\sigma}_b} \in \left[0, \frac{\bar{M}}{\varepsilon(\bar{M} + \varepsilon)}\right].$$

By reversing the transformation,

$$m_3 = \frac{\bar{M} - m}{(\bar{M} + \varepsilon)(m + \varepsilon)}, \qquad m \triangleq \frac{f(\sigma_a) - f(\hat{\sigma}_a)}{\sigma_a - \hat{\sigma}_a},$$

which, since $m \in [0, \bar{M}]$, establishes the result.

The condition for absolute stability in the finite sector $[0, \bar{M}/(\varepsilon(\bar{M} + \varepsilon))]$ is applied to the third transformed system to yield the frequency domain restriction that $Z(s) \in \{Z_{\mathrm{RL}}(s)\}$ must exist such that

$$[W_3 + \varepsilon(\bar{M} + \varepsilon)/\bar{M}]Z(s) = [(\bar{M} + \varepsilon)/\bar{M}]^2[W(s) + \bar{M}^{-1}]^{-1}Z(s) \in \{\mathrm{PR}\},$$

which is equivalent to the condition

$$[W(s) + \bar{M}^{-1}]Z^{-1}(s) \in \{\mathrm{PR}\},$$

that is, an RC multiplier may be used in guaranteeing absolute stability.

The closed interval $[0, \bar{M}]$ has been considered so that the frequency domain criterion using either $Z(s)$ or $Z^{-1}(s)$ may be stated in a unified manner. If the interval $[0, \bar{M})$ were considered, then it would only be necessary that $A \in \{A_1\}$ for $Z^{+1}(s)$ to be used in Eq. (5-12), while $A \in \{A_1\}$ and $A_{\bar{M}} \in \{A_1\}$ are required if $Z^{-1}(s)$ is used.

Criterion 2a (Monotonic gains, principal case). The system described by Eqs. (5-1) and (5-13) is absolutely $\{F_m\}$ stable if there exists some $Z(s)$ such that $Z^{\pm 1}(s) \in \{Z_{\mathrm{RL}}\}$ and

$$[W(s) + \bar{M}^{-1}]Z(s) \in \{\mathrm{PR}\}. \ \square \tag{5-12}$$

The particular case, $A \in \{A_0\}$, indicated in Eq. (5-26) with $f(\sigma_0) \in \{F_m\}$ and $\Delta f(\sigma_0)/\Delta\sigma_0 \in (0, \bar{M}]$, may be treated in essentially the same manner with only a few modifications to take into account the use of Lemma 5 instead of the MKY lemma.

The class of multipliers $\{Z_{\mathrm{RL}}\}$ has the property that if $Z(s) \in \{Z_{\mathrm{RL}}\}$, then

$$0 \leqslant \arg Z(i\omega) < \pi/2,$$

as does the Popov multiplier $(\beta_0 s + \gamma_0)$; however, the phase angle no longer needs to increase monotonically from 0 to $\pi/2$. For this reason there are many situations when an RL multiplier may be used and $(\beta_0 s + \gamma_0)$ cannot. Several examples of $Z \in \{Z_{\mathrm{RL}}\}$ are shown in Fig. 5-2. The same comment applies vis-à-vis $\{Z_{\mathrm{RC}}\}$ and $(\beta_0 s + \gamma_0)^{-1}$, except that $\arg Z(i\omega) \in (-\pi/2, 0]$.

Frequency domain conditions sufficient to guarantee the stability of systems with a single monotonic gain were first sought by Zames [1, 2] and Brockett and Willems [1]. The criterion derived in the latter work is the same as Criterion 2a except that $df/d\sigma$ is assumed to exist everywhere, whereas an equivalent condition restricting $(f(\sigma_1) - f(\sigma_2))/(\sigma_1 - \sigma_2)$ is sufficient. The result of Zames is also similar to Criterion 2a. While the LTI portion of the system may be more general, that is, $W(s)$ is not necessarily rational, the nonlinearity is never allowed to occupy the infinite sector, as $0 \leqslant f(\sigma_0)/\sigma_0$

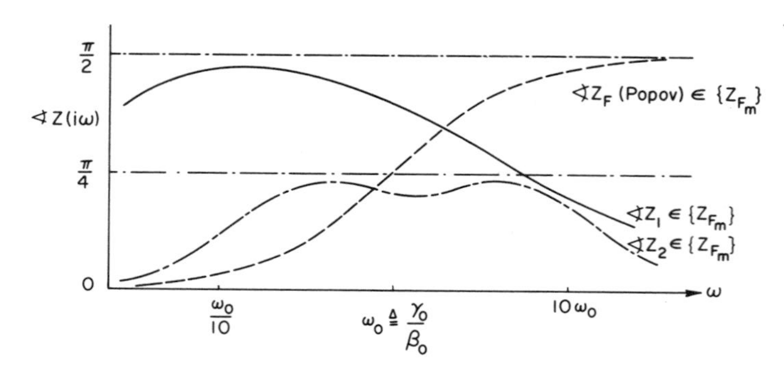

Fig. 5-2. Typical RL multipliers.

$< \bar{F} < \infty$ is always demanded. This restriction precludes the formal treatment of nonlinear gains such as $f(\sigma_0) = \kappa\sigma_0{}^3$. At about the same time Yakubovich [2, Part II], Dewey [1], Dewey and Jury [2], and Willems [1] reported stability criteria for slope restricted nonlinearities which are not expressed in terms of frequency domain multipliers, but are more nearly related to the analytic formulation of the off-axis circle criterion [Eq. (7-24)]. In 1966 O'Shea [1, 2] introduced a new approach to the study of stability of systems with a monotonic gain by use of the bounds of cross correlation functions; this approach was corrected and refined in Falb and Zames [1, 2], Zames and Falb [1], and further extended by Baker and Desoer [1].

The treatment presented here combines the multiple-integral Lyapunov function approach of Narendra and Neuman [1] with the augmentation method of Brockett and Willems [1] and the zero-shifting or finite sector–infinite sector transformation of Rekasius and Gibson [1].

If the nonlinearity lies in a sector $[0, \bar{F}]$ which is much smaller than the range of the slope, that is, if $\Delta f(\sigma_0)/\Delta\sigma_0 \in [0, \bar{M}]$ where $\bar{M} \gg \bar{F}$, then a less strict result due to Srinath and Thathachar [1] may be used.

CRITERION 2b (Monotonic gains, sector and slope restrictions). The system described by Eq. (5-1) with

$$0 \leqslant f(\sigma_0)/\sigma_0 \leqslant \bar{F}, \qquad 0 \leqslant \frac{f(\sigma_1) - f(\sigma_2)}{\sigma_1 - \sigma_2} \leqslant \bar{M} \gg \bar{F}, \qquad A \in \{A_1\}, \qquad A_{\bar{F}} \in \{A_1\}, \tag{5-41}$$

is absolutely stable if $Z(s) \in \{Z_{\mathrm{RL}}\}$ [Eq. (5-34)] exists such that

$$[W(s) + \bar{F}^{-1}]Z(s) - \sum_{i=1}^{m} \gamma_i(\bar{F}^{-1} - \bar{M}^{-1}) \in \{\mathrm{PR}\}, \tag{5-42}$$

where the parameters γ_i are defined in Eq. (5-34). □

A similar criterion for use for unrestricted slopes ($\bar{M} = \infty$) was proved by Narendra and Neuman [1]. Since the frequency domain condition is not of the standard type [Eq. (5-12)], we do not consider this result in detail.

The inversion transformation used in deriving Criterion 2a is found to be invalid in this case, so $Z^{-1}(s) \in \{Z_{RL}\}$ cannot be used as a multiplier.

3. Linear Systems

The preceding section demonstrates that as the class of nonlinear functions allowed in the feedback path is made more restrictive, the conditions on the linear part of the system that ensure absolute stability may be relaxed. The question naturally arises, in considering the limiting case when the feedback gain is linear, whether this approach can yield conditions equivalent to the Nyquist criterion, that is, conditions which are both necessary and sufficient for stability. In Chapter IV it is mentioned that the Nyquist condition to be satisfied for asymptotic stability with $\kappa \in [0, \bar{K})$ guarantees the existence of a shifted LC multiplier $Z(s) \in \{Z_\alpha(s)\}$ such that $[W(s) + \bar{K}^{-1}]Z(s) \in \{\text{PR}\}$. This in turn implies, as is briefly shown in this section, that a Lyapunov function that is quadratic in the state variables and τ, and linear in the gain parameter, namely

$$v(x) = \tfrac{1}{2}x^{\mathrm{T}}[P + \kappa M]x + \tfrac{1}{2}\beta_0 \rho \tau^2$$

exists to assure the stability of the system for $\kappa \in [0, \bar{K})$. This fundamental result relating the Nyquist criterion to the Lyapunov approach consequently provides an additional justification for the search for frequency domain multipliers discussed in this chapter, since the application of this technique to the LTI case is no more restrictive than the Nyquist criterion.

The frequency domain multipliers used in Section 2 for monotonic nonlinearities contained only real poles and zeros. For more restricted types of nonlinearities, we are interested in obtaining frequency domain multipliers with complex poles, for example $Z(s) \in \{Z_\alpha(s)\}$. While the techniques used in dealing with such systems are conceptually simple extensions of those presented in Section 2, the generation of multipliers with complex poles adds to the computational work involved in the analysis.

Consider

$$\tau = -\kappa\sigma_0, \qquad \kappa \in [0, \bar{K}), \quad A \in \{A_1\}, \tag{5-43}$$

where $[0, \bar{K})$ is one of the Nyquist ranges of the system described by Eqs. (5-1) and (5-43).

The canonical form of the class of impedance functions that may be realized using only inductance (L) and capacitance (C) is given in Eq. (4-42). The shifted LC function is identical, except that $(s + \alpha)$ is substituted for s:

$$\{Z_\alpha(s)\} \triangleq \Big\{Z = \beta_0(s+\alpha) + \frac{\gamma_1}{s+\alpha} + \sum_{i=1}^{k} \frac{\delta_i(s+\alpha)}{(s+\alpha)^2 + \mu_i^2}; \quad \alpha > 0,\ \beta_0 \geqslant 0,\ \gamma_1 \geqslant 0,\ \delta_i > 0,\ i = 1, 2, \ldots, k\Big\}. \tag{5-44}$$

By choosing $\alpha > 0$, it is guaranteed that $\{Z_\alpha\} \subset \{\text{SPR}\}$.

In deriving the frequency domain criterion, it is necessary to augment $W(s)$ by creating a matrix A_a which has both the eigenvalues of the original A matrix and additional eigenvalues corresponding to the poles of the multiplier $Z(s)$. If $Z(s)$ must have poles at $s = -\alpha$ and $s = -\alpha \pm i\mu_i$, $i = 1, 2, \ldots, k$, this augmentation procedure can be implemented by successively adding poles and zeros. For simplicity, we only consider one set of complex poles and zeros in all that follows, since the addition of a real pole has been treated. Let

$$W_a(s) = W(s)\frac{(s+\alpha)^2+\mu^2}{(s+\alpha)^2+\mu^2} = \rho_a + h_a^{\mathrm{T}}(sI - A_a)^{-1}b_a;$$

as in Eq. (5-21), we expand directly to obtain

$$\rho_a = \rho, \qquad b_a = \begin{bmatrix} 0 \\ 0 \\ \hline b \end{bmatrix}, \qquad h_a = \begin{bmatrix} 0 \\ 0 \\ \hline h \end{bmatrix} + 2\alpha \begin{bmatrix} 0 \\ \hline h \\ \hline 0 \end{bmatrix} + (\alpha^2+\mu^2)\begin{bmatrix} h \\ \hline 0 \\ 0 \end{bmatrix}, \tag{5-45}$$

and denoting the last row of A_a as $-\hat{a}_1, -\hat{a}_2, \ldots, -\hat{a}_{n+2}$, we have

$$\begin{bmatrix} \hat{a}_1 \\ \hat{a}_2 \\ \\ \cdot \\ \cdot \\ \cdot \\ \\ \hat{a}_{n+1} \\ \hat{a}_{n+2} \end{bmatrix} = \begin{bmatrix} 0 \\ 0 \\ a_1 \\ a_2 \\ \cdot \\ \cdot \\ \cdot \\ a_{n-1} \\ a_n \end{bmatrix} + 2\alpha \begin{bmatrix} 0 \\ a_1 \\ a_2 \\ \cdot \\ \cdot \\ \cdot \\ a_n \\ 1 \end{bmatrix} + (\alpha^2+\mu^2)\begin{bmatrix} a_1 \\ a_2 \\ \cdot \\ \cdot \\ \cdot \\ a_n \\ 1 \\ 0 \end{bmatrix}. \tag{5-46}$$

In order to obtain a general term in a multiplier $Z(s)$ that has these two poles, it is necessary to define two vectors that correspond to the vector c of Section 2.

$$d \triangleq \begin{bmatrix} h_1 \\ h_2 \\ \cdot \\ \cdot \\ \cdot \\ h_n \\ 0 \\ 0 \end{bmatrix} + \rho \begin{bmatrix} a_1 \\ a_2 \\ \cdot \\ \cdot \\ \cdot \\ a_n \\ 1 \\ 0 \end{bmatrix}, \qquad e \triangleq A_a^{\mathrm{T}} d = \begin{bmatrix} 0 \\ h_1 \\ h_2 \\ \cdot \\ \cdot \\ \cdot \\ h_n \\ 0 \end{bmatrix} + \rho \begin{bmatrix} 0 \\ a_1 \\ a_2 \\ \cdot \\ \cdot \\ \cdot \\ a_n \\ 1 \end{bmatrix}. \tag{5-47}$$

The last expression for e may be obtained by performing the indicated matrix multiplication. These vectors may be shown by expansion to satisfy

(i) $d^{\mathrm{T}}b_a = 0, \qquad e^{\mathrm{T}}b_a = \rho,$ (5-48)

(ii) $$e^{\mathrm{T}}(sI - A_a)^{-1}b_a = s[d^{\mathrm{T}}(sI - A_a)^{-1}b_a] = \frac{sW(s)}{s^2 + 2\alpha s + (\alpha^2 + \mu^2)}, \tag{5-49}$$

(iii) $$d^{\mathrm{T}}A_a = e^{\mathrm{T}}, \qquad e^{\mathrm{T}}A_a = h_a^{\mathrm{T}} - 2\alpha e^{\mathrm{T}} - (\alpha^2 + \mu^2)d^{\mathrm{T}}. \tag{5-50}$$

LEMMA 5-3. Given that the augmented system $W_a(s)$ has poles at $s = -\alpha \pm i\mu$, vectors d and e exist such that Eqs. (5-48)–(5-50) are satisfied. □

We now choose as a Lyapunov function candidate

$$v(z) = \tfrac{1}{2}\left[z^{\mathrm{T}}Pz + \beta_0(\kappa\sigma_a^2 + \rho\tau_a^2) + \sum_{i=1}^{3} \beta_i\kappa\sigma_i^2\right], \tag{5-51}$$

where

$$\begin{aligned} \sigma_1 &\triangleq r_1^{\mathrm{T}}z \triangleq (\gamma_1/\beta_1)c^{\mathrm{T}}z, \\ \sigma_2 &\triangleq r_2^{\mathrm{T}}z \triangleq \gamma_2 d^{\mathrm{T}}z + e^{\mathrm{T}}z, \\ \sigma_3 &\triangleq r_3^{\mathrm{T}}z \triangleq \gamma_3 d^{\mathrm{T}}z + e^{\mathrm{T}}z \end{aligned} \tag{5-52}$$

and according to Lemma 5-1b (with $\rho + \bar{K}^{-1}$ substituted for ρ), the vector c satisfies

$$\begin{aligned} &\text{(i)} && c^{\mathrm{T}}b_a = \rho + \bar{K}^{-1}, \\ &\text{(ii)} && c^{\mathrm{T}}(sI - A_a)^{-1}b_a = (W_a(s) + \bar{K}^{-1})/(s + \alpha), \\ &\text{(iii)} && c^{\mathrm{T}}A_a z = h_a^{\mathrm{T}}z - \alpha c^{\mathrm{T}}z. \end{aligned} \tag{5-53}$$

The vectors d and e satisfy Eqs. (5-48)–(5-50) with ρ replaced by $(\rho + \bar{K}^{-1})$, so that the finite sector problem may be treated directly.

As in the previous sections, the validity of $v(z)$ as an absolute Lyapunov function candidate is easily verified. Considering the total time derivative of $v(z)$ along the trajectories of the system and using algebraic operations similar to those described in earlier sections† we obtain

$$\begin{aligned} \dot{v} = {} & \tfrac{1}{2}z^{\mathrm{T}}[A_a^{\mathrm{T}}P + PA_a]z - \kappa\sigma_a z^{\mathrm{T}}[Pb_a - \beta_0(A_a^{\mathrm{T}}h_a + \alpha h_a) - \gamma_1 c \\ & - \delta_1(\alpha d + e)] - \beta_0[h_a^{\mathrm{T}}b_a + \alpha(\rho + \bar{K}^{-1})](\kappa\sigma_a)^2 \\ & - \beta_0\alpha\kappa\sigma_a^2[1 - \kappa/\bar{K}] - \alpha\kappa\sum_{i=1}^{3}\beta_i\sigma_i^2. \end{aligned} \tag{5-54}$$

The last two terms of Eq. (5-54) are negative semidefinite, and using the MKY lemma, the first three terms may be made negative semidefinite by satisfying a frequency domain condition of the form

$$[W(s) + \bar{K}^{-1}]Z(s) \in \{\mathrm{PR}\}, \tag{5-12}$$

† This procedure involves the following interrelations between the variables γ_i and β_i:

$$\beta_1 = \gamma_1(1 - \kappa/\bar{K}) > 0, \qquad \beta_2 = \beta_3 = \tfrac{1}{2}\delta_1/(1 - \kappa/\bar{K}) > 0,$$
$$\gamma_2 = \alpha + \mu, \qquad \gamma_3 = \alpha - \mu,$$

where $\mu \neq 0$ is arbitrary.

where $Z(s) \in \{Z_\alpha(s)\}$. Hence we have the result that $\dot{v} \leqslant 0$ if the above frequency domain condition is satisfied. The argument that $\dot{v} \not\equiv 0$ unless $z \equiv 0$ proceeds as in previous cases. Since $Z(s) \in \{Z_{LC}\}$ implies $Z^{-1}(s) \in \{Z_{LC}(s)\}$, and similarly for $Z(s) \in \{Z_\alpha(s)\}$, a separate development is not needed for inverse multipliers. The final criterion for linear time-invariant systems may be stated as follows.

CRITERION 3 (Linear systems). The system described by Eqs. (5-1) and (5-43) is asymptotically stable if and only if $Z(s) \in \{Z_\alpha(s)\}$ [Eq. (5-44)] exists such that

$$[W(s) + \bar{K}^{-1}]Z(s) \in \{\mathrm{PR}\}. \quad \square \tag{5-12}$$

As mentioned earlier, this result is formally equivalent to the Nyquist criterion, thus supplying an alternative necessary and sufficient condition for asymptotic stability. The derivation presented here follows closely that of Thathachar and Srinath [1], who proved a conjecture of Narendra and Neuman [2]. The class of multipliers $\{Z_\alpha(s)\}$ is not the entire class of multipliers that may be used for linear time-invariant systems. A simple argument demonstrates that the existence of any multiplier $Z(s) \in \{\mathrm{SPR}\}$ such that

$$H(s) = [W(s) + \bar{K}^{-1}]Z(s) \in \{\mathrm{PR}\}$$

is a necessary and sufficient condition for asymptotic stability (see Chapter IV, Section 3).

4. Odd Monotonic Gains

The derivation of stability criteria for other classes of nonlinear gain functions proceeds as in Sections 2 and 3. By constraining the nonlinearity class $\{N\}$ to be less general than $\{F_m\}$, a frequency domain condition that is intermediate in strictness between those of Criterion 2 and Criterion 3 may be obtained. There are, however, two compelling reasons for considering only relatively simple nonlinearity classes: If $\{N\}$ is closely specified, then it may require extensive measurement of the input–output characteristics of the nonlinear device ($f(\sigma_0)$ versus σ_0), which runs counter to the philosophy of absolute $\{N\}$ stability, and if $\{N\}$ consists of a complicated class of functions, then $\{Z_N(s)\}$ is likewise apt to be unwieldy and the application of such stability criteria becomes extremely difficult at best. Even the stability criterion outlined briefly in this section suffers from this defect, and is therefore found to be primarily of theoretical interest.

If the class of odd monotonic functions is defined by

$$\{F_{mo}\} \triangleq \{f(\sigma_0): \quad f(\sigma_0) \in \{F_m\}, \quad f(-\sigma_0) = -f(\sigma_0) \quad \text{for all } \sigma_0\}, \tag{5-55}$$

the following lemma is found to be useful in rearranging the terms of the derivative of the absolute Lyapunov function candidate prior to an application of the MKY lemma.

LEMMA 5-4. If $f(\sigma_0) \in \{F_{mo}\}$ and $\Delta f(\sigma_0)/\Delta\sigma_0 \in [0, \infty)$, then $\sigma_1 f(\sigma_1) + \sigma_2 f(\sigma_2) \pm [\sigma_1 f(\sigma_2) \pm \sigma_2 f(\sigma_1)] \geqslant 0$ for all σ_1 and σ_2. □

The Lyapunov function candidate chosen has the same form as that for the monotonic case [Eq. (5-28)] except that the upper limits of the integrals σ_i are defined in terms of $d^{\mathrm{T}}x$ and $e^{\mathrm{T}}x$ as well as $c^{\mathrm{T}}x$, as in Eq. (5-52) for the linear case. The time derivative $\dot{v}$ contains terms of the form $\sigma_i f(\sigma_j)$, and Lemma 5-4 can be used to incorporate these terms into negative semidefinite forms as in the monotonic case. This procedure, in conjunction with Lemma 5-4, permits σ_i composed of $d^{\mathrm{T}}x$ and $e^{\mathrm{T}}x$ to be used in the absolute Lyapunov function candidate and consequently allows the use of complex poles in the multiplier $Z(s)$. The resulting class of multipliers that may be used in establishing absolute $\{F_{mo}\}$ stability corresponds to a special class of RLC functions, that is, $\{Z_{F_{mo}}\} \subset \{Z_{\mathrm{RLC}}\}$.

LEMMA 5-5.

$$\begin{aligned} Z(s) \triangleq (s + \lambda_0)\Big[\beta_0 + \sum_{i=1}^{n_1} \frac{\gamma_i}{s + \eta_i}\Big] + \sum_{i=n_1+1}^{m_1} \gamma_i \frac{s + \rho_i\eta_i}{s + \eta_i} \\ + \sum_{i=1}^{n_2} \delta_i \frac{s + \zeta_i}{s^2 + 2\lambda_i s + (\lambda_i^2 + \mu_i^2)} \\ + \sum_{i=n_2+1}^{m_2} \kappa_i \frac{s^2 + \phi_i s + \psi_i}{s^2 + 2\lambda_i s + (\lambda_i^2 + \mu_i^2)} \end{aligned} \tag{5-56}$$

is a member of $\{Z_{F_{mo}}(s)\}$ if each term of $Z(s)$ is strictly positive real, if the parameter constraints $1 < \rho_i \leqslant 2$, $\zeta_i < 2\lambda_i$, $\phi_i < 2\lambda_i \triangleq \pi_i$ and $\psi_i < (\lambda_i^2 + \mu_i^2)$ are satisfied, and if the auxiliary parameters ν_i and ξ_i defined by

$$0 < \nu_i \triangleq \begin{cases} \left[\dfrac{\delta_i}{2\beta_i}\left\{\left(1 + \dfrac{(\zeta_i - \lambda_i)^2}{\mu_i^2}\right)^{1/2} + 1\right\}\right]^{1/2}, & i = 1, 2, \ldots, n_2 \\ \left[\dfrac{\kappa_i(\pi_i - \phi_i)}{2\beta_i}\left\{\left(1 + \dfrac{[(\rho_i - \psi_i)/(\pi_i - \phi_i) - \lambda_i]^2}{\mu_i^2}\right)^{1/2} + 1\right\}\right]^{1/2}, & \\ & i = (n_2 + 1), \ldots, m_2; \end{cases}$$

$$0 \leqslant \xi_i \triangleq \begin{cases} \left[\dfrac{\delta_i}{2\beta_i}\left\{\left(1 + \dfrac{(\zeta_i - \lambda_i)^2}{\mu_i^2}\right)^{1/2} - 1\right\}\right]^{1/2}, & i = 1, 2, \ldots, n_2 \\ \left[\dfrac{\kappa_i(\pi_i - \phi_i)}{2\beta_i}\left\{\left(1 + \dfrac{[(\rho_i - \psi_i)/(\pi_i - \phi_i) - \lambda_i]^2}{\mu_i^2}\right)^{1/2} - 1\right\}\right]^{1/2}, & \\ & i = (n_2 + 1), \ldots, m_2, \end{cases} \tag{5-57}$$

satisfy

$$\begin{aligned} &\text{(i)} \quad \Big(\gamma_0 + \sum_{i=n_2+1}^{m_2} \kappa_i\Big) - \sum_{i=1}^{m_2} \beta_i(\nu_i + \xi_i) \triangleq \varepsilon_1 \geqslant 0, \\ &\text{(ii)} \quad (\lambda_i - \mu_i) - \nu_i \triangleq \varepsilon_{2i} \geqslant 0, \qquad i = 1, 2, \ldots, m_2. \;\square \end{aligned} \tag{5-58}$$

One important feature is the position of singularities of $Z(s)$. Denoting the singularities of $Z(s)$ by $s_i(Z)$, we define the sector S_1 in the s-plane:

$$S_1 \triangleq \{3\pi/4 < \arg[s_i(Z)] < 5\pi/4\}.$$

The condition that $1 < \rho_i \leqslant 2$ restricts the terms 1 and 2 of Eq. (5-56) to have their zeros in sector S_1, and the second part of condition (5-58) requires that poles of $Z(s)$ must lie in the same sector. That the singularities of $Z(s)$ must lie in S_1 is a necessary but not sufficient condition for $Z(s) \in \{Z_{F_{mo}}\}$.

The first two terms of $\{Z_{F_{mo}}\}$ were obtained by Narendra and Neuman [1]. The multiplier class defined in Lemma 5-5 was first obtained in its entirety by Thathachar, Srinath, and Ramapriyan [1], using the same approach outlined here. A similar result using functional analytic techniques is due to Narendra and Cho [1].

Some of the complexity of the parameter constraints detailed above is avoided by using the shifted LC function [Eq. (5-44)] whenever possible. The chief simplification occurs in the auxiliary parameters ν_i and ξ_i; they become $\nu_i = (\delta_i/\beta_i)^{1/2}$, $\xi_i = 0$.

COROLLARY TO LEMMA 5-5. A shifted LC function $Z(s) \in \{Z_\alpha(s)\}$ [Eq. (5-44)] is a member of $\{Z_{F_{mo}}(s)\}$ if $\gamma_1 < \alpha^2\beta_0$ where $\beta_0 > 0$, and if

$$\text{(i)} \qquad \beta_0\alpha - \sum_{i=1}^{k} (\beta_i\delta_i)^{1/2} \triangleq \varepsilon_1 \geq 0, \tag{5-59}$$

$$\text{(ii)} \qquad \alpha - \mu_i - (\delta_i/\beta_i)^{1/2} \triangleq \varepsilon_{2i} \geqq 0, \qquad i = 1, 2, \ldots, k. \square$$

Again, these constraints guarantee that $s_i(Z) \in S_1$.

The obvious complexity of even the latter conditions [for $Z(s) \in \{Z_\alpha(s)\}$] indicates that they are not particularly useful for most practical purposes, unless a multiplier containing a few terms is adequate to satisfy the frequency domain condition. In some circumstances it may prove to be useful to consider other, more complicated classes of nonlinear gains, however; this would certainly be the case if a specified type of nonlinearity were intentionally introduced in the design of a system in order to achieve a particular type of response. Results for power law nonlinearities are reported by Thathachar [1], and for nonlinearities with restricted nonmonotonicity by Thathachar and Srinath [3].

5. The General Finite Sector Problem

A single elementary transformation may be used to extend all of the previous results to the general finite sector case. Thus, it is possible to state all of the earlier criteria in one succinct result.

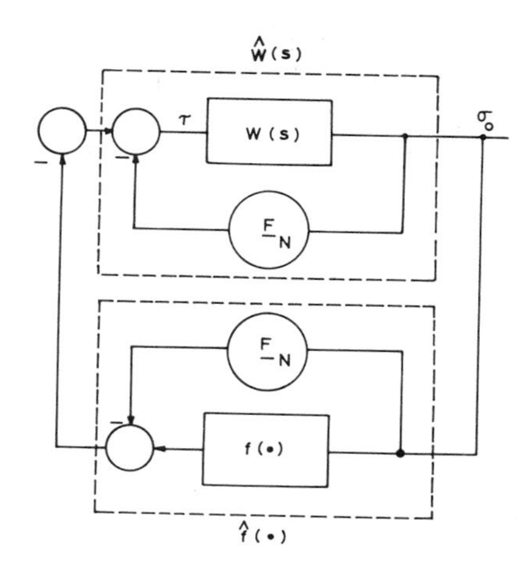

Fig. 5-3. The general finite sector/finite sector transformation.

The general finite sector–finite sector transformation of Rekasius and Gibson [1],

$$\hat{W}(s) \triangleq W(s)/(1 + \underline{F}_N W(s)), \qquad \hat{f}(\sigma_0) \triangleq f(\sigma_0) - \underline{F}_N \sigma_0, \tag{5-60}$$

may be demonstrated not to affect the stability properties of the system; in Fig. 5-3 we see that it is the complement of the transformation shown in Fig. 5-1. Thus the absolute stability of the system represented by $\{\hat{W}(s), \hat{f}(\sigma_0)\}$ guarantees the absolute stability of the solutions of Eq. (5-1) constrained by Eq. (5-2a) or (5-2b). Thus, it is only necessary to show that the transformed system is in the correct form to fall within the ambit of the earlier work.

(i) $\hat{W}(s)$ is by definition asymptotically stable if Eq. (5-2a) holds, since $A_{\underline{F}N} \in \{A_1\}$, and it has no more zeros than poles. If condition (5-2b) is considered, then $A_{\underline{F}N} \in \{A_0\}$ yields a similar result.

(ii) $\hat{f}(\sigma_0) \in \{N\}$, $\hat{f}(\sigma_0)/\sigma_0 \in [0, \bar{F}_N - \underline{F}_N]$ or $\in (0, \bar{F}_N - \underline{F}_N]$; this transformation guarantees these conditions by inspection.

Being satisfied that the previous conditions for absolute stability are applicable, we inspect condition (5-12) which restricts $\hat{W}(s)$: by direct substitution, $Z(s)$ must exist such that $Z^{\pm 1}(s) \in \{Z_N(s)\}$ and

$$H(s) \triangleq \left[\frac{1 + \bar{F}_N W(s)}{1 + \underline{F}_N W(s)}\right] Z(s) \tag{5-61}$$

satisfies $H(s) \in \{\text{PR}\}$ if Eq. (5-2a) is considered, or $H(s) \in \{\text{SPR}_0\}$ if $f(\sigma_0)$ is constrained by Eq. (5-2b).

General Stability Criterion 1. The system described by Eq. (5-1), constrained by Eq. (5-2a) or (5-2b), is absolutely $\{N\}$ stable if $Z(s)$ exists such that $Z^{\pm 1}(s) \in \{Z_N(s)\}$ and $H(s)$ [Eq. (5-61)] satisfies $H(s) \in \{\text{PR}\}$ for Eq.

(5-2a) or $H(s) \in \{\mathrm{SPR}_0\}$ for Eq. (5-2b), where the nonlinear function classes $\{N\}$ and the corresponding multiplier classes $Z_N(s)$ are given by:

DEFINITION 1:

$$\{N\} = \{F\}; \quad \{Z_F(s)\} \triangleq \{Z(s) = (\beta_0 s + \gamma_0), \quad \beta_0 \geqslant 0, \ \gamma_0 > 0\}. \tag{5-11}$$

DEFINITION 2:

$$\{N\} = \{F_m\}; \quad \{Z_{F_m}(s)\} \triangleq \{Z_{\mathrm{RL}}(s)\}. \tag{5-34}$$

DEFINITION 3:

$$\{N\} = \{F_{mo}\}; \quad \{Z_{F_{mo}}(s)\} \triangleq \{Z(s) \text{ satisfying Lemma 5-5 or its corollary}\}.$$

DEFINITION 4:

$$\{N\} = \{L\}; \ \{Z_L(s)\} \triangleq \{\mathrm{SPR}\} \supset \{Z_\alpha(s)\}. \ \square \tag{5-44}$$

General Stability Criterion 1 summarizes most of the results presented in this chapter. For the problem considered by Popov, the multiplier consists of a single term $(\beta_0 s + \gamma_0)$ or its inverse. When the class of nonlinear functions includes only monotonic nonlinearities, the multiplier can have an arbitrary number of alternating real poles and zeros. For odd monotonic functions in the feedback path, the multiplier belongs to a restricted class of Z_{RLC} functions. Finally, for a linear gain, the multiplier can be any arbitrary rational strictly positive real function, and the stability criterion is equivalent to the Nyquist criterion.

VI

STABILITY OF NONLINEAR TIME-VARYING SYSTEMS

The first absolute stability criterion that was developed for an extension of the Lur'e model of Chapter V to the time-varying case [$g(\sigma_0, t)$ or $f(\sigma_0)k(t)$ replacing $f(\sigma_0)$ in Eq. (5-1)] is the circle criterion. The derivation of this sufficient condition for system stability utilized the common quadratic Lyapunov function $v = \frac{1}{2}x^T Px$ rather than an extended Lur'e–Postnikov form with integral terms based on the NLTV gain. Since v does not contain $g(\sigma_0, t)$, the continuity of $g(\sigma_0, t)$ with respect to time is not required in order to satisfy the conditions necessary to establish the validity of v as a candidate (Chapter III, Section 2). Similarly, as $\dot{v}$ does not contain $\partial g/\partial t$, the rate of time variation is completely unconstrained.

It is shown [Eq. (6-11)] that if the circle criterion is applied to the system described by Eq. (5-1) for $f(\sigma_0)/\sigma_0 \in [0, \bar{F}]$, it is required that

$$H(s) = [W(s) + \bar{F}^{-1}] \in \{\text{SPR}\}.$$

This is essentially a special case of the Popov condition (see Criterion 1b, Chapter V, Section 1), where $\beta_0 = 0$ and $\gamma_0 = 1$, except that the frequency domain condition is somewhat more strict. The same closed range $[0, \bar{F}]$ is valid for any nonlinear time-varying gain $g(\sigma_0, t)$ by the circle criterion, regardless of its rate of time variation. The fact that a special case of the Popov criterion can be extended to an NLTV situation makes it reasonable to expect that the general form of the Popov condition ($\beta_0 \neq 0$) may be extended in some way to render it applicable to nonlinear time-varying systems. The Lyapunov function candidate that makes this generalization possible is a straightforward extension of the Lur'e–Postnikov form where $f(\sigma_0)$ is replaced with $g(\sigma_0, t)$ or $f(\sigma_0)k(t)$ in the integral term. This form is related to the concept of the κ-dependent Lyapunov function candidate introduced in Chapter IV; in particular, for LTV systems with $\rho = 0$, this candidate is linear in $k(t)$:

$$v = \tfrac{1}{2}[x^{\mathrm{T}}Px + k(t)x^{\mathrm{T}}hh^{\mathrm{T}}x].$$

Since $\dot{v} \triangleq (\nabla v)^{\mathrm{T}}\dot{x} + \partial v/\partial t$ now must contain dk/dt or $\partial g/\partial t$, it is to be anticipated that the rate of variation of the gain with respect to time must be restricted in some manner.

All of the stability criteria of Chapter V for nonlinear systems may be generalized using the same arguments applied to the Popov criterion to allow their application in time-varying situations. The resultant conditions that the rate of time variation must satisfy are found to depend on a certain index of the nonlinearity of the system ($\underline{\Phi}$, defined in Chapter II, Section 1) and a property of the frequency domain multiplier $Z(s)$, defined in Section 2 of this chapter to be the multiplier margin $\bar{\Lambda}$.

The direct utilization of Lyapunov's stability theorem yields a point condition that $\partial g/\partial t$ must satisfy, that is, $\partial g/\partial t$ must be constrained at each instant in order to guarantee that $\dot{v} < 0$ for all t (see General Stability Criterion 2, Section 4). The application of a corollary of the theorem of Corduneanu, on the other hand, yields the requirement that $\partial g/\partial t$ be bounded only in a time-averaged sense. This latter criterion (General Stability Criterion 3, Section 6) is more general than General Stability Criterion 2 in this respect, and in fact contains General Stability Criterion 2 for the infinite sector problem as a special case.

Since the systems under investigation in this chapter are no longer autonomous, Theorem 1 used in obtaining point constraints on $\partial g/\partial t$ necessitates showing that $\dot{v}$ is negative definite (except in the case of periodic systems, Section 5); for this reason, the Lefschetz form of the Kalman–Yakubovich lemma (the LKY lemma) is used in lieu of the MKY lemma in Sections 2 and 3. The application of the LKY lemma results in a frequency domain requirement that

$$H(s) \triangleq \tfrac{1}{2}\psi + k^{\mathrm{T}}(sI - A)^{-1}b \in \{\mathrm{SPR}\}.$$

The negative definiteness of $\dot{v}$ is thus obtained at the expense of requiring that $H(s)$ satisfy a more restrictive frequency domain condition than that obtained in Chapter V.

All stability problems treated in this chapter pertain to a system described by the state vector differential equation

$$\dot{x} = Ax + b\tau, \qquad \sigma_0 = h^{\mathrm{T}}x + \rho\tau, \qquad \tau = -g(\sigma_0, t). \tag{6-1}$$

We note in Section 2 that there is no advantage to be achieved in treating the particular case. Thus we generally assume that

$$g(\sigma_0, t) \in \{G_i[N, T]\}, \qquad g(\sigma_0, t)/\sigma_0 \in [\underline{G}_N, \bar{G}_N],$$
$$A_{\underline{G}_N} \in \{A_1\}, \qquad A_{\bar{G}_N} \in \{A_1\}, \tag{6-2}$$

where, as in the NLTI case, we usually take $\underline{G}_N$ to be zero (except in the derivation of the circle criterion) until we treat the general finite sector problem by the elementary transformation of Chapter V, Section 5. The same initial assumption that $A_\kappa \in \{A_1\}$, $\kappa \in [\underline{G}_N, \bar{G}_N]$, is made as in the NLTI case and it is subsequently discarded except at the extrema [Eq. (6-2)], since the frequency domain condition for absolute stability subsumes this condition for $\kappa \in (\underline{G}_N, \bar{G}_N)$.

1. The Circle Criterion

The system whose stability is to be determined is described by the vector differential equation (6-1), where the NLTV gain and the A matrix are constrained by

$$g(\sigma_0, t) \in \{G_i[F, K_0]\}, \qquad g(\sigma_0, t)/\sigma_0 \in [\underline{G}, \bar{G}], \qquad \underline{G} \leqslant 0 \leqslant \bar{G},$$
$$A_{\underline{G}} \in \{A_1\}, \qquad A_{\bar{G}} \in \{A_1\}. \tag{6-3}$$

The nonlinear time-varying gain function g need not be separable, and this is the only case considered in this book where $g(\sigma_0, t)$ may have a discontinuous time variation. From the specified range, the useful inequality

$$(\underline{G}\sigma_0 + \tau)(\bar{G}\sigma_0 + \tau) \leqslant 0 \tag{6-4}$$

follows directly.

It is possible in this case that $\underline{G} < 0$, that is, that $g(\sigma_0, t)$ may pass outside the first and third quadrants; however, in this sector g must still behave like a member of $\{G_i[F, K_0]\}$, in that Eq. (6-3) is the only restriction. We do not consider $\underline{G} > 0$ at this time, since this case may be simply treated by the transformation of Section 4.

It is desired to find the range $[\underline{G}, \bar{G}]$ such that $v(x) = \frac{1}{2}x^{\mathrm{T}}Px$ exists as a common quadratic Lyapunov function for the dynamic system described by Eq. (6-1); directly,

$$\dot{v} = \tfrac{1}{2}x^{\mathrm{T}}(A^{\mathrm{T}}P + PA)x - g(\sigma_0, t)x^{\mathrm{T}}Pb.$$

Using the practice established in Chapter V, this derivative may be expanded to yield

$$\dot{v} = \tfrac{1}{2}\{x^{\mathrm{T}}(A^{\mathrm{T}}P + PA)x + 2\tau x^{\mathrm{T}}[Pb - \tfrac{1}{2}(\underline{G} + \bar{G} + 2\rho\underline{G}\bar{G})h]$$
$$- (1 + \rho\underline{G})(1 + \rho\bar{G})\tau^2 - \underline{G}\bar{G}(h^{\mathrm{T}}x)^2 + (\bar{G}\sigma_0 + \tau)(\underline{G}\sigma_0 + \tau)\}.$$

This expression is very much like that obtained in each development of Chapter V just prior to the application of the MKY lemma (Lemma 3, Chapter III, Section 4). This lemma gives a necessary and sufficient frequency domain condition that the equations

$$A^{\mathrm{T}}P + PA = -qq^{\mathrm{T}} - M, \qquad Pb - k = \sqrt{\psi}\, q$$

have appropriate solutions; the first three terms of $\dot{v}$ [Eq. (5-4)] then form a negative perfect square and a negative semidefinite quadratic form. The development presented by Narendra and Goldwyn [1] provides a similar reduction for the first four terms of the above equation; however, the conditions which ensure that $\dot{v}$ is negative definite as required in treating time-varying systems (Theorem 1, Chapter III, Section 3) are not given explicitly. This problem is circumvented by using a lemma of the Lefschetz form which yields $\dot{v}$ having a nonpositive perfect square and a negative definite quadratic form. A general statement of suitable form is adapted from the result of Rekasius and Rowland [1].

LEMMA 6. Given $A \in \{A_1\}$, (A, b) completely controllable, a real vector k, scalars $\psi \geqslant 0$ and $\varepsilon > 0$, and arbitrary matrices $R = R^{\mathrm{T}} > 0$ and $S = S^{\mathrm{T}} \geqslant 0$, then the matrix $P = P^{\mathrm{T}} > 0$ and a real vector q satisfying

(a) $$A^{\mathrm{T}}P + PA = -qq^{\mathrm{T}} - \varepsilon R - S, \tag{6-5}$$

(b) $$Pb - k = \sqrt{\psi}\, q, \tag{6-6}$$

exist if and only if ε is sufficiently small and

(c) $$H(s) \triangleq \psi + 2k^{\mathrm{T}}m(s) - m^*(s)Sm(s) \in \{\mathrm{SPR}\}, \tag{6-7}$$

where $m(s) \triangleq (sI - A)^{-1}b$, $m^*(s) = m^{\mathrm{T}}(s^*)$ and $s^* = \sigma - i\omega$. □

This result differs from the lemma of Rekasius and Rowland in that S is more general (not specified to be proportional to hh^{T}) and a term $-\varepsilon R$ is included in Eq. (6-5) to ensure that $\dot{v}$ is negative definite. To achieve this end, a more strict frequency domain condition (6-7) has to be satisfied.

The steps in the proof of this lemma are almost identical to those taken in demonstrating the validity of Lemma 5 in Chapter III, Section 4. A great deal of simplification arises from the assumption that $A \in \{A_1\}$.

In deriving the desired stability criterion, we identify

$$S \triangleq -\underline{G}\bar{G}hh^{\mathrm{T}} \geqslant 0, \qquad k \triangleq \tfrac{1}{2}(\underline{G} + \bar{G} + 2\rho\underline{G}\bar{G})h, \qquad \psi \triangleq (1 + \rho\underline{G})(1 + \rho\bar{G}) \tag{6-8}$$

so that $\dot{v}$ becomes

$$\dot{v} = -\tfrac{1}{2}\{[x^{\mathrm{T}}q - \sqrt{\psi}\,\tau]^2 - (\underline{G}\sigma_0 + \tau)(\bar{G}\sigma_0 + \tau) + \varepsilon x^{\mathrm{T}}Rx\} \leqslant -\tfrac{1}{2}\varepsilon x^{\mathrm{T}}Rx, \tag{6-9}$$

which is negative definite due to inequality (6-4) and Lemma 6. Substituting Eq. (6-8) into Eq. (6-7) and taking the real part of $H(i\omega)$ gives us the relation

$$\operatorname{Re} H(i\omega) = 1 + (\underline{G} + \bar{G}) \operatorname{Re} W(i\omega) + \underline{G}\bar{G} \,|\, W(i\omega) \,|^2 > 0,$$

where we recall that $W(s) = \rho + h^{\mathrm{T}}(sI - A)^{-1}b$. Note that this is the numerator of

$$\operatorname{Re} \left\{ \frac{1 + \bar{G}W(i\omega)}{1 + \underline{G}W(i\omega)} \right\} = \frac{1 + (\underline{G} + \bar{G})U + \underline{G}\bar{G}(U^2 + V^2)}{(1 + \underline{G}U)^2 + (\underline{G}V)^2},$$

where $W(i\omega) \triangleq U(\omega) + iV(\omega)$, so Lemma 6 subject to Eq. (6-8) is satisfied if

$$[1 + \bar{G}W(s)]/[1 + \underline{G}W(s)] \in \{\mathrm{SPR}\}. \tag{6-10}$$

CRITERION 4 (The circle criterion). The system defined by Eqs. (6-1) and (6-3) is absolutely $\{G_i[F, K_0]\}$ stable if condition (6-10) is satisfied. □

This criterion is of great utility owing to its simple Nyquist-like geometric interpretation in the polar plot (Im $W(i\omega)$ versus Re $W(i\omega)$) as detailed in Chapter VII.

The genesis of this criterion is quite complex. The first incidence appears to be due to Rozenvasser [1], who proved a special case of the above corresponding to a nonlinear time-varying function $g(\sigma_0, t)$ in the sector $[0, \bar{G}]$; his proof involved the use of a common quadratic Lyapunov function. Bongiorno [1, 2] used an appeal to Floquet theory and Fourier transform techniques to prove a special case for an LTV system with $|k(t)| \leqslant \bar{G}$.

The criterion for nonlinear time-varying gains in the statement above with $\rho = 0$ is due to Narendra and Goldwyn [1], who again used a common quadratic Lyapunov function approach. An independent result of Kudrewicz [1] gives the same frequency domain constraint (6-10) as a sufficient condition for energetic stability, that is, if a system input exists (see Fig. 2-1) satisfying

$$\| v \| \triangleq \lim_{T \to \infty} \left[T^{-1} \int_0^T | v(t) |^2 \, dt \right]^{1/2} = 0,$$

then the system output σ_0 satisfies $\| \sigma_0 \| = 0$ if condition (6-10) is satisfied. Shortly thereafter, more general forms of the circle criterion were obtained by Sandberg [2] and Zames [2], using functional analytic techniques. The generality of the latter results devolves from an integral formulation of the LTI portion of the system, which allows the representation of plants with pure time delays [e^{Ts} in $W(s)$] and distributed systems; $\rho = 0$ is still assumed in these derivations, however.

If the lower bound on $g(\sigma_0, t)/\sigma_0$ is $\underline{G} = 0$, then the condition of Criterion 4 simplifies to

$$[\bar{G}^{-1} + W(s)] \in \{\mathrm{SPR}\}, \tag{6-11}$$

which is essentially a special case of the Popov condition for the finite sector problem [Criterion 1b, refer to Eq. (5-9)] with $\beta_0 = 0$, $\gamma_0 = 1$. The full importance of this case also lies in its geometric interpretation.

2. An Extension of the Popov Criterion—Point Conditions

The stability criteria developed in this section represent a straightforward extension of the results of Chapter V for NLTI systems with first and third quadrant gains. The application of Lyapunov's stability theorem results in a condition that $\partial g/\partial t$ must satisfy at each instant of time (a point condition) in order to render $\dot{v}$ negative definite for all t. Since the rate of variation of $g(\sigma_0, t)$ with respect to time is being restricted in general, it is not at all surprising that the condition to be satisfied by $W(s)$ may often be less strict than that imposed by the circle criterion. This in turn may mean that the upper bound on $g(\sigma_0, t)/\sigma_0$ may be allowed to be higher for a given $W(s)$ if $\partial g/\partial t$ is restricted. As $\partial g/\partial t$ becomes less restricted, the constraint on $W(s)$ approaches that imposed by the circle criterion for the range $[0, \bar{G}]$. These relationships, which provide the main motivation for this study, may be most clearly appreciated by considering the corresponding geometric criteria (Chapter VII) and applications (Chapter VIII).

The system under consideration is represented by the vector differential equation (6-1) with the NLTV gain and the matrix A specified by

$$g(\sigma_0, t) \in \{G_i[F, K_1]\}, \qquad g(\sigma_0, t)/\sigma_0 \in [0, \bar{G}], \qquad A \in \{A_1\}, \qquad A_{\bar{G}} \in \{A_1\}. \tag{6-12}$$

This formulation is similar to that considered in deriving the circle criterion, except that the gain g must be a continuous function of time, that is, K_1 replaces K_0 in the specification of the behavior of g with respect to time, as required for the validity of the absolute Lyapunov function candidate. Only separable gains are considered in detail; the result for inseparable nonlinear time-varying functions is obtained directly at the conclusion.

The absolute Lyapunov function candidate is an obvious generalization of the Lur'e–Postnikov form as modified by Popov and used in Chapter V, Section 1:

$$v(x, t) = \tfrac{1}{2}x^{\mathrm{T}}Px + \hat{\beta}_0 k(t)\int_0^{\sigma_0} f(\zeta)\, d\zeta + \tfrac{1}{2}\hat{\beta}_0 \rho\tau^2. \tag{6-13}$$

As in previous developments $P = P^{\mathrm{T}} > 0$ and, initially, $\hat{\beta}_0 \geqslant 0$ and $\rho \geqslant 0$ are assumed to guarantee its validity according to Lemma V3, Chapter III, Section 2. There are several reasons for this choice of Lyapunov function candidate, aside from the obvious one that Eq. (6-13) arises from a direct substitution of $k(t)f(\sigma_0)$ in lieu of $f(\sigma_0)$ in the original Lyapunov function candidate for NLTI systems (Chapter V, Section 1), as indicated earlier (Chapter III, Section 2, and Chapter IV, Sections 1 and 3).

In considering the finite sector problem for separable gains, we have $\bar{G} \triangleq \bar{F}\bar{K}$, where the upper bounds $\bar{F}$ and $\bar{K}$ are specified only within an arbitrary constant multiplier;

$$0 \leqslant (\sigma_0)^{-1}f(\sigma_0)k(t) = (\sigma_0)^{-1}[\alpha f(\sigma_0)][\alpha^{-1}k(t)] \leqslant \bar{G} \tag{6-14}$$

yields upper bounds $\alpha\bar{F}$ and $\bar{K}/\alpha$ for any $\alpha > 0$. The total time derivative of $v(x, t)$ along trajectories of the system described by Eqs. (6-1) and (6-12) is

$$\begin{aligned}\dot{v} = &\tfrac{1}{2}x^{\mathrm{T}}(A^TP + PA)x - k(t)f(\sigma_0)x^{\mathrm{T}}[Pb - \hat{\beta}_0A^{\mathrm{T}}h - \gamma_0 h]\\ &-[\hat{\beta}_0h^{\mathrm{T}}b + \gamma_0(\rho + \bar{G}^{-1})][k(t)f(\sigma_0)]^2\\ &-(\gamma_0/\bar{K})\sigma_0 f(\sigma_0)k^2(t)[1 - f(\sigma_0)/(\bar{F}\sigma_0)]\\ &-k(t)\int_0^{\sigma_0} f(\zeta)\,d\zeta\,[\gamma_0\phi(\sigma_0)\{1 - k(t)/\bar{K}\} - \hat{\beta}_0k^{-1}\,dk/dt].\end{aligned} \tag{6-15}$$

We arrive at this point using manipulations that are analogous to those used in deriving Eq. (5-4); the principal departure is the appearance of the term involving dk/dt. The first three terms of this expression may be guaranteed to be negative definite by stipulating that the frequency domain condition imposed by the LKY lemma is satisfied. The fourth term is never positive if $\gamma_0 \geqslant 0$, since $f(\sigma_0)/\sigma_0 \in [0, \bar{F}]$. The fifth term contains the ratio

$$\phi(\sigma_0) \triangleq [\sigma_0 f(\sigma_0)/\textstyle\int_0^{\sigma_0} f(\zeta)\,d\zeta] \geqslant \underline{\Phi} \triangleq \min_{\sigma_0}\{\phi(\sigma_0)\} \geqslant 0; \tag{6-16}$$

the lower bound $\underline{\Phi}$ provides an effective measure or index of the nonlinear behavior of the gain. This last term of $\dot{v}$ leads to the constraint on dk/dt.

To be more explicit, two conditions suffice to ensure that the system described by Eqs. (6-1) and (6-12) is absolutely stable:

(a) $H(s) \triangleq [W(s) + \bar{G}^{-1}](\beta_0 s + \gamma_0)^{\pm 1} \in \{\mathrm{SPR}\}; \qquad \gamma_0 > 0, \quad \beta_0 \geqslant 0,$ (6-17)

(b) $$\gamma_0\underline{\Phi}k(t)[1 - k(t)/\bar{K}] - \hat{\beta}_0\,dk/dt \geqslant 0. \tag{6-18}$$

In obtaining the first or frequency domain condition, the application of the LKY lemma to obtain what is essentially the Popov constraint on $W(s)$ proceeds as in Chapter V. Define the parameters k and ψ according to Eq. (5-5) with $\bar{G}$ replacing $\bar{F}$; then $\hat{H}(s) \in \{\mathrm{SPR}\}$ is demanded where

$$\hat{H}(s) \triangleq \hat{\beta}_0 s[h^{\mathrm{T}}(sI - A)^{-1}b] + \gamma_0[W(s) + \bar{G}^{-1}]$$

in order to render the first three terms of $\dot{v}$ negative definite [see Eq. (6-19)]. Then we extend the argument of Chapter V, Section 1 to relax the requirements that $\hat{\beta}_0 \geqslant 0$ and $\rho \geqslant 0$ for $v(x, t)$ to be a valid absolute Lyapunov function candidate. This entails replacing $k(t)f(\sigma_0)$ by $\hat{\kappa}(\sigma_0, t)\sigma_0$ and utilizing the fact that $\hat{\kappa} \in [0, \bar{G}]$ in proving that $v(x, t)$ is equal in value at every instant to a positive definite quadratic form. As in Chapter V, the frequency domain condition $\hat{H}(s) \in \{\text{SPR}\}$ with $\hat{\beta}_0$ allowed to be positive or negative is equivalent to condition (a), where $\beta_0 \triangleq |\hat{\beta}_0|$. We again express this restriction in terms of the class of multipliers $\{Z_F(s)\}$,

$$\{Z_F(s)\} \triangleq \{Z(s) = (\beta_0 s + \gamma_0), \quad \beta_0 \geqslant 0, \ \gamma_0 > 0\}; \tag{5-11}$$

thus condition (a) may be expressed as the equivalent requirement that $Z(s)$ must exist such that $Z^{\pm 1}(s) \in \{Z_F(s)\}$ and $[W(s) + \bar{G}^{-1}]Z(s) \in \{\text{SPR}\}$. This condition again permits the removal of the implicit assumption that $A_\kappa \in \{A_1\}$, $\kappa \in (0, \bar{G})$ as in the NLTI case.

The second or instantaneous time-domain restriction on dk/dt directly guarantees that $\dot{v}$ is bounded above by the negative definite quadratic form provided by the application of the LKY lemma. To demonstrate this, we substitute for k and ψ in $\dot{v}$, eliminate the fourth term of Eq. (6-15) (which is never positive due to the finite sector condition), and arrive at

$$\begin{aligned} \dot{v} \leqslant & -\tfrac{1}{2}[x^{\mathrm{T}}q + \sqrt{\psi}\, k(t) f(\sigma_0)]^2 - \tfrac{1}{2}\varepsilon x^{\mathrm{T}} L x \\ & - k(t) \int_0^{\sigma_0} f(\zeta)\, d\zeta\, [\gamma_0 \phi(\sigma_0)(1 - k(t)/\bar{K}) - \hat{\beta}_0 k^{-1}\, dk/dt] \leqslant -\tfrac{1}{2}\varepsilon x^{\mathrm{T}} L x, \end{aligned} \tag{6-19}$$

where the right-hand inequality is valid by virtue of condition (b), Eq. (6-18). If $(\beta_0 s + \gamma_0)^{+1}$ is used in the frequency domain constraint (a), then condition (b) is simply

$$dk/dt \leqslant (\gamma_0/\beta_0)\underline{\Phi} k(t)[1 - k(t)/\bar{K}], \tag{6-20$^+$}$$

whereas if the use of $(\beta_0 s + \gamma_0)^{-1}$ is required by the form of $W(s)$, then taking $\hat{\beta}_0 = -\beta_0 < 0$ in condition (b) yields the constraint

$$dk/dt \geqslant -(\gamma_0/\beta_0)\underline{\Phi} k(t)[1 - k(t)/\bar{K}]. \tag{6-20$^-$}$$

The above definition of $\{Z_F(s)\}$ may be used to refine the time domain restriction on dk/dt as well. If a frequency domain shift of λ units is made for some specific member of $\{Z_F(s)\}$, we have

$$Z(s - \lambda) = [\beta_0(s - \lambda) + \gamma_0].$$

If we constrain $Z(s - \lambda)$ to be a member of $\{Z_F(s)\}$ and define the multiplier margin of $Z(s)$ to be the upper bound on the frequency domain shift allowed under this restriction, we have

$$\bar{\Lambda} \triangleq \max \Lambda\colon\ Z(s - \lambda) \in \{Z_F(s)\} \qquad \text{for all} \quad \lambda \in [0, \Lambda). \tag{6-21}$$

In this case, $Z(s - \lambda) \in \{Z_F(s)\}$ if $(\gamma_0 - \beta_0\lambda) > 0$ or if $\lambda < \gamma_0/\beta_0 \triangleq \bar{\Lambda}$; this ratio appears in condition (6-20$^\pm$).

The above derivations, definitions and comments complete the proof of the first extension of the Popov criterion for the absolute stability of NLTV systems.

CRITERION 5a (A point criterion, first and third quadrant gains). The system defined by Eqs. (6-1) and (6-12) is absolutely $\{G_1[F, K_1]\}$ stable if

(1) there exists some $Z(s)$ such that $Z^{\pm 1}(s) \in \{Z_F(s)\}$ [Eq. (5-11)] and

$$[W(s) + \bar{G}^{-1}]Z(s) \in \{\mathrm{SPR}\}; \tag{6-22}$$

(2) dk/dt is restricted by

$$dk/dt \leqslant \underline{\Phi}\bar{\Lambda}k(t)[1 - k(t)/\bar{K}] \quad \text{if } Z^{+1} \in \{Z_F(s)\}, \tag{6-23$^+$}$$

or

$$dk/dt \geqslant -\underline{\Phi}\bar{\Lambda}k(t)[1 - k(t)/\bar{K}] \quad \text{if } Z^{-1} \in \{Z_F(s)\}, \tag{6-23$^-$}$$

where $\bar{\Lambda} = \gamma_0/\beta_0$ is the multiplier margin of $Z(s)$ [Eq. (6-21)] and $\underline{\Phi}$ is the index of nonlinearity [Eq. (6-16)]. □

It is important to emphasize that this theorem enables us to completely separate the constraint on dk/dt from the values of σ_0 and $f(\sigma_0)$ at each instant. This was accomplished by two artifices: discarding the fourth term of Eq. (6-15) and defining $\underline{\Phi}$ as an index of nonlinearity. Without taking these steps, the time domain condition would be

$$\frac{dk}{dt} \leqslant \frac{\gamma_0}{\beta_0} \frac{\sigma_0 f(\sigma_0)}{\int_0^{\sigma_0} f(\zeta)\, d\zeta} k(t) \left[1 - \frac{f(\sigma_0)k(t)}{\bar{G}\sigma_0}\right]$$

(for $Z^{+1}(s) \in \{Z_F(s)\}$), which is difficult, if not impossible, to apply directly. On the other hand, the above might conceivably be less strict, since condition (6-23$^+$) is only a sufficient condition for this constraint to be satisfied. The comparative simplicity of Criterion 5a would seem to far outweigh any such considerations. The parameter $\underline{\Phi}$ was introduced by Narendra and Taylor [1]; since it is important to be able to make a liberal estimate of this parameter so as to allow dk/dt as large a variation as possible, a detailed discussion of $\underline{\Phi}$ is given in Chapter II, Section 1.

It is noted [Eq. (6-11)] that the circle criterion for $g(\sigma_0, t)/\sigma_0 \in [0, \bar{G}]$ requires that condition (6-22) be satisfied for $Z(s) = 1$. Since $\gamma_0 > 0$, $\bar{\Lambda} = \gamma_0/\beta_0$ becomes infinite as $\beta_0 \longrightarrow 0$, so the same result (unbounded time variation) is allowed by (6-23$^\pm$). Thus the extended Popov condition agrees with the result obtained by an application of the circle criterion; in this sense, the circle criterion may be viewed as a special case of Criterion 5a.

In application, this theorem lends itself to a quite direct interpretation. First, the class of multipliers $\{Z_F(s)\}$ provides an explicit constraint on the type of behavior that $[W(s) + \bar{G}^{-1}]$ can exhibit. This condition is equivalent to a simple geometric restriction on $W(i\omega)$, considered in detail in Sections 2 and 5 of Chapter VII. Once this part of the absolute stability criterion is known to be satisfied, it is necessary to extract two pieces of information, one from the LTI portion of the system ($\bar{\Lambda}$, or the multiplier margin of $Z(s)$), and one from the behavior of the nonlinearity ($\underline{\Phi}$, or the index of nonlinearity). Based on these parameters, it is then possible to constraint dk/dt in such a manner as to guarantee absolute stability by using condition (6-23$^{\pm}$). The parameter $\bar{\Lambda}$ is simply obtained from the geometric constraint on $W(i\omega)$, and condition (6-23$^{\pm}$) also permits a simple graphical interpretation in the phase plane (dk/dt versus k) (Chapter VII, Section 5A).

Considering the form of $\dot{v}$ under the substitution of $g(\sigma_0, t)$ for $k(t) f(\sigma_0)$ (that is, under the assumption that the nonlinear time-varying gain is not separable), the only condition of Criterion 5a that must be altered is (2) [Eq. (6-23$^{\pm}$)]. Since $g(\sigma_0, t)$ is not separable, it is no longer possible to express the restriction on the rate of time variation in such simple terms. As all other details in considering $g \in \{G_0\}$ are identical, the stability criterion may be written by inspection.

CRITERION 5b (A point criterion, inseparable gains). The system defined by Eqs. (6-1) and (6-12) is absolutely $\{G_0 [F, K_1]\}$ stable if

(1) condition (1) of Criterion 5a is satisfied;
(2) $\partial g/\partial t$ is restricted by

$$\int_0^{\sigma_0} \partial g(\zeta, t)/\partial t \, d\zeta \leqslant \bar{\Lambda}\sigma_0 g(\sigma_0, t)[1 - g(\sigma_0, t)/(\bar{G}\sigma_0)] \qquad \text{(6-24}^{+}\text{)}$$

for all σ_0 and t if $Z^{+1}(s) \in \{Z_F(s)\}$, or

$$\int_0^{\sigma_0} \partial g(\zeta, t)/\partial t \, d\zeta \geqslant -\bar{\Lambda}\sigma_0 g(\sigma_0, t)[1 - g(\sigma_0, t)/(\bar{G}\sigma_0)] \qquad \text{(6-24}^{-}\text{)}$$

for all σ_0 and t if $Z^{-1} \in \{Z_F(s)\}$; $\bar{\Lambda} = \gamma_0/\beta_0$ is the multiplier margin [Eq. (6-21)] of $Z(s)$. □

The stability condition (2) in Criterion 5b is difficult to apply in general. In the infinite sector case, however, this criterion may be tractable.

Criteria for the particular case of one zero eigenvalue may be obtained by making suitable modifications in the derivations of Criterion 5a and Criterion 5b. We find, however, that in order to constrain $\dot{v}$ to be negative definite as required in time-varying situations, it is necessary that $g(\sigma_0, t)/\sigma_0 \geqslant \varepsilon > 0$ for all σ_0 and t. Ranges of the form $[\varepsilon, \bar{G}]$ may be treated by the general finite sector transformation (Section 4), since stability in the limit (see Chapter V, Section 1) guarantees that $A_\varepsilon \in \{A_1\}$. The rationale for considering the particular case for NLTI systems is that we can permit $f(\sigma_0)/\sigma_0$ to approach zero asymptotically, that is, $f(\sigma_0)/\sigma_0 \in (0, \bar{F}]$; since this cannot be allowed in NLTV situations, there is no point in treating this problem separately.

The first instance of point constraints on dk/dt is due to Narendra and Goldwyn [3]. Brockett and Forys [1] derived the special case of Criterion 5a for LTV systems ($\Phi = 2$). Rekasius and Rowland [1] obtained a number of possible constraints on $\int_0^{\sigma_0} \partial g(\zeta, t)/\partial t \, d\zeta$, one of which corresponds to Criterion 5b. Sandberg [5] extended the result of Brockett and Forys by using a more general integral formulation and functional analytic techniques.

3. Stability Criteria for Restricted Nonlinear Behavior—Point Conditions

The derivation of stability criteria for all NLTV systems with restricted nonlinear behavior is sufficiently similar to the extension of Popov's theorem that only the class $\{G_1[F_m, K_1]\}$ requires complete analysis. Even this case is made relatively simple in some respects, given the result of Chapter V for monotonic gains. In particular, the extensive preliminaries regarding system augmentation and the validity of treating the augmented system to determine the stability of the original system (Lemma 5-2) are identical, as is the frequency domain analysis leading to the use of the RL multiplier in the stability criteria. We denote the state vector by z so that it remains evident that the system under consideration has been suitably augmented, but the system of subscripting (A_a, b_a, h_a) is eliminated. This reduction in preparatory considerations is counterbalanced to some extent by the greater complexity of the various transformations due to the time variation of the gain.

We should again stress that the motivation of this analysis is the understanding that as the nonlinear behavior is restricted, we would expect in general to be able to relax the conditions to be met by the remainder of the system. This usually leads to either an increase in the range permitted for the NLTV gain, or a less stringent restriction on dk/dt or $\partial g/\partial t$, or both.

The stability properties of the system described by the vector differential equation (6-1) with the NLTV gain and matrix A specified by

$$\begin{aligned} &g(\sigma_0, t) \in \{G_i[F_m, K_1]\}, \qquad \Delta g(\sigma_0, t)/\Delta\sigma_0 \in [0, \bar{G}_M], \\ &A \in \{A_1\}, \qquad A_{\bar{G}_M} \in \{A_1\}, \end{aligned} \tag{6-25}$$

are to be ascertained. For the finite sector case we have $k(t) \in [0, \bar{K}]$ and

$$0 \leqslant \frac{f(\sigma_1) - f(\sigma_2)}{\sigma_1 - \sigma_2} \leqslant \bar{M} \qquad \text{for all} \quad \sigma_1 \quad \text{and} \quad \sigma_2 \neq \sigma_1; \qquad \bar{G}_M \triangleq \bar{M}\bar{K}$$

in the separable case, or

$$0 \leqslant \frac{g(\sigma_1, t) - g(\sigma_2, t)}{\sigma_1 - \sigma_2} \leqslant \bar{G}_M \qquad \text{for all} \quad \sigma_1 \quad \text{and} \quad \sigma_2 \neq \sigma_1 \quad \text{and} \quad t.$$

In the developments that follow, the subscript M on the upper bound is suppressed.

The multiplier class $\{Z_{F_m}(s)\}$ consists of arbitrary RL functions. The class $\{Z_{RL}\}$ is defined in Chapter V by the expansion

$$\begin{aligned} &\{Z_{\mathrm{RL}}(s)\} \triangleq \{Z = (s + \lambda_1)\Big[\beta_0 + \sum_{i=1}^{m} \gamma_i/(s + \eta_i)\Big]; \\ &0 < \lambda_1 < \min_i(\eta_i),\ \beta_0 \geqslant 0,\ \gamma_i > 0\} \subset \{\mathrm{SPR}\}. \end{aligned} \tag{5-34}$$

The reciprocal multiplier, $Z(s) \in \{Z_{\mathrm{RC}}(s)\} \triangleq \{Z_{\mathrm{RL}}^{-1}(s)\}$ also may be used for monotonic gains. Since the singularity of $Z(s)$ that is closest to the origin in the s-plane is at $s = -\lambda_1$, we have the result that $Z(s - \lambda) \in \{Z_{RL}\}$ for any value of $\lambda \in [0, \lambda_1)$, so

$$\bar{\Lambda} \triangleq \lambda_1 \tag{6-26}$$

establishes the multiplier margin [as defined in Eq. (6-21)] of any frequency domain multiplier belonging to $\{Z_{RL}\}$.

A. *The Infinite Sector Problem*

The steps for determining the conditions for stability in this case closely resemble those that have been presented for the previous case. The time-varying absolute Lyapunov function candidate has the form

$$v(z, t) = \tfrac{1}{2} z^T P z + \sum_{i=0}^{m} \beta_i k(t) \int_0^{\sigma_i} f(\zeta)\, d\zeta + \tfrac{1}{2} \beta_0 \rho \tau^2, \tag{6-27}$$

where the usual parameter constraints hold and σ_i, $i = 1, 2, \ldots, m$ are chosen, as in Section 2 of Chapter V, to give the desired poles in $Z(s)$. After application of the LKY lemma, the corresponding derivative $\dot{v}$ contains terms of the form $-\sum_{i=0}^{m} \beta_i k(t) \int_0^{\sigma_i} f(\zeta)\, d\zeta [\lambda_1 \underline{\Phi} - k^{-1}\, dk/dt]$ in addition to the negative definite term $-\frac{1}{2}\varepsilon z^T L z$ and other negative semidefinite forms. As in the previous case, dk/dt has to be bounded by $\bar{\Lambda}\underline{\Phi} k(t)$ to yield a negative definite $\dot{v}$. For the infinite sector problem, the absolute stability of the system is thus guaranteed if $Z(s) \in \{Z_{RL}\}$ exists such that

$$W(s)Z(s) \in \{\mathrm{SPR}\}$$

and (in the separable case)

$$dk/dt \leqslant \bar{\Lambda} \underline{\Phi}\, k(t). \tag{6-28a}$$

When the nonlinear gain is not separable, the latter condition must be changed to

$$\partial/\partial t \int_0^{\sigma_i} g(\zeta, t)\, d\zeta \leqslant \bar{\Lambda} \sigma_i g(\sigma_i, t). \tag{6-28b}$$

B. *The Finite Sector Problem*

While considering the extension of the Popov problem in Section 2, the stability conditions are investigated directly for the finite sector case. For the case when the feedback gain is monotonic, the finite sector criterion is obtained using the results for the infinite sector case and a transformation procedure similar to that of Chapter V, Section 2. In Chapter V, we must consider the range $\Delta f(\sigma_0)/\Delta\sigma_0 \in [0, \bar{M})$ in order to guarantee the validity of the transformation; here we see that it is possible to treat the case $\Delta g(\sigma_0, t)/\Delta\sigma_0 \in [0, \bar{G}]$ directly by making minor modifications in this procedure. This is because the final frequency domain condition is in the form $H(s) \in \{\text{SPR}\}$. Define

$$\hat{W}(s) \triangleq W(s) + (\bar{G} + \varepsilon)^{-1}, \tag{6-29}$$

$$\hat{\sigma}_0 \triangleq \sigma_0 - k(t)f(\sigma_0)/(\bar{G} + \varepsilon), \tag{6-30}$$

$$\hat{g}(\hat{\sigma}_0, t) \triangleq k(t)f(\sigma_0); \ \Delta\hat{g}/\Delta\hat{\sigma}_0 \in [0, \bar{G}(\bar{G} + \varepsilon)/\varepsilon], \tag{6-31}$$

where $0 \leqslant k(t)\Delta f(\sigma_0)/\Delta\sigma_0 \leqslant \bar{G}$. Applying the previous criterion to the transformed system $\{\hat{W}(s), \hat{g}(\hat{\sigma}_0, t)\}$, it is found that the system $\{\hat{W}, \hat{g}\}$ is absolutely stable if the conditions

$$\begin{gathered} [W(s) + (\bar{G} + \varepsilon)^{-1}]Z(s) \in \{\text{SPR}\}, \qquad Z(s) \in \{Z_{\text{RL}}\} \\ (\hat{\sigma}_i \hat{g}(\hat{\sigma}_i, t))^{-1}\, \partial/\partial t \int_0^{\hat{\sigma}_i} \hat{g}(\zeta, t)\, d\zeta \leqslant \lambda_1 \triangleq \bar{\Lambda} \qquad \text{for all } \hat{\sigma}_i \text{ and } t \end{gathered} \tag{6-32}$$

are satisfied. Interpreting the second condition in terms of the original system gain function $k(t)f(\sigma_0)$ requires inversion of the transformation [Eqs. (6-29) to (6-31)]; we ultimately obtain

$$dk/dt \leqslant \bar{\Lambda}\underline{\Phi}k(t)[1 - k(t)/\bar{K}].\dagger \tag{6-23$^+$}$$

The frequency domain condition in Eq. (6-32) can be shown to be equivalent to the standard condition that $Z(s)$ must exist such that $Z(s) \in \{Z_{\text{RL}}\}$ and

$$[W(s) + \bar{G}^{-1}]Z(s) \in \{\text{SPR}\}.\ddagger \tag{6-33}$$

Thus, conditions (6-23$^+$) and (6-33) may be demonstrated to be sufficient conditions for absolute stability in the finite sector $[0, \bar{G}]$. For inseparable gains, the condition (6-23$^+$) must be altered to the equivalent constraint (6-24$^+$).

C. *The Inverse Multiplier*

The derivation of a general stability criterion of the form of Criterion 5a may be completed by observing that for all monotonic gains, the system

† In terms of the transformed variables $\hat{\sigma} = \sigma - k(t)f(\sigma)/(\bar{G} + \varepsilon)$, we have

$$\int_0^{\hat{\sigma}} \hat{g}(\zeta, t)\, d\zeta = k(t) \int_0^{\sigma} f(\xi)\, d\xi - [k(t)f(\sigma)]^2/(\bar{G} + \varepsilon),$$

$$\partial/\partial t \int_0^{\hat{\sigma}} \hat{g}(\zeta, t)\, d\zeta = dk/dt \int_0^{\sigma} f(\xi)\, d\xi.$$

‡ By making the standard approximation $(\bar{G} + \varepsilon)^{-1} \approx \bar{G}^{-1} - \varepsilon/\bar{G}^2$, the condition given in (6-32) is essentially

$$[W(s) + \bar{G}^{-1}]Z(s) - (\varepsilon/\bar{G}^2)Z(s) \in \{\text{SPR}\}.$$

If condition (6-33) is satisfied and $Z(s) \in \{\text{SPR}\}$, then ε can be chosen to be sufficiently small that $(\varepsilon/\bar{G}^2)Z(s)$ is dominated by the first term for all finite s in the closed right half plane. This is also true as $|s| \to \infty$ if $1 + \rho\bar{G} > 0$, that is, if $A_{\bar{G}} \in \{A_1\}$, as assumed.

differential equation may be inverted as in Chapter V, Section 2. The relation (5-39) and the subsidiary transformations used in this procedure are more complex as we are considering a time-varying situation.

By successively using the transformations

(i) $W_1(s) \triangleq \dfrac{W(s)}{1 - \varepsilon W(s)}$; $\quad \dfrac{g_1(\sigma_0, t)}{\sigma_0} \triangleq \left[\dfrac{k(t)f(\sigma_0)}{\sigma_0} + \varepsilon\right] \triangleq \dfrac{\hat{\sigma}_0}{\sigma_0} \in [\varepsilon, \bar{G} + \varepsilon]$,

(ii) $W_2(s) \triangleq (W_1(s))^{-1}$; $\quad \dfrac{g_2(\hat{\sigma}_0, t)}{\hat{\sigma}_0} \triangleq \dfrac{\sigma_0}{\hat{\sigma}_0} \in [(\bar{G} + \varepsilon)^{-1}, \varepsilon^{-1}]$,

(iii) $W_3(s) \triangleq \dfrac{W_2(s)}{1 + W_2(s)/(\bar{G} + \varepsilon)}$;

$$\frac{g_3(\hat{\sigma}_0, t)}{\hat{\sigma}_0} \triangleq \frac{g_2(\hat{\sigma}_0, t)}{\hat{\sigma}_0} - (\bar{G} + \varepsilon)^{-1} \in \left[0, \frac{\bar{G}}{\varepsilon(\bar{G} + \varepsilon)}\right],$$

the final transformed system $\{W_3(s), g_3(\hat{\sigma}_0, t)\}$ is of the correct form so that the preceding finite sector result can be applied. This in turn can be used to show that conditions (6-23$^-$) or (6-24$^-$), with $Z(s) \in \{Z_{\mathrm{RC}}(s)\}$ satisfying the frequency domain condition, are also sufficient to prove the absolute stability of the system under consideration. Although several transformations are again required in various parts of the proof, this result is seen to be identical in form to the extension of the Popov result.

CRITERION 6 (A point criterion, monotonic separable gains). The system defined by Eqs. (6-1) and (6-25) is absolutely $\{G_1[F_m, K_1]\}$ stable if

(1) there exists some $Z(s)$ such that $Z^{\pm 1}(s) \in \{Z_{F_m}(s)\}$ [Eq. (5-34)] and

$$[W(s) + \bar{G}_M^{-1}]Z(s) \in \{\mathrm{SPR}\}; \tag{6-22}$$

(2) dk/dt is restricted by

$$dk/dt \leqslant \underline{\Phi}\bar{\Lambda}k(t)[1 - k(t)/\bar{K}] \qquad \text{if} \quad Z^{+1} \in \{Z_{F_m}(s)\}, \tag{6-23$^+$}$$

or

$$dk/dt \geqslant -\underline{\Phi}\bar{\Lambda}k(t)[1 - k(t)/\bar{K}] \qquad \text{if} \quad Z^{-1} \in \{Z_{F_m}(s)\}, \tag{6-23$^-$}$$

where $\bar{\Lambda} = \lambda_1$ is the multiplier margin of $Z(s)$ [Eq. (6-21)] and $\underline{\Phi}$ is the index of nonlinearity of $f(\sigma_0)$ [Eq. (6-16)]. □

The criterion for the principal case with inseparable gains is derived from Criterion 6 directly as in the extension of the Popov criterion (Section 2), so it is not repeated here.

Zames [3] obtained this result with the conservative index of nonlinearity $\underline{\Phi} \equiv 1$. Narendra and Taylor [1] obtained Criterion 6 in the infinite sector case; more general results are reported in Cho and Narendra [1] and Narendra and Taylor [2]. Independent work by Srinath, Thathachar, and Ramapriyan [1] established identical results for the inseparable case.

The derivation of stability criteria for systems with other classes of restricted nonlinearities and for LTV systems is identical to that given above for monotonically nondecreasing functions. In all cases, the stability of the system is assured if a frequency domain condition is satisfied by the linear part of the system and dk/dt or $\partial g/\partial t$ is constrained in the fashion indicated above. For a given LTI plant $W(s)$, it may be appreciated that the constraint on the rate of time variation of the NLTV gain generally becomes less stringent as the allowed type of nonlinear behavior specified by $\{N\}$ is made more restrictive. This arises from two phenomena: as $\{N\}$ is restricted, the class $\{Z_N(s)\}$ becomes more general, so it may be possible to find a multiplier $Z(s)$ satisfying the frequency domain condition (6-22) which has a larger multiplier margin $\bar{\Lambda}$, and also as $\{N\}$ is restricted it may be possible to arrive at a higher estimated value of the nonlinearity index Φ (see Chapter II, Section 1). This is discussed from the point of view of geometric interpretations in Chapter VII, Section 5.

Results for various classes of nonlinear behavior may be found in Cho and Narendra [1], Narendra and Taylor [1, 2] and in Srinath, Thathachar, and Ramapriyan [1]. Brockett and Forys [1] obtained the condition corresponding to $Z(s) = (1 + \alpha s)/(1 + \beta s)$ and $\bar{\Lambda} = \min(\alpha^{-1}, \beta^{-1})$ for LTV systems. A notable result for LTV systems was obtained by Gruber and Willems [1]: a system is absolutely $\{K_1\}$ stable if $k(t) \in [0, \bar{K})$, $Z(s)[W(s) + \bar{K}^{-1}] \in \{\text{SPR}\}$, $Z(s - \lambda) \in \{\text{SPR}\}$ for $0 \leqslant \lambda < \bar{\Lambda}$ and $dk/dt \leqslant 2\bar{\Lambda}k(t)[1 - k(t)/\bar{K}]$.

4. The General Finite Sector Problem

The stability criteria given in the preceding sections apply to nonlinear time-varying gains which lie in the interval $[0, \bar{G}_N]$. Using the simple transformations discussed in Chapter V, Section 5, these results can be easily extended to the case where the NLTV gains lie in a range $[\underline{G}_N, \bar{G}_N]$. Define the transformed system $\{\hat{W}(s), \hat{g}(\sigma_0, t)\}$ by

$$\hat{W}(s) = W(s)/(1 + \underline{G}_N W(s)), \qquad \hat{g}(\sigma_0, t) = g(\sigma_0, t) - \underline{G}_N \sigma_0;$$

if the original system described by Eq. (6-1) is constrained by Eq. (6-2), then $\{\hat{W}, \hat{g}\}$ falls within the ambit of one of the preceding absolute stability criteria. Hence we can summarize all of these results as follows.

General Stability Criterion 2. The system described by Eq. (6-1) constrained by Eq. (6-2) is absolutely $\{G_i[N, K_1]\}$ stable if

(1) $Z(s)$ exists such that $Z^{\pm 1}(s - \lambda) \in \{Z_N(s)\}$, $\lambda \in [0, \bar{\Lambda})$ where the classes $\{Z_N(s)\}$ are defined in Definitions 1–4, General Stability Criterion 1,

and

$$\frac{1 + \bar{G}_N W(s)}{1 + \underline{G}_N W(s)} Z(s) \in \{\text{SPR}\};$$

(2) dk/dt or $\partial g/\partial t$ are restricted by (a) or (b):

(a) *Separable gains:*

$$\frac{dk}{dt} \leqslant \underline{\Phi}\bar{\Lambda}[k(t) - \underline{K}][1 - k(t)/\bar{K}]\frac{\bar{G}_N}{\bar{G}_N - \underline{G}_N}, \qquad Z^{+1} \in \{Z_N\},$$

or (6-34±)

$$\frac{dk}{dt} \geqslant -\underline{\Phi}\bar{\Lambda}[k(t) - \underline{K}][1 - k(t)/\bar{K}]\frac{\bar{G}_N}{\bar{G}_N - \underline{G}_N}, \qquad Z^{-1} \in \{Z_N\},$$

(b) *Inseparable gains:*

$$\int_0^\sigma \partial/\partial t\, g(\zeta, t)\, d\zeta \leqslant \bar{\Lambda}\sigma^2 \left[\frac{g(\sigma, t)}{\sigma} - \underline{G}_N\right]\left[1 - \frac{g(\sigma, t)}{\bar{G}_N \sigma}\right]\frac{\bar{G}_N}{\bar{G}_N - \underline{G}_N}, \qquad Z^{+1} \in \{Z_N\},$$

or (6-35±)

$$\int_0^\sigma \partial/\partial t\, g(\zeta, t)\, d\zeta \geqslant -\bar{\Lambda}\sigma^2 \left[\frac{g(\sigma, t)}{\sigma} - \underline{G}_N\right]\left[1 - \frac{g(\sigma, t)}{\bar{G}_N \sigma}\right]\frac{\bar{G}_N}{\bar{G}_N - \underline{G}_N}, \qquad Z^{-1} \in \{Z_N\},$$

where $\underline{\Phi}$, the index of nonlinearity for separable gains, is defined by

$$\underline{\Phi} \triangleq \min_\sigma \sigma f(\sigma) / \textstyle\int_0^\sigma f(\zeta)\, d\zeta; \qquad (6\text{-}16)$$

the multiplier margin for each class $\{Z_N\}$ is

DEFINITION 1: $\bar{\Lambda} = \gamma_0/\beta_0$,

DEFINITION 2: $\bar{\Lambda} = \lambda_1$,

DEFINITION 3: $\bar{\Lambda} = \min\{\lambda_0; \eta_i(2 - \rho_i),\ i = n_1 + 1, \ldots, m_1;\ \varepsilon_1;\ \varepsilon_{2i},\ i = 1, 2, \ldots, m_2\}$

if $Z(s)$ in Lemma 5-5 is used, or

$$\bar{\Lambda} = \min\{[\alpha - (\gamma_1/\beta_0)^{1/2}];\ \varepsilon_1;\ \varepsilon_{2i}, i = 1, 2, \ldots, k\}$$

if $Z(s) \in \{Z_\alpha(s)\}$ is used according to the corollary to Lemma 5-5,

DEFINITION 4: $\bar{\Lambda} = \alpha$. □

This result clearly subsumes all of the earlier criteria derived in Sections 1 to 3. In the case $Z(s) = 1$, we have $\bar{\Lambda} = \infty$ and the condition $H(s) \in \{\text{SPR}\}$ is identical to Eq. (6-10); this situation corresponds to $v(x) = \frac{1}{2}x^T Px$, so we have an exact duplication of the circle criterion in that the NLTV gain may be discontinuous with respect to time; in fact, the result here is more general than that given in Section 1 in that the range $[\underline{G}, \bar{G}]$ need not include zero.

If the gain is separable the restriction on dk/dt is easily interpreted in the phase plane (a plot of dk/dt versus k), as shown in Chapter VII, Section 5. For inseparable gains, however, condition ($6\text{-}35^{\pm}$) would most probably prove to be very difficult to apply.

5. Periodic Nonlinear Time-Varying Gains

The constraints derived in the previous sections may be significantly relaxed if the system is periodic. In this case, we may use the theorem of LaSalle, which provides the same conditions for stability as in the NLTI case.

COROLLARY A2 (LaSalle [1]). If a function $v(x, t) = v(x, t + T)$ for all t and fixed x satisfies conditions (i)–(iv) of Theorem A, then a sufficient condition for the uniform asymptotic stability in the whole of the solutions to

$$\begin{aligned} \dot{x} &= f(x, t) \in \{S\}, \\ f(x, t) &= f(x, t + T) \qquad \text{for all} \quad t \quad \text{and fixed} \quad x \end{aligned} \tag{1-3e}$$

is that condition (A1 : v) is satisfied. □

Thus, we can permit $\dot{v}$ to be equal to zero as long as $\dot{v} \not\equiv 0$ unless $x \equiv 0$; this in turn allows the use of the MKY lemma rather than the LKY lemma, which results in the less strict frequency domain condition $H(s) \in \{\text{PR}\}$.

To see that this theorem is directly applicable, we consider only the point constraint on dk/dt obtained in extending the Popov criterion to the separable NLTV case. Assume that $H(s)$ [Eq. (6-17)] satisfies the condition $H(s) \in \{\text{PR}\}$; then $\dot{v} \leqslant 0$ and $\dot{v} = 0$ only when

(i) $x^{\mathrm{T}}Mx = 0 \quad$ where $\quad M = M^{\mathrm{T}} \geqslant 0$,

(ii) $[x^T q + \sqrt{\psi}\, k(t) f(\sigma_0)] = 0$,

[by the application of the MKY lemma to Eq. (6-15)];

(iii) $k(t) f(\sigma_0) = 0 \quad$ or $\quad f(\sigma_0) = \bar{F}\sigma_0$,

(iv) $\gamma_0 k \sigma_0 f(\sigma_0)[1 - k(t)/\bar{K}]$

$$= \hat{\beta}_0\, dk/dt \int_0^{\sigma_0} f(\xi)\, d\xi \qquad \text{or} \quad k(t) f(\sigma_0) = 0$$

[from the fourth and fifth terms of Eq. (6-15)]. If $\dot{v} \equiv 0$ along a trajectory, then from (iii), either $k(t) f(\sigma_0) \equiv 0$ (in which case $\dot{v} \equiv -\frac{1}{2}(x^{\mathrm{T}}q)^2 - \frac{1}{2}x^{\mathrm{T}}Mx$, which is identically zero only if $x \equiv 0$), or $f(\sigma_0) \equiv \bar{F}\sigma_0$. In the second case, condition (iv) requires that $2\gamma_0 k[1 - k/\bar{K}] \equiv \hat{\beta}_0\, dk/dt$. From a practical viewpoint, very little loss of generality results by assuming that $k(t)$ does not satisfy this relation.

COROLLARY TO GENERAL STABILITY CRITERION 2. (Periodic separable systems). If $g(\sigma_0, t) = k(t)f(\sigma_0) = k(t + T)f(\sigma_0)$ for all t in Eq. (6-1), then the frequency domain condition of General Stability Criterion 2 can be replaced by

$$\frac{1 + \bar{G}_N W(s)}{1 + \underline{G}_N W(s)} Z(s) \in \{\mathrm{PR}\},$$

provided that $dk/dt \not\equiv \pm 2\bar{\Lambda}[k(t) - \underline{K}][1 - k(t)/\bar{K}]\bar{G}_N/(\bar{G}_N - \underline{G}_N)$. □

Similar relaxed frequency domain conditions may be stated for the absolute stability of inseparable periodic systems. Since for periodic gains the constraint $g(\sigma_0, t)/\sigma_0 \in (\underline{G}_N, \bar{G}_N)$ is equivalent to $g(\sigma_0, t)/\sigma_0 \in [\underline{G}_N + \varepsilon, \bar{G}_N - \varepsilon]$, there is again no point in treating the particular case, $A_{\underline{G}_N} \in \{A_0\}$, for the reason discussed in Section 2.

6. Extension of the Popov Criterion—Integral Conditions

In deriving less restrictive results for time-varying systems, it has been found to be beneficial to weaken the requirement that $\dot{v}$ be negative definite. The standard conditions $v > 0$, $\dot{v} < 0$ used heretofore are actually rather strict in view of the ultimate goal of Lyapunov's direct method: proving the asymptotic stability of a system by showing that $v(x(t), t) \longrightarrow 0$ uniformly as t goes to infinity. Theorem 2, which is a special case of the noteworthy result of Corduneanu [1], provides the desired relaxed conditions for stability. To recapitulate this stability theorem, a system is uniformly asymptotically stable in the whole if $v(x, t)$ is a valid absolute Lyapunov function candidate (Lemma V3) such that there exists a real-valued continuous function $p(t)$ that satisfies

(i) $$\dot{v} \leqslant p(t)v \tag{6-36}$$

and

(ii) $$\lim_{t\to\infty} \int_{t_0}^{t} p(\tau)\, d\tau = -\infty \tag{6-37a}$$

uniformly with respect to t_0 if $p(t)$ is aperiodic, or

$$\int_0^T p(t)\, dt < 0 \tag{6-37b}$$

if $p(t + T) = p(t)$ for all t.

As previously mentioned, this theorem demonstrates the direct relationship between the stability properties of x and $v(x, t)$.

Due to increased complexity, we only consider the separable gain case. Ultimately, $p(t)$ in condition (6-36) contains the term $k^{-1}\, dk/dt$, so in order to satisfy the continuity requirement of this theorem it is necessary that $k(t) \in \{K_2\}$, that is, that $k(t)$ possesses a continuous derivative, and that $k(t) > 0$. Hence consider the behavior of the system described by the differential equation (6-1) constrained by

$$\begin{aligned} &g(\sigma_0, t) \in \{G_1[F, K_2]\}, \quad k(t) \in (0, \bar{K}], \quad f(\sigma_0)/\sigma_0 \in [0, \bar{F}], \\ &\rho \geqslant 0, \quad A \in \{A_1\}, \quad A_{\bar{F}\bar{K}} \triangleq A_{\bar{G}_F} \in \{A_1\}. \end{aligned} \tag{6-38}$$

In order to apply this theorem, it is necessary to have a term in $\dot{v}$ that is proportional to each term in v; that is, since we use the absolute Lyapunov function candidate indicated in Eq. (6-13), we require in $\dot{v}$ terms proportional to $x^{\mathrm{T}}Px$, to $k(t) \int_0^{\sigma_0} f(\zeta)\, d\zeta$, and to τ^2. It has been demonstrated in Section 2 that the derivative of this candidate evaluated along trajectories of a system with a separable gain in the finite sector is

$$\begin{aligned} \dot{v} = {} & \tfrac{1}{2}x^{\mathrm{T}}(A^{\mathrm{T}}P + PA)x - k(t)f(\sigma_0)x^{\mathrm{T}}[Pb - \beta_0 A^{\mathrm{T}}h - \gamma_0 h] \\ & - [\beta_0 h^{\mathrm{T}}b + \gamma_0(\rho + (\bar{G}_F)^{-1})]\tau^2 - (\gamma_0/\bar{K})\sigma_0 k^2(t) f(\sigma_0)[1 - f(\sigma_0)/(\bar{F}\sigma_0)] \\ & + \beta_0 k(t) \int_0^{\sigma_0} f(\zeta)\, d\zeta [k^{-1}\, dk/dt - (\gamma_0/\beta_0)\phi(\sigma_0)\{1 - k(t)/\bar{K}\}]. \end{aligned} \tag{6-15}$$

One term proportional to $k(t) \int_0^{\sigma_0} f(\zeta)\, d\zeta$ and another that varies as τ^2 exist, but none that varies as $x^{\mathrm{T}}Px$ is in evidence.

At this point, as in Chapter V, Section 1, we might directly apply the MKY lemma, which provides a condition that guarantees that the first three terms of $\dot{v}$ are negative semidefinite. If, however, $\hat{A} \triangleq A + \mu I \in \{A_1\}$, then we may apply the following modification of the MKY lemma.

LEMMA 4 (Chapter III, Section 4). Given $\hat{A} \in \{A_1\}$, real vectors b and k, a real scalar ψ, then a real vector q and matrices P, $P = P^{\mathrm{T}} > 0$, and N, $N = N^{\mathrm{T}} \geqslant 0$, exist satisfying

(a) $$A^{\mathrm{T}}P + PA = -qq^{\mathrm{T}} - 2\mu P - N$$

(b) $$Pb - k = \sqrt{\psi}\, q$$

if and only if

(c) $$\hat{H}(s) \triangleq \tfrac{1}{2}\psi + k^{\mathrm{T}}[(s - \mu)I - A]^{-1}b \in \{\mathrm{PR}\}. \ \square$$

In applying this lemma, we define

$$\begin{aligned} k &\triangleq \beta_0 A^{\mathrm{T}}h + \gamma_0 h, \\ \tfrac{1}{2}\psi &\triangleq \beta_0 h^{\mathrm{T}}b + (\gamma_0 - \beta_0\mu)(\rho + (\bar{G}_F)^{-1}); \end{aligned} \tag{6-39}$$

then the terms of $\dot{v}$ are reordered so that an application of Lemma 4 yields

$$\begin{aligned} \dot{v} \leqslant {} & p(t)v - [2\mu + p(t)][\tfrac{1}{2}x^{\mathrm{T}}Px + \tfrac{1}{2}\beta_0\rho\tau^2] \\ & - \beta_0 k(t) \int_0^{\sigma_0} f(\zeta)\, d\zeta [p(t) + \bar{\Lambda}\phi(\sigma_0) - ((\bar{\Lambda} - \mu)/\bar{K})k(t)\phi(\sigma_0) - k^{-1}\, dk/dt] \\ & - \beta_0\sigma_0 k^2(t) f(\sigma_0)[(\bar{\Lambda} - \mu)/\bar{K}][1 - f(\sigma_0)/(\bar{F}\sigma_0)] \end{aligned} \tag{6-40}$$

(where the nonpositive terms $-\frac{1}{2}[x^T q + \sqrt{\psi}\, k(t) f(\sigma_0)]^2$ and $-\frac{1}{2}x^T N x$ have been discarded in view of the inequality sign and $p(t)v$ has been added and subtracted), if and only if $W(s - \mu)$ is asymptotically stable $[\hat{A} \in \{A_1\}]$ and $\hat{H}(s) \in \{\mathrm{PR}\}$. By substituting for k and ψ into the latter condition, the equivalent constraint is that $H(s - \mu) \in \{\mathrm{PR}\}$, where

$$H(s) \triangleq [W(s) + (\bar{G}_F)^{-1}](\beta_0 s + \gamma_0),$$

as in the earlier derivations.

The last term of $\dot{v}$ is negative semidefinite if

$$\bar{\Lambda} \triangleq \gamma_0/\beta_0 \geqslant \mu > 0; \tag{6-41}$$

the first inequality is satisfied if $H(s - \mu) \in \{\mathrm{PR}\}$, since the factor $[\beta_0 s + (\gamma_0 - \beta_0 \mu)]$ must represent a closed left hand plane zero of $H(s - \mu)$.

In order to satisfy condition (6-36), it is necessary that

(i) $\quad p(t) \geqslant -2\mu \quad$ and $\quad [\frac{1}{2}x^T P x + \frac{1}{2}\beta_0 \rho \tau^2] \geqslant 0,$

(ii) $\quad p(t) \geqslant -[\{\bar{\Lambda} - ((\bar{\Lambda} - \mu)/\bar{K})k(t)\}\underline{\Phi} - k^{-1}\, dk/dt]$

$\quad$ and $\quad \beta_0 k(t) \int_0^{\sigma_0} f(\zeta)\, d\zeta \geqslant 0.$

On the other hand, it is desired that $p(t)$ be as negative as possible so that the integrals in condition (6-37) may be negative; thus it is most advantageous to choose

$$p(t) \triangleq \sup\{-2\mu; -[\{\bar{\Lambda} - (\bar{\Lambda} - \mu)k(t)/\bar{K}\}\underline{\Phi} - k^{-1}\, dk/dt]\}. \tag{6-42}$$

Here we see the necessity of assuming that $\beta_0 \geqslant 0$ and $\rho \geqslant 0$: this definition of $p(t)$ guarantees that $\dot{v} \leqslant p(t)v$ only if $v(x, t)$ has been divided into two parts satisfying $v_1(x, t) \geqslant 0$ and $v_2(x, t) \geqslant 0$; if $\beta_0 < 0$ is allowed, it cannot be guaranteed that $v_1 \triangleq \frac{1}{2}x^T P x + \frac{1}{2}\beta_0 \rho \tau^2$ is nonnegative, for example.

The corresponding result that permits the use of $[\beta_0(s - \mu) + \gamma_0]^{-1}$ in condition (b) presents some difficulties that are not resolved at this time. The results of the derivation above are collected to give us the first example of a stability criterion which does not constrain dk/dt at every instant.

Criterion 7 (A time-averaged criterion, first and third quadrant gains). The system defined by Eqs. (6-1) and (6-38) is absolutely $\{G_1[F, K_2]\}$ stable if

(1) $W(s - \mu)$ is asymptotically stable for some $\mu > 0$;
(2) $H(s - \mu)$ is positive real where

$$H(s) \triangleq [W(s) + (\bar{G}_F)^{-1}](\beta_0 s + \gamma_0); \qquad \beta_0 \geqslant 0, \quad \gamma_0 > 0; \tag{6-43}$$

(3) $p(t)$ [Eq. (6-42)] with $\bar{\Lambda} = \gamma_0/\beta_0$ and $\underline{\Phi}$ as defined previously [Eq. (6-16)] satisfies condition (6-37). □

As $\mu \longrightarrow 0$, this criterion clearly approaches Criterion 5a except that the inverse multiplier cannot be used. In some instances it is advantageous to allow

μ to be as large as possible, while in other circumstances the most lenient conditions for absolute stability may be obtained by taking intermediate values of μ [$\mu \in (0, \bar{\Lambda})$] or by using Criterion 5a. This is demonstrated in Chapter VIII, Section 5 in treating a nonlinear differential equation related to the damped Mathieu equation. In treating linear time-varying situations, experience has so far indicated that taking $\mu = \bar{\Lambda}$ or using Criterion 5a yields the least stringent conditions for absolute $\{K_2\}$ stability (see Chapter VIII, Section 4, and Section 8 of this chapter), so it does not appear to be necessary to keep μ as a free parameter and formally determine $\hat{\mu} \in [0, \bar{\Lambda}]$ by using standard techniques of calculus to find the greatest permitted range of $k(t)$ or its maximum frequency if $k(t)$ is periodic. In NLTV situations, however, any attempt to find such optimal conditions for absolute stability does require these techniques, as demonstrated in Chapter VIII, Section 5. In general, Criterion 7 with $\mu = \bar{\Lambda}$ is the most lenient if the phase plane plot of $k(t)$ is symmetric about the k-axis, while condition (6-23$^+$) is apt to be less strict if dk/dt exhibits large negative excursions but takes on only small positive values, as shown in the example given in Section 8.

7. Integral Conditions for Restricted Nonlinear and Linear Gains

The procedures used in the development of the previous section may be applied to any criterion for NLTV gains of restricted nonlinear behavior (see Section 3, especially Criterion 6) for the infinite sector case essentially by inspection. In order to obtain the stability criterion for the finite sector case, however, we would be forced to resort to rather extensive transformational methods, so we do not consider this case.

General Stability Criterion 3. The system described by Eq. (6-1) constrained by

$$g(\sigma_0, t) \in \{G_1[N, K_2]\}, \qquad k(t) \in (0, \bar{K} < \infty)$$
$$f(\sigma_0)/\sigma_0 \in [0, \infty), \qquad (A + \mu I) \in \{A_1\} \tag{6-44}$$

is absolutely $\{G_1[N, K_2]\}$ stable if

(1) $Z(s)$ exists such that $Z(s - \lambda) \in \{Z_N(s)\}$, $\lambda \in [0, \bar{\Lambda})$, where the classes $\{Z_N(s)\}$ are defined in Definitions 1–4, General Stability Criterion 1, and for some $\mu > 0$

$$W(s - \mu)Z(s - \mu) \in \{\mathrm{PR}\};$$

(2) dk/dt is restricted by demanding that

$$p(t) = \sup\{-2\mu; -[\underline{\Phi}\bar{\Lambda} - k^{-1}\, dk/dt]\} \tag{6-45}$$

satisfy condition (6-37), where the index of nonlinearity $\underline{\Phi}$ is defined by Eq. (6-16). □

It is quite clear that there is little point in attempting to derive any directly analogous result for inseparable NLTV gains; the time domain condition would be virtually intractable. It is not possible to derive an analogous result for $A \in \{A_0\}$ and $k(t)f(\sigma_0) \in (0, \bar{G}_N]$, because under these circumstances $W(s)$ has a pole at $s = 0$ and thus the condition that $W(s - \mu)$ must be asymptotically stable $[A + \mu I \in \{A_1\}]$ cannot be met for any $\mu > 0$.

The same comments made in reference to Criterion 7 apply to this result also; in particular, the stability criterion corresponding to $Z^{-1} \in \{Z_N\}$ has not been obtained in a final form.

The interpretation of the various steps taken in the application of this criterion is similar to that given for the point criteria. The two principal differences are that dk/dt is only constrained in an average sense and that two parameters must be evaluated on the basis of the properties of the LTI plant; in addition to the multiplier margin $\bar{\Lambda}$ used in General Stability Criterion 2, we take μ as a measure of the passivity of $H(s) = W(s)Z(s)$. Obviously μ and $\bar{\Lambda}$ are interrelated; usually $\bar{\Lambda}$ must be reduced if μ is increased as shown in the applications indicated in Chapter VIII.

The application of General Stability Criterion 3, as well as the circle criterion and General Stability Criterion 2, to linear and nonlinear systems is demonstrated in Chapter VIII. Several conclusions may be drawn in light of this analysis of two forms of the Mathieu equation (see Chapter VIII, Sections 4 and 5); we particularly stress that the degree to which each theorem approximates the necessary and sufficient condition for stability (a "figure of merit" of each criterion) is nearly in direct proportion to the sophistication of the criterion and hence the complexity of its application.

The above result with $\rho = 0$ was first reported by Taylor and Narendra [2].

8. Integral Conditions for Linear Time-Varying Systems

Since the transformation procedures required for the generalization of General Stability Criterion 3 to the general finite sector case for LTV systems are quite straightforward, we treat this case to obtain the most general result for this class of systems. For periodic gains, we then obtain the result of Freedman and Zames [1] as a special case. Consider Eq. (6-1) constrained by

$$\begin{aligned} \tau = -k(t)\sigma_0, \qquad k(t) \in \{K_2\}, \qquad k(t) \in [\underline{K} + \varepsilon, \bar{K} - \varepsilon], \\ (A_{\underline{K}} + \mu I) \in \{A_1\}, \qquad (A_{\bar{K}} + \mu I) \in \{A_1\}. \end{aligned} \tag{6-46}$$

The two transformations used extensively in previous developments are defined by the relations (5-36) and (5-60). Applying these successively, we consider the equivalent system defined by

$$W_1(s) \triangleq \frac{1 + \bar{K}W(s)}{1 + \underline{K}W(s)},$$

$$k_1(t) \triangleq \frac{k(t) - \underline{K}}{\bar{K} - k(t)} \in \left[\frac{\varepsilon}{\bar{K} - \underline{K} - \varepsilon}, \frac{\bar{K} - \underline{K} - \varepsilon}{\varepsilon}\right]. \tag{6-47}$$

Applying General Stability Criterion 3 to this transformed system $\{W_1, k_1\}$, we obtain the conditions that $Z(s)$ must exist such that $Z(s) \in \{Z_\alpha(s)\}$ [Eq. (5-44)] and $W_1(s - \mu)Z(s - \mu) \in \{PR\}$, and that dk/dt must be restricted by demanding that

$$p(t) \triangleq \sup\left\{-2\mu; -\left[2\alpha - \frac{\bar{K} - \underline{K}}{(k - \underline{K})(\bar{K} - k)}\frac{dk}{dt}\right]\right\} \tag{6-48$^+$}$$

satisfies condition (6-37).

Although the class $\{Z_\alpha(s)\}$ contains its own inverse as noted previously, the time domain condition is different if we consider the equivalent system

$$W_2(s) \triangleq -W(s)/(1 + \bar{K}W(s)),$$

$$k_2(t) \triangleq \bar{K} - k(t) \in [\varepsilon, \bar{K} - \underline{K} - \varepsilon]$$

obtained by a simple transformation similar to that discussed in Chapter V. Since this is in the form considered in Eq. (6-46) with $\underline{K}_2 = 0$ and $\bar{K}_2 = \bar{K} - \underline{K}$, we have to satisfy

$$[W_2(s - \mu) + (\bar{K}_2)^{-1}]Z(s - \mu)$$
$$= \frac{1 + \underline{K}W(s - \mu)}{(\bar{K} - \underline{K})[1 + \bar{K}W(s - \mu)]} \cdot Z(s - \mu) \in \{PR\},$$

where $Z(s) \in \{Z_\alpha(s)\}$, while $p(t)$ [Eq. (6-48$^+$)] becomes

$$p(t) = \sup\left\{-2\mu; -\left[2\alpha - \frac{\bar{K} - \underline{K}}{(\bar{K} - k)(k - \underline{K})}\left(-\frac{dk}{dt}\right)\right]\right\}. \tag{6-48$^-$}$$

The ability to choose the sign of the term in dk/dt in Eq. (6-48$^\pm$) is useful in some applications.

CRITERION 8 (Integral conditions, linear time-varying systems). The system described by Eqs. (6-1) and (6-46) is absolutely $\{K_2\}$ stable if $Z(s)$ exists such that $Z(s) \in \{Z_\alpha(s)\}$ [Eq. (5-44)] and

$$\frac{1 + \bar{K}W(s - \mu)}{1 + \underline{K}W(s - \mu)} Z(s - \mu) \in \{PR\}, \tag{6-49}$$

where $\mu > 0$, and $p(t)$ [Eq. (6-48$^\pm$)] satisfies condition (6-37). □

Note that the specification $k(t) \in [\underline{K} + \varepsilon, \bar{K} - \varepsilon]$ guarantees that $k_1(t)$ and $k_2(t)$ are always bounded (General Stability Criterion 3). For periodic gains, the distinction between $k(t) \in (\underline{K}, \bar{K})$ and $k(t) \in [\underline{K}+\varepsilon, \bar{K}-\varepsilon]$ disappears since these conditions are equivalent for finite t.

To arrive at a special case of Criterion 8, we consider periodic gains and define

$$\hat{W}_\mu(s) \triangleq \frac{1 + \bar{K}W(s-\mu)}{1 + \underline{K}W(s-\mu)}. \tag{6-50}$$

If $\hat{W}_\mu(s)$ satisfies the condition

$$-\pi < \arg\{\hat{W}_\mu(i\omega)\} < +\pi \tag{6-51}$$

as required by the constraint (6-49), then we note in Chapter IV, Section 3 that there must exist some $Z(s) \in \{Z_{\mathrm{LC}}\}$ such that $\hat{W}_\mu(s)Z(s) \in \{\mathrm{PR}\}$. This is equivalent to the statement that $\hat{Z}(s) \in \{Z_\mu(s)\}$ exists satisfying Eq. (6-49), so we know that if Eq. (6-51) is satisfied, we can take $\alpha = \mu$ in $p(t)$.

If we take the period $[0, T]$ and define the two sets of subintervals τ_p and τ_n by

$$\tau_p \triangleq t \in [0, T]: \quad dk/dt \geqslant 0,$$
$$\tau_n \triangleq t \in [0, T]: \quad dk/dt < 0$$

then according to one definition [Eq. (6-48$^+$)], we have

$$p(t) = \begin{cases} \dfrac{(\bar{K}-\underline{K})\,dk/dt}{(\bar{K}-k)(k-\underline{K})} - 2\mu, & t \in \tau_p, \\ -2\mu, & t \in \tau_n, \end{cases}$$

and condition (6-37) for periodic gains is satisfied if

$$\int_{\tau_p} \frac{(\bar{K}-\underline{K})\dot{k}}{(\bar{K}-k)(k-\underline{K})}\,dt < 2\mu T. \tag{6-52}$$

We recall that the integrand is $(k_1)^{-1}\,dk_1/dt$ or $(d/dt)(\log k_1)$. Since we are assuming that k is periodic, the same must be true of $\log k_1(t)$; thus

$$\int_0^T ((k_1)^{-1}\,dk_1/dt)\,dt = \log[k_1(T)/k_1(0)] = 0,$$

and in terms of the instants of τ_p and τ_n we consequently have

$$\int_{\tau_p} ((k_1)^{-1}\,dk_1/dt)\,dt = \int_{\tau_n} (k_1)^{-1}|dk_1/dt|\,dt = \tfrac{1}{2}\int_0^T (k_1)^{-1}|dk_1/dt|\,dt.$$

The constraint (6-52) is thus equivalent to

$$T^{-1}\int_0^T \frac{(\bar{K}-\underline{K})|\dot{k}|}{(\bar{K}-k)(k-\underline{K})}\,dt < 4\mu, \tag{6-53}$$

and we see that considering dk/dt only when $dk/dt \geqslant 0$ as in Eq. (6-52) is entirely equivalent to constraining $|\dot{k}|$ over the entire period.

CRITERION 9 (Integral conditions, periodic LTV systems). The system described by Eqs. (6-1) and (6-46) with $k(t) = k(t + T)$ for all t is absolutely $\{K_2\}$ stable if

$$-\pi < \arg\left\{\frac{1 + \bar{K}W(i\omega - \mu)}{1 + \underline{K}W(i\omega - \mu)}\right\} < \pi, \tag{6-54}$$

and $k(t)$ satisfies the constraint (6-53). □

While this result is quite easily applied, it should be noted that it is weaker than Criterion 8 in several respects. In the first instance, although condition (6-54) guarantees that $Z(s) \in \{Z_{\mathrm{LC}}\}$ exists satisfying $\hat{W}_\mu(s)Z(s) \in \{\mathrm{PR}\}$, the existence of $\tilde{Z}(s) \in \{Z_\alpha(s)\}$ satisfying the same condition is not precluded. If $\tilde{Z}(s) \in \{Z_\alpha\}$ in this constraint, then $Z(s)$ in Eq. (6-49) satisfies $Z(s) \in \{Z_{\alpha+\mu}\}$, that is, $Z(s)$ is an LC multiplier shifted by $(\alpha + \mu)$ units and the condition restricting dk/dt is less restrictive than Eq. (6-53). The second shortcoming found in Criterion 9 arises from the implicit assumption that $k(t)$ is least restricted by taking $\bar{\Lambda} = \mu$, which is not always the case. This latter point may be clarified by considering the following situation. Given

$$W(s) = (s^2 + 2\zeta s + \zeta^2)^{-1}$$

and

$$k(t) = \begin{cases} \exp(\delta(t - nT)), & nT \leqslant t < (nT + t_1) \triangleq nT + \delta^{-1}\ln(1 + \beta) \\ \exp(4\delta[(n+1)T - t]), & (nT + t_1) \leqslant t < (n+1)T \triangleq nT + 5\ln(1+\beta)/(4\delta) \end{cases}$$
$$n = 0, 1, 2, \ldots .$$

This time-varying gain is an exponential "sawtooth" function varying from 1 to $(1 + \beta)$. $k^{-1}\, dk/dt$ is given by the discontinuous function

$$k^{-1}\, dk/dt = \begin{cases} \delta, & nT \leqslant t < (nT + t_1), \\ -4\delta, & (nT + t_1) \leqslant t < (n + 1)T; \end{cases}$$

to render the criterion valid we assume that $k(t)$ is actually a continuously differentiable function that approximates this form as closely as desired.

First we apply Criterion 9: $W(s - \mu)$ is asymptotically stable for $0 \leqslant \mu < \zeta$, and the Nyquist plot of $W(i\omega - \mu)$ does not intersect the negative real axis. Thus only the integral condition remains to be satisfied:

$$I_1 \triangleq T^{-1}\int_0^T k^{-1}|dk/dt|\, dt = T^{-1}[\delta t_1 + 4\delta(T - t_1)] = \tfrac{8}{5}\delta.$$

Choosing μ to be as large as possible yields the final stability condition $\delta < \hat{\delta}_1 \triangleq \frac{5}{2}\zeta$.

Now we consider Criterion 8 applied to the above problem: again $W(s - \mu)$ is asymptotically stable for all $\mu < \zeta$. The Popov multiplier $(s + \alpha)$ is used as a special case of $Z(s) \in \{Z_\alpha\}$; the function $W(s - \mu)\cdot(s + \alpha - \mu)$

is positive real for all $\alpha \leqslant 2\zeta - \mu$. Hence these first two conditions are satisfied if for any $\rho \in (0, 1)$ the parameters μ and α satisfy

$$\mu = (1 - \rho)\zeta, \qquad \alpha = (1 + \rho)\zeta.$$

The time-averaged constraint that $I_2 \triangleq \int_0^T p(t)\, dt < 0$, where

$$p(t) \triangleq \begin{cases} \delta - 2(1 + \rho)\zeta, & 0 \leqslant t < t_1, \\ -2(1 - \rho)\zeta, & t_1 \leqslant t < T, \end{cases}$$

yields $I_2 = \frac{4}{5}[\delta - (\frac{5}{2} + \frac{3}{2}\rho)\zeta] < 0$, which is satisfied if $\delta < (\frac{5}{2} + \frac{3}{2}\rho)\zeta$. If $\rho \approx 0$ is chosen, then the result corresponding to Criterion 9 is obtained; if, however, $\rho \approx 1$, the stability condition [as dictated by the point condition (6-23$^+$)] is

$$\delta < \hat{\delta}_2 = 4\zeta.$$

In this simple example, it could be seen by inspection that the point condition would yield the less restrictive upper bound on δ. If $k(t)$ were more complex, however, it would not be clear which criterion to apply. As demonstrated above, Criterion 8 takes this problem into account.

Criterion 9 is essentially identical to a theorem of Freedman and Zames [1], except that they used a more general system formulation allowing the theorem to be applied to nonrational transfer functions $W(s)$. The relation between General Stability Criterion 3 and the result of Freeman and Zames was noted in the infinite sector case by Taylor and Narendra [2].

VII

GEOMETRIC STABILITY CRITERIA

In the preceding chapters, stability criteria are derived for systems containing a single gain that may be nonlinear, time-varying, or both over a specified range. In the application of such criteria, the primary system description is given in terms of the LTI plant $W(s)$ and the NLTV gain $g(\sigma_0, t)$ in the closed loop negative feedback configuration specified by

$$W(s) = L(\sigma_0)/L(\tau) = \rho + h^{\mathrm{T}}(sI - A)^{-1}b,$$
$$\tau = -g(\sigma_0, t), \qquad g(\sigma_0, t) \in \{G_i[N, T]\}, \tag{7-1}$$

as depicted in Fig. 2-1a. The state variable formulation (involving ρ, h, A, b) is suppressed here, although it is recalled that $W(s)$ is rational and that it may not have more zeros than poles.

If the NLTV gain lies in the closed range $[\underline{G}_N, \bar{G}_N]$ as we generally assume, then the subsidiary conditions $A_{\underline{G}_N} \in \{A_1\}$ and $A_{\bar{G}_N} \in \{A_1\}$ must be imposed (see Eq. (4-2)); the same type of constraint is required in NLTI situations. We have also discussed the case $f(\sigma_0)/\sigma_0 \in (0, \bar{F}_N]$ if the matrix A has a single zero eigenvalue, that is, if $W(s)$ has a single pole at $s = 0$; we consider that eventuality in this chapter only in passing.

In all cases, the primary condition for absolute stability is expressed in terms of the transfer function $W(s)$, the range of $g(\sigma_0, t)/\sigma_0$, and a multiplier function $Z(s)$ such that $Z^{\pm 1}(s) \in \{Z_N(s)\}$, where the class $\{Z_N(s)\}$ is determined by $\{N\}$ or $\{G_i[N, T]\}$. The function

$$H(s) \triangleq \frac{1 + \bar{G}_N W(s)}{1 + \underline{G}_N W(s)} Z(s) \tag{7-2}$$

is always constrained to be at least positive real; for NLTV systems, $H(s)$ must be strictly positive real (Chapter III, Section 5) unless the system is periodic (Chapter VI, Section 5). For an NLTI system, a frequency domain condition of this form is all that is required for absolute stability, whereas in NLTV situations the time variation of the gain must generally be constrained. Consequently, the investigation of the stability properties of a given NLTI system, using the frequency domain approach, generally reduces to a search for specific members of classes of multipliers. Whereas the multiplier $Z(s)$ has a simple form for the problem of Lur'e and Postnikov as treated by Popov, that is, $Z(s) = (\beta_0 s + \gamma_0)^{\pm 1}$, it becomes increasingly more complex as the class of nonlinear functions in the feedback path is made more restrictive. For a given $W(s)$, particularly when the order of the system is high, it is seldom apparent how such multipliers can be found to satisfy the frequency domain criteria.

It is shown in Chapter IV that the existence of a shifted LC function $Z(s) \in \{Z_\alpha(s)\}$, Eq. (5-44), where $\alpha > 0$ may be arbitrarily small, which satisfies the condition $H(s) \in \{PR\}$ (Eq. (7-2), with $\underline{G}_N = 0$, $\bar{G}_N = \bar{K}$) is both necessary and sufficient to ensure that the closed loop system is asymptotically stable for all linear gains $\tau = -\kappa\sigma_0$ in the interval $0 \leqslant \kappa < \bar{K}$. For stability in this range, Nyquist's criterion may be stated in terms of the argument of $W(i\omega)$ as

$$-\pi < \phi_{\bar{K}}(\omega) \triangleq \arg[W(i\omega) + \bar{K}^{-1}] < \pi. \tag{7-3}$$

In the above multiplier condition, the argument of $Z(i\omega)$ lies in the range $(-\pi/2, \pi/2)$. By making α arbitrarily small, the derivative of the argument with respect to ω can be increased without bound, and in the limit when $\alpha \to 0$ we obtain the LC function, which has a discontinuous argument. It is not surprising that this condition does not impose any constraints on the rate at which the argument of $W(i\omega)$ can vary with ω as long as $\phi_{\bar{K}}(\omega)$ is continuous, since it reduces to the Nyquist criterion (7-3) for LTI systems. In the following sections, it is seen that the rate at which the phase of $W(i\omega)$ varies with ω is related to the stability properties of the system with nonlinear functions in the feedback path. In general terms, it is found that as the rate of variation of $\phi_{\bar{K}}(\omega)$ decreases, a member of a more restricted class of multiplier functions $\{Z_N(s)\}$ may be chosen to ensure that $Z(s)[W(s) + \bar{K}^{-1}] \in \{PR\}$; this in turn assures that the closed loop system containing $W(s)$ is stable for a larger class of nonlinear functions $f(\sigma_0)$. This correspondence is only implicit in the geometric criteria that follow, however; there is generally no direct relationship that can be used in the determination of frequency domain multipliers for a given plant $W(s)$.

In the case of NLTV systems, the rate of time variation of the gain must usually be constrained. The severity of this restriction is determined both by

$Z(s)$ (specifically, by $\bar{\Lambda}$, the multiplier margin, Eq. (6-21)) and the nonlinear behavior of the gain (as quantified by the index of nonlinearity $\underline{\Phi}$, Chapter VI, Section 2). Thus, in the application of absolute stability criteria for time-varying systems, the problem of determining frequency domain multipliers is compounded by the desire to maximize $\bar{\Lambda}$ so as to obtain the least restrictive constraint on the rate of time variation of the gain. The examples treated in Chapter VIII demonstrate the degree of complexity that may be encountered even in considering relatively simple systems. In dealing with more complicated systems, this may lead to the necessity of resorting to tedious trial-and-error methods or to the use of a digital computer.

In this chapter, some criteria are derived which completely by-pass the need to determine multipliers to prove stability. The criteria are geometric interpretations of the frequency domain criteria for absolute stability developed previously, and are, in general, sufficient conditions for the existence of specific classes of multipliers. In situations where the geometric condition is not necessary (not in one-to-one correspondence with the existence of $Z(s) \in \{Z_N(s)\}$), the result may be more stringent than the stability theorems stated in Chapters V and VI. The ease with which such criteria can be applied to specific problems makes them particularly attractive, whether or not they are as general as the criteria developed earlier.

1. Linear Time-Invariant Systems

The intimate relation that exists between the stability properties of LTI systems and the corresponding stability problems for NLTI and NLTV systems is indicated in Chapter IV. The forms of the Lyapunov functions chosen to establish stability and the corresponding frequency domain conditions are motivated by the LTI case. It is therefore not surprising that the two approaches used to obtain geometric criteria for NLTI and NLTV systems are both closely related to corresponding criteria for LTI systems.

The first and most widely used approach entails constructions on the Nyquist plot of $W(i\omega)$ or on some modified frequency response diagram, and thus is related to the Nyquist criterion. A significant amount of information regarding the stability of a system comprised of a linear plant represented by $W(i\omega)$ and a nonlinear and time-varying gain in a closed loop configuration [Eq. (7-1)] can be extracted from such a plot; for example, the geometric criteria might be used to aid in the design of compensators in the frequency domain to suitably modify the stability characteristics of the feedback system by changing the frequency response of the LTI part of the system. Such frequency domain compensation has been extensively treated in the control theory literature, although only with the goal of achieving some performance specification for LTI systems and not as a solution to the problem of NLTV system stabilization.

The second approach deals with the poles and zeros of the overall transfer function $W(s)/(1 + \kappa W(s))$ as determined by the poles and zeros of $W(s)$ for constant values of the feedback gain κ, and hence is linked with the root locus technique.

A. *The Nyquist Criterion*

The criterion of Nyquist permits the use of the frequency response $W(i\omega)$ of LTI plant and the behavior of $W(s)$ as $|s| \longrightarrow \infty$ in the right half-plane in making precise inferences about the stability of the closed loop system specified by $\tau = -\kappa\sigma_0$ in Eq. (7-1). The condition for asymptotic stability is stated in terms of the phase-amplitude characteristic (Nyquist plot) of $W(i\omega)$, that is, the curve determined by the parametric equations

$$U(\omega) \triangleq \text{Re}\, W(i\omega), \qquad V(\omega) \triangleq \text{Im}\, W(i\omega)$$

in the U, V plane as ω takes on values $\omega \in [-R, R]$, and the knowledge of the behavior of $W(s)$ on the semicircle $s = \text{Re}^{i\theta}$, $-\pi/2 \leqslant \theta \leqslant +\pi/2$, where R is arbitrarily large.

As detailed in Chapter IV, Section 3A, if $\mathbf{\Gamma}_R$ represents the above contour in the complex s-plane and $\mathcal{R}_R$ denotes the finite RHP or the region specified by $\mathbf{\Gamma}_R$ and its interior, the Nyquist plot $\mathbf{\Gamma}_W$ can be considered to be a mapping of the contour $\mathbf{\Gamma}_R$ under the transformation $W(s) = U(s) + iV(s)$, $s \in \mathbf{\Gamma}_R$. According to the criterion, the closed-loop system (7-1) with $\tau = -\kappa\sigma_0$ is asymptotically stable for all values of the gain in the interval $0 \leqslant \kappa < \bar{K}$, if the LTI plant $W(s)$ is asymptotically stable ($A \in \{A_1\}$) and the Nyquist plot $\mathbf{\Gamma}_W$ does not intersect the negative real axis to the left of the point $-\bar{K}^{-1}$, as shown in Fig. 7-3a for the value $\bar{K}_1$. If the Nyquist plot intersects the negative real axis at several points, several ranges $\underline{K}_j < \kappa < \bar{K}_j$ exist within which the system is asymptotically stable. These ranges can be determined by traversing the plot $\mathbf{\Gamma}_W$ in the direction of increasing ω; if the interval $(-\underline{K}_j^{-1}, -\bar{K}_j^{-1})$ on the real axis lies to the left of the curve, then the closed loop system is asymptotically stable for all values of the gain κ in the interval $\underline{K}_j < \kappa < \bar{K}_j$. This is the technique used to determine $\mathcal{W}(\mathcal{R}_R)$ or the mapping of $\mathcal{R}_R$; only those values of κ such that the point $(-\kappa^{-1}, 0)$ is outside $\mathcal{W}(\mathcal{R}_R)$ lead to asymptotically stable systems. If $W(s)$ goes to a constant value or zero as $s \longrightarrow \infty$, then we generally locate $\mathcal{W}(\mathcal{R}_\infty)$, as in Fig. 7-3a. Since it is always clear which region is under consideration, the subscript R or ∞ is dispensed with.

While treating various absolute stability criteria, it is found beneficial to consider ranges of the nonlinear gain other than $[0, \bar{G}]$; in particular the range may be $[\underline{G}, \bar{G}]$, which does not necessarily include zero. This implies that $W(s)$ is not required to be stable for the closed loop system to be asymptotically stable. This is found to be the case for the circle criterion considered in the next section, for example. In order to discuss the absolute stability criteria in this general form, we first consider the stability of LTI systems using the Nyquist criterion under the same conditions.

Assume that $W(s)$ has m poles λ_i such that $\text{Re}\,\lambda_i \geqslant 0$. It is necessary to modify Γ_R, the contour in the s-plane previously discussed, to include all of these m poles. This curve is thus indented into the left half-plane to enclose any of these poles on the imaginary axis ($\text{Re}\,\lambda_i = 0$), as shown in the case of a single imaginary axis pole at $s = 0$ in Fig. 7-1a. The region $\mathcal{R}$ is again made up of the extended curve and its interior. The region $\mathcal{W}(\mathcal{R})$ is located as described previously; the Nyquist criterion in terms of this mapping is a direct extension of the special case given in Chapter IV, Section 3A.

DEFINITION OF REGION $\mathcal{R}$. For any $W(s)$ having poles at $s = \lambda_i$ satisfying $\text{Re}\,\lambda_i \geqslant 0$, $i = 1, 2, \ldots, m$; $\text{Re}\,\lambda_i < 0$, $i = (m + 1), \ldots, n$, the region $\mathcal{R}$ in the s-plane consists of the union of all points satisfying

(i) $|s| \in [0, R]$ and $\measuredangle s \in [-\pi/2, \pi/2]$, where R is arbitrarily large (or $R \to \infty$ if $W(\infty) < \infty$), and

(ii) $|s - \lambda_i| \leqslant \rho_0$, $i = 1, 2, \ldots, m$, where $\rho_0 > 0$ may be arbitrarily small.† □

THEOREM (Nyquist). The closed loop system described by Eq. (7-1) with $\tau = -\kappa\sigma_0$ is asymptotically stable if the point $(U = -\kappa^{-1}, V = 0)$ does not lie in $\mathcal{W}(\mathcal{R})$. □

In the example $W_1(s) = (s + \beta)/(s(s - \alpha))$ shown in Fig. 7-1b, the region corresponding to $\kappa > \alpha$ is outside $\mathcal{W}(\mathcal{R})$, which guarantees that the overall transfer function

$$W_\kappa(s) = (s^2 + (\kappa - \alpha)s + \kappa\beta)^{-1}$$

has its poles in the open left half plane for $\kappa > \alpha$.

This formulation of the criterion of Nyquist is equivalent to the more common "encirclement" statement: two poles of $W_1(s)$ are encircled by the extension of Γ_R in the clockwise sense, so stability of the closed loop system for some value of κ is ensured if the point $(-\kappa^{-1}, 0)$ is encircled twice in the counterclockwise sense. This implies by the principle of the argument that $[\kappa^{-1} + W(s)]$ has no zeros in the closed right half-plane; the only singularities

† None of the remaining poles at $s = \lambda_i$, $i = (m + 1), \ldots, n$ may lie in $\mathcal{R}$.

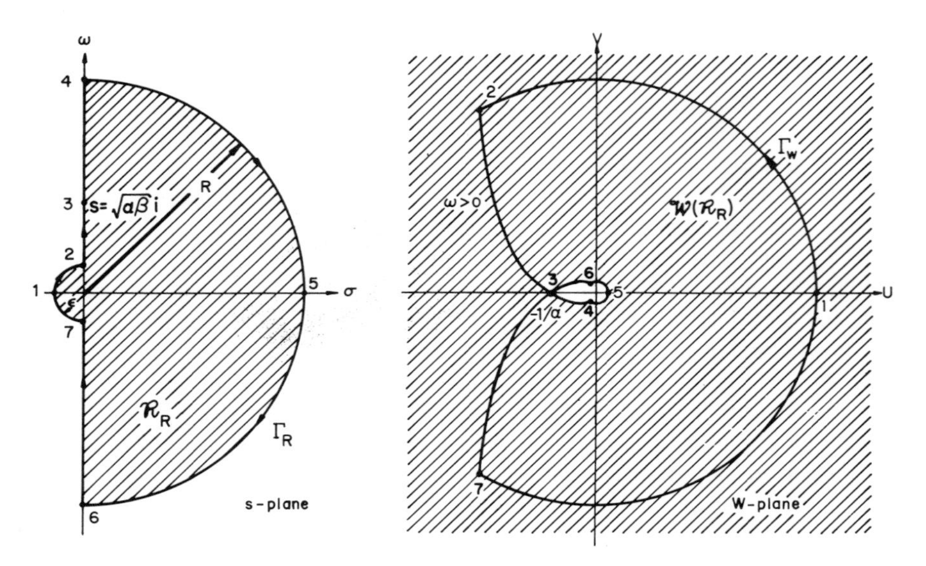

Fig. 7-1. The Nyquist criterion applied to $W_1(s) = (s + \beta)/(s(s - \alpha))$.

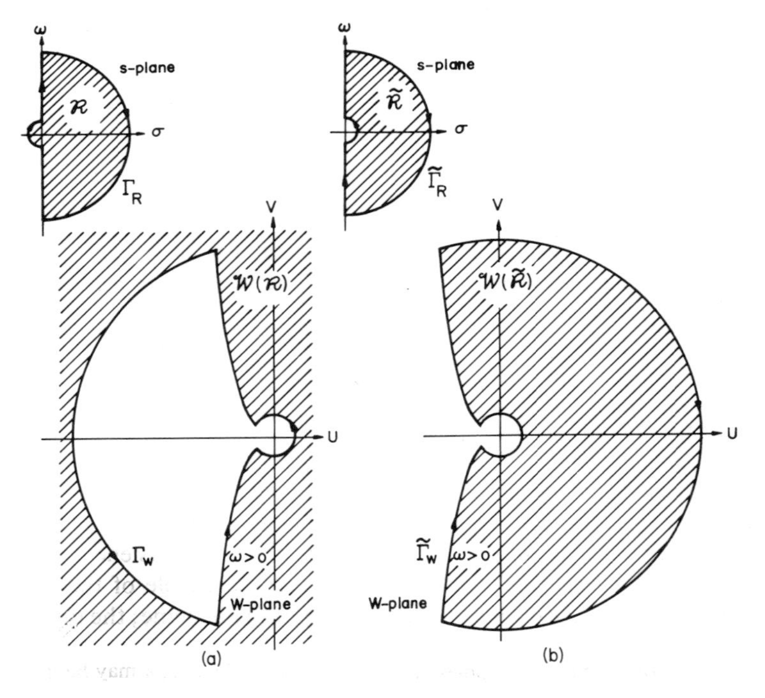

Fig. 7-2. Γ_W for $W(s) = (s(s + \alpha))^{-1}$ corresponding to two s-plane contours: (a) Γ_R enclosing $s = 0$; (b) $\tilde{\Gamma}_R$ excluding $s = 0$.

in $\mathcal{R}$ are the prespecified poles. Using the mapping properties of $\mathcal{R} \xrightarrow{W(s)} \mathcal{W}(\mathcal{R})$, a somewhat more unified and direct statement of the criterion is obtained.

The reason for enclosing all of the poles of $W(s)$ in the closed right half-plane is demonstrated by taking $W_2(s) = (s(s + \alpha))^{-1}$; as shown in Fig. 7-2a, this contour correctly gives the asymptotic stability range as $(0, \infty)$, whereas the polar diagram of $W_2(s)$ for s on a contour that does not enclose the pole at $s = 0$ would seem to imply asymptotic stability for $\kappa \in [0, \infty)$ (Fig. 7-2b).

B. The Root Locus Technique

A second approach to the stability problem for LTI systems uses the root locus plot. If the poles and zeros of the open loop transfer function $W(s)$ are specified, the root locus plot of the LTI system is defined as the loci of the poles of the closed loop system [zeros of $1 + \kappa W(s)$] as a function of the parameter κ. The constructional methods used to generate these loci are treated in most introductory texts dealing with the theory of feedback control systems. Since an LTI system is asymptotically stable when all of its poles are in the open left half of the complex s-plane, the asymptotic stability of the system in the range $K_1 < \kappa < K_2$ of the parameter κ is assured if the root locus corresponding to this range of parameter value lies in the open left half-plane. Any Hurwitz range $\underline{K}_j < \kappa < \bar{K}_j$, therefore, is determined by finding those values of the gain κ at which the root locus intersects the imaginary axis, provided that the root locus for $\kappa \in (\underline{K}_j, \bar{K}_j)$ is in the open left half-plane. A root locus plot corresponding to the plant $W(s)$ whose Nyquist plot is shown in Fig. 4-4 demonstrates these ideas in Fig. 7-3b.

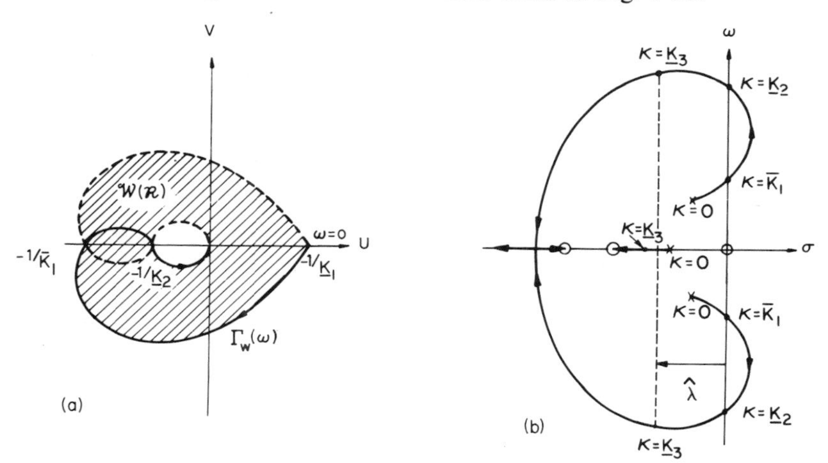

Fig. 7-3. The Nyquist and root locus techniques: (a) typical Nyquist plot, (b) corresponding root locus plot; ⊙: open loop zeros, x: open loop poles.

If a specified range $K_1 \leqslant \kappa \leqslant K_2$ lies within some Hurwitz range, that is, $\underline{K}_j < K_1 \leqslant \kappa \leqslant K_2 < \bar{K}_j$, the poles of the closed loop λ_i as determined by κ satisfy the inequality

$$\operatorname{Re}\{\lambda_i(\kappa)\} < 0, \qquad \kappa \in [K_1, K_2]. \tag{7-4}$$

If we define the parameter $\hat{\lambda}$ by the relation

$$\max_{K_1 \leqslant \kappa \leqslant K_2} \operatorname{Re}\{\lambda_i(\kappa)\} \triangleq -\hat{\lambda} < 0, \tag{7-5}$$

then $\hat{\lambda} > 0$ represents the minimum distance between the loci of the poles of the closed loop system and the imaginary axis as shown in Fig. 7-3b for $\kappa \geqslant \underline{K}_3 > \underline{K}_2$. This parameter is generally used as a measure of the stability of the closed loop system in the given range $[K_1, K_2]$ of the gain, in this case, $\kappa \in [\underline{K}_3, \infty)$. The impulse response

$$w_\kappa(t) \triangleq L^{-1}[W_\kappa(s)] \triangleq L^{-1}[W(s)/(1 + \kappa W(s))]$$

of the closed loop system with $\kappa \in [\underline{K}_3, \infty)$ decays at least as fast as $\phi \exp(-\hat{\lambda}t)$, where ϕ is an arbitrary constant. If the feedback path contains a time-varying gain $k(t)$, the stability of the system may be expressed in terms of $\hat{\lambda}$; quantitatively, the maximum rate of variation of $k(t)$ that may be tolerated in order to maintain stability is directly proportional to $\hat{\lambda}$.

While the root locus plot which yields the stability range of the parameter κ of the closed loop system can be drawn from a knowledge of the open loop pole and zero locations (as determined directly by the differential equation), the Nyquist criterion yields the same Hurwitz range or ranges $\underline{K}_j < \kappa < \bar{K}_j$ from the frequency response of the open loop, which can be obtained experimentally. The practical implications of this fact are evident: If stability criteria are established in terms of the frequency response of the open loop, stability analysis and compensation of nonlinear systems can be attempted (as in the linear case) using experimental data rather than the exact mathematical model of the open loop plant that is required for a root locus analysis. This not only saves effort, but leads to an avoidance of the inaccuracies that are inherent in the modeling procedure and the measurement of system parameters. If, on the other hand, the plant transfer function $W(s)$ is known, it is not necessary to reverse this procedure to determine $\mathcal{W}(\mathcal{R})$, since the root locus technique may be used directly to obtain those values of κ which lead to imaginary axis crossings by the root loci.

The importance of LTI systems in the study of the stability properties of NLTV systems is reemphasized by again considering the role played by the Hurwitz range of κ in the absolute stability problem. Although Aizerman's conjecture was disproved by Pliss and other workers, which implies that the stability of the LTI system in a Hurwitz range $\underline{K}_j < \kappa < \bar{K}_j$ does not guarantee the stability of a nonlinear system with the nonlinear gain in the same interval, the Hurwitz range is important since it represents the maximum range for which one can hope to prove absolute stability.

2. The Circle Criterion

Among the geometric criteria for absolute stability that exist at the present time, perhaps the most well known is the circle criterion, which provides a sufficient condition for the stability of a system with a single nonlinear time-varying gain. In Chapter VI, Section 4, it is shown that if $W(s)$ is the transfer function in the forward path and the nonlinear time-varying gain in the feedback path $g(\sigma_0, t)$ lies in the sector $[\underline{G}, \bar{G}]$, that is,

$$\underline{G} \leqslant g(\sigma_0, t)/\sigma_0 \leqslant \bar{G},$$

where zero need not lie in the range $[\underline{G}, \bar{G}]$ as assumed in Chapter VI, Section 1, absolute stability of the null solution is assured if a common quadratic Lyapunov function of the form $\frac{1}{2}x^{\mathrm{T}}Px$ exists over the entire range $\underline{G} \leqslant \kappa \leqslant \bar{G}$ for the corresponding LTI system. This in turn reduces to the frequency domain condition

$$H(s) \triangleq (1 + \bar{G}W(s))/(1 + \underline{G}W(s)) \in \{\mathrm{SPR}\}$$

if $g(\sigma_0, t)$ is aperiodic or $H(s) \in \{\mathrm{PR}\}$ if $g(\sigma_0, t) = g(\sigma_0, t + T)$ for fixed σ_0; we first consider the aperiodic case. This condition guarantees that

$$J(\omega^2) \triangleq 1 + (\underline{G} + \bar{G}) \operatorname{Re} W(i\omega) + \underline{G}\bar{G}\,|W(i\omega)|^2 > 0, \omega \in (-\infty, \infty). \quad (7\text{-}6)$$

This result may be interpreted in the U, V plane by noting that if neither bound is zero, condition (7-6) may be rewritten in the form

$$\begin{aligned} J(\omega^2) &= \underline{G}\bar{G}\{[U + \tfrac{1}{2}(\underline{G}^{-1} + \bar{G}^{-1})]^2 + V^2 - \tfrac{1}{4}(\underline{G}^{-1} - \bar{G}^{-1})^2\} \\ &\triangleq \underline{G}\bar{G}\{[U - \upsilon_0]^2 + V^2 - \rho_0^2\} > 0. \end{aligned} \quad (7\text{-}7)$$

The relation $J(\omega^2) = 0$ thus corresponds to the circle $(U - \upsilon_0)^2 + V^2 = \rho_0^2$. The region in the U, V plane that corresponds to $J(\omega^2) > 0$ depends upon the signs of $\underline{G}$ and $\bar{G}$; if $\operatorname{sgn} \underline{G} = \operatorname{sgn} \bar{G}$ ($\underline{G}\bar{G} > 0$), then $J(\omega^2) > 0$ outside the circle, while if $\operatorname{sgn} \underline{G} = -\operatorname{sgn} \bar{G}$, then $J(\omega^2) > 0$ inside the specified circle. The U-axis intercepts of this circle are of particular interest; they are $\upsilon_0 \pm \rho_0$, or

$$U_1 \triangleq \upsilon_0 + \rho_0 = -\underline{G}^{-1}, \qquad U_2 \triangleq \upsilon_0 - \rho_0 = -\bar{G}^{-1}. \quad (7\text{-}8)$$

Two special cases arise when $g(\sigma_0, t)$ lies in a range having an extremal value of zero.

(i) $g \in [0, \bar{G}]$ requires that $J(\omega^2) = 1 + \bar{G} \operatorname{Re} W(i\omega) > 0$;
(ii) $g \in [\underline{G}, 0]$ requires that $J(\omega^2) = 1 - |\underline{G}| \operatorname{Re} W(i\omega) > 0$.

In situation (i), the Nyquist diagram of $W(i\omega)$ must lie strictly to the right of a vertical line passing through $U_2 = -\bar{G}^{-1} < 0$, while in (ii) the polar plot must lie strictly to the left of a vertical line passing through $U_1 = -\underline{G}^{-1} > 0$; in both of these eventualities the boundary is a degenerate circle.

The statement of the circle criterion in all of the cases considered heretofore can be simplified using the following definition.

Definition. In the U, V plane $C[\underline{G}, \bar{G}]$ denotes the generalized circle

$$C[\underline{G}, \bar{G}] \triangleq \begin{cases} \{(U, V)\colon & [U + \frac{1}{2}(\bar{G}^{-1} + \underline{G}^{-1})]^2 + V^2 = \frac{1}{4}(\underline{G}^{-1} - \bar{G}^{-1})^2\} \\ \{(U, V)\colon & U = -\bar{G}^{-1}\}, \qquad \underline{G} = 0 \\ \{(U, V)\colon & U = |\underline{G}|^{-1}\}, \qquad \bar{G} = 0 \end{cases} \tag{7-9}$$

and the closed interior of $C[\underline{G}, \bar{G}]$ is the region in the U, V plane defined by

$$I[\underline{G}, \bar{G}] \triangleq \begin{cases} \{(U, V)\colon & \underline{G}\bar{G}[(U + (2\bar{G})^{-1} + (2\underline{G})^{-1})^2 + V^2 \\ & \quad - \frac{1}{4}(\underline{G}^{-1} - \bar{G}^{-1})^2] \leqslant 0\} \\ \{(U, V)\colon & U \leqslant -\bar{G}^{-1}\}, \qquad \underline{G} = 0 \\ \{(U, V)\colon & U \geqslant -\underline{G}^{-1}\}, \qquad \bar{G} = 0. \ \square \end{cases} \tag{7-10}$$

In essence, the circle $C[\underline{G}, \bar{G}]$ excludes the origin on the U, V plane in every case, that is, $0 \notin I[\underline{G}, \bar{G}]$.

The three cases, (a) $0 \leqslant \underline{G}_a < \bar{G}_a$, (b) $\underline{G}_b < 0 < \bar{G}_b$, and (c) $\underline{G}_c < \bar{G}_c \leqslant 0$, are shown in Fig. 7-4 along with representative Nyquist diagrams $W_a(i\omega)$, $W_b(i\omega)$ and $W_c(i\omega)$, for which a closed loop NLTV system [Eq. (7-1)] with $g(\sigma_0, t)/\sigma_0 \in [\underline{G}, \bar{G}]$ is absolutely stable. In each case the circle criterion simply reduces to the condition that $\mathcal{W}(\mathcal{R})$ or the mapping of the s-plane region $\mathcal{R}$ (Section 2) defined by $W = W(s \in \mathcal{R})$ must lie outside $I[\underline{G}, \bar{G}]$: $\mathcal{W}(\mathcal{R}) \cap I[\underline{G}, \bar{G}] = \phi$ (the empty set).

In Fig. 7-4, it has been assumed that $W(s)$ is asymptotically stable [$A \in \{A_1\}$ in Eq. (7-1)]. If $\underline{G}\bar{G} > 0$, or if $0 \notin [\underline{G}, \bar{G}]$, then $W(s)$ may have poles in the closed right half of the complex s-plane, that is, Re $\lambda_i \geqslant 0$. This situa-

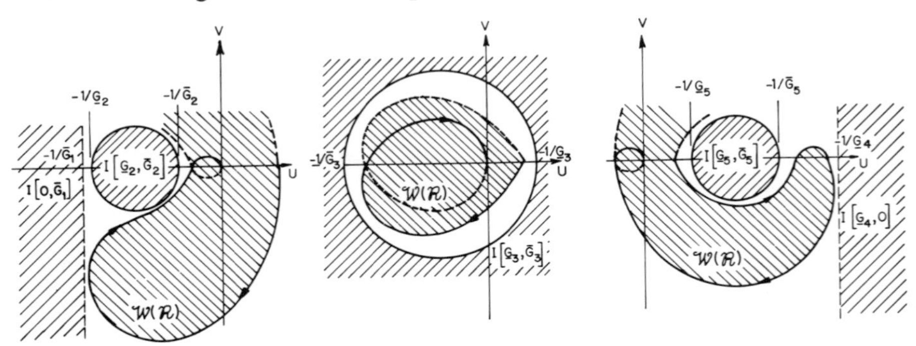

Fig. 7-4. Applications of the circle criterion for aperiodic gains.

tion corresponds to the more general statement of the Nyquist condition for LTI systems stated in Section 1A; the condition for absolute stability is still that no point of $\mathcal{W}(\mathcal{R})$ can be permitted to lie in $I[\underline{G}, \bar{G}]$.

In treating periodic systems, we can allow $\mathcal{W}(\mathcal{R})$ and $C[\underline{G}, \bar{G}]$ to have points in common subject to the standard constraint $A_{\underline{G}} \in \{A_1\}$, $A_{\bar{G}} \in \{A_1\}$.

Geometric Criterion 1 (The circle criterion). The system described by Eq. (7-1) with

$$g(\sigma_0, t) \in \{G_i[F, K_0]\}, \qquad g(\sigma_0, t)/\sigma_0 \in [\underline{G}, \bar{G}];$$
$$A_{\underline{G}} \in \{A_1\}, \qquad A_{\bar{G}} \in \{A_1\} \tag{7-11}$$

is absolutely $\{G_i[F, K_0]\}$ stable if

(i) $\mathcal{W}(\mathcal{R}) \cap I[\underline{G}, \bar{G}] = \phi$, $g(\sigma_0, t)$ aperiodic, or
(ii) $\mathcal{W}(\mathcal{R}) \cap (I[\underline{G}, \bar{G}] - C[\underline{G}, \bar{G}]) = \phi$, $g(\sigma_0, t) = g(\sigma_0, t + T)$ for all t and σ_0 fixed. □

For a given transfer function $W(s)$, an infinite number of ranges of the nonlinear time-varying gain can result in absolute stability. In general, either $\underline{G}$, or $\bar{G}$, or the arithmetic mean of $\sigma_0/g(\sigma_0, t)$ must be specified to determine the largest possible circle uniquely. These parameters determine one of the real axis intercepts or the center of the circle respectively; given this information, the least restrictive range of $g(\sigma_0, t)$ is obtained by drawing the largest circle in the U, V plane which fails to have a common point with $\mathcal{W}(\mathcal{R})$ if the system is aperiodic, or which contacts $\mathcal{W}(\mathcal{R})$ at points other than on the real axis if $g(\sigma_0, t) = g(\sigma_0, t + T)$.

In general, the range $[\underline{G}, \bar{G}]$ obtained by an application of the circle criterion for NLTV gains may be much smaller than the range permitted by the Nyquist criterion for $g(\sigma_0, t) = \kappa\sigma_0$, as is the case for the plant $W(s)$ whose Nyquist diagram is shown in Fig. 7-5a. For the special case shown in Fig. 7-5b, where the circle through $(-\underline{G}^{-1}, 0)$ and $(-\bar{G}^{-1}, 0)$ passes arbitrarily close to the curve $\mathbf{\Gamma}_W$ at those points, the two ranges coincide almost exactly. This is a rather exceptional situation in which a restricted form of Aizerman's conjecture (where the gain may approach the extreme values of the Nyquist range as closely as desired for finite σ_0 and t but it may not approach the Nyquist bounds asymptotically) is valid even for NLTV gains.

The circle criterion as stated above was derived from conditions guaranteeing the existence of a common quadratic Lyapunov function; it is thus a sufficient condition for the stability of a nonlinear time-varying system. The question naturally arises as to whether the condition is also necessary. For the case of a linear time-varying system, it has been shown that the condition is not necessary. If a second-order system is represented by the differential equation $\ddot{x} + 2\zeta\dot{x} + k(t)x = 0$, then the circle criterion guarantees absolute $\{K_0\}$ stability in the range $0 < \varepsilon \leqslant k(t) \leqslant 4\zeta^2 - \varepsilon$. Variational analysis

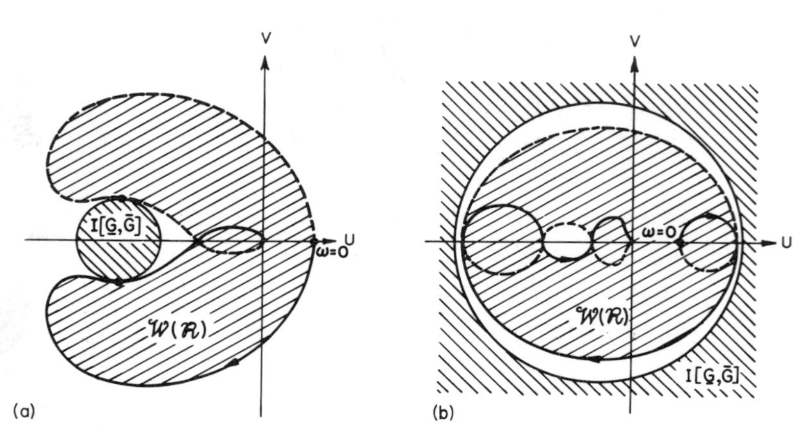

Fig. 7-5. A comparison of stability ranges: (a) $[\underline{G}, \bar{G}]$ much smaller than the Nyquist range $[g(\sigma_0, t)$ periodic]; (b) $[\underline{G}, \bar{G}]$ coinciding with one Nyquist range.

indicates that stability is guaranteed for larger upper bounds on $k(t)$. Specific examples are treated by Taylor and Narendra [1] and Brockett [1]; the analysis of Chapter VIII indicates this same point. For NLTI systems with $\tau = -f(\sigma_0) \in \{F\}$, $f(\sigma_0)/\sigma_0 \in [0, \bar{F}]$, the application of the circle criterion also often fails to yield the largest range possible, as is demonstrated in Section 3 by a graphical interpretation of the absolute stability condition of Popov. However, for nonlinear time-varying systems, Geometric Criterion 1 may indeed prove to be a necessary condition for absolute stability.

3. The Popov Criterion

In Chapter V, it is first shown that when $W(s)$ is asymptotically stable and a multiplier of the form $Z(s) = (\beta_0 s + \gamma_0)^{\pm 1}$ where $\gamma_0 > 0$ and $\beta_0 \geqslant 0$ exists such that $[W(s) + \bar{F}^{-1}]Z(s) \in \{\mathrm{PR}\}$, the system described by Eq. (5-1) is absolutely stable for all first and third quadrant nonlinearities restricted to the finite sector $[0, \bar{F})$. Criterion 1b allows the range $[0, \bar{F}]$ if another constraint is imposed, and particular cases may be treated, as we consider subsequently. In his original paper, Popov [3] suggested interesting algebraic and geometric interpretations of Criterion 1a.

A. Algebraic Conditions for Absolute $\{F\}$ Stability

In terms of the real and imaginary parts of the asymptotically stable transfer function $W(i\omega)$, Popov's criterion may be expressed as

$$\gamma_0 J(\omega^2) \triangleq \gamma_0[U + \bar{F}^{-1}] - \hat{\beta}_0 \omega V \geqslant 0, \qquad \omega \in (-\infty, \infty) \qquad (7\text{-}12)$$

(where we can allow $\hat{\beta}_0$ to be negative), provided that $\rho \geqslant -1/\bar{F}$. This constraint corresponds to condition (5-7), which necessitates the additional restriction $\rho \geqslant -1/\bar{F}$, as we note subsequently.

Dividing by γ_0 and defining α as $\alpha \triangleq \hat{\beta}_0/\gamma_0$, this reduces to

$$J(\omega^2) = (U + \bar{F}^{-1}) - \alpha\omega V \geqslant 0, \qquad \omega \in (-\infty, \infty). \tag{7-13}$$

The numerator of $J(\omega^2)$ is a polynomial in the variable ω^2 and can be expressed as $h(\theta, \alpha)$ where $\theta \triangleq \omega^2$. Condition (7-13) consequently reduces to the requirement that a polynomial in θ must be nonnegative for all $\theta \geqslant 0$. The choice of the arbitrary parameter α to satisfy the condition for any specific case can be carried out using standard algebraic methods. Generally this method is useful only if $W(s)$ is given analytically and is of low order.

EXAMPLE. To demonstrate the application of this method to a system with a single zero eigenvalue, consider

$$W(s) = (s^3 + a_3 s^2 + a_2 s)^{-1}.$$

Since the Nyquist range is $(0, a_2 a_3)$, we apply Criterion 1c with $\bar{F} = (a_2 a_3 - \varepsilon)$. Absolute stability for $f(\sigma_0)/\sigma_0 \in (0, a_2 a_3 - \varepsilon]$ is guaranteed since

$$h(\theta, \alpha) = \theta\{(a_2 - \theta)^2 + (a_2 - \theta)[\alpha(a_2 a_3 - \varepsilon) - a_3{}^2] + \varepsilon a_3\} > 0$$

is satisfied for all $\theta > 0$ by choosing $\alpha = a_3{}^2/(a_2 a_3 - \varepsilon)$. □

B. *Geometric Conditions for Absolute* $\{F\}$ *Stability*

A geometric interpretation of the criterion can also be derived from the Popov inequality (7-13). In the U, $\hat{V}$ plane the equation

$$[U(\omega^2) + \bar{F}^{-1}] - \alpha\hat{V}(\omega^2) = 0 \tag{7-14}$$

represents a straight line with slope α^{-1} passing through the point $(-\bar{F}^{-1}, 0)$. Thus, according to the inequality (7-13) the imaginary axis of the s-plane transformed by $\mathrm{Re}\, W(i\omega) \triangleq U$, $\omega\, \mathrm{Im}\, W(i\omega) \triangleq \hat{V}$ must lie on or to the right of this straight line as shown in Fig. 7-6. The plot of $\omega\, \mathrm{Im}\, W(i\omega)$ versus $\mathrm{Re}\, W(i\omega)$ is referred to as the modified phase amplitude characteristic, the modified Nyquist plot, or the Popov plot; note that since $U(\omega^2)$ and $\hat{V}(\omega^2) = \omega V(\omega)$ are both even functions of ω, the diagram for $\omega < 0$ is identical to that for $\omega > 0$. According to the Popov criterion the modified Nyquist plot of $W(i\omega)$ must lie in a half plane for absolute stability.

GEOMETRIC CRITERION 2 (Popov, Criterion 1a). Given a system described by Eq. (7-1) with $\rho \geqslant -\bar{F}^{-1}$ and

$$g(\sigma_0, t) = f(\sigma_0) \in \{F\}, \qquad f(\sigma_0)/\sigma_0 \in [0, \bar{F}), \tag{7-15}$$

then if in the U, $\hat{V}$ plane, the Popov plot $\hat{\Gamma}_W$ of W touches or lies to the right of a straight line passing through the point $(-\bar{F}^{-1}, 0)$ with nonzero slope, the system is absolutely $\{F\}$ stable. □

Since the slope of the line is α^{-1}, it is observed that the specific multiplier used in the Popov criterion can also be directly obtained from the modified Nyquist plot. If $\alpha > 0$, then the multiplier $(1 + \alpha s)$ may be used in Criterion 1a, while if $\alpha < 0$, the multiplier $Z(s) = (1 + |\alpha| s)^{-1}$ satisfies this criterion.

Since the lower bound is specified to be zero, the largest range of $f(\sigma_0)$ that guarantees absolute stability is uniquely determined. This maximum upper limit $\bar{F}$ is determined by the intersection with the negative real axis of a line tangent to the modified Nyquist plot $\hat{\Gamma}_W$ of $W(i\omega)$. Any system which permits the use of a straight line tangent to $\hat{\Gamma}_W$ at a point on the real axis (Fig. 7-6a) provides an example of a system for which the Aizerman conjecture is valid: one Nyquist range of the system is $[0, \bar{F})$, as can be appreciated from the fact that any real-axis crossings of $\hat{\Gamma}_W$ are also real axis crossings of the Nyquist diagram and vice versa for ω finite. In case the Popov plot does not lie to the right of the line tangent to $\hat{\Gamma}_W$ at a point on the real axis, the upper limit $\bar{F}$ is determined by drawing a straight line twice tangent to the plot (once above the real axis, once below it) as shown in Fig. 7-6b.

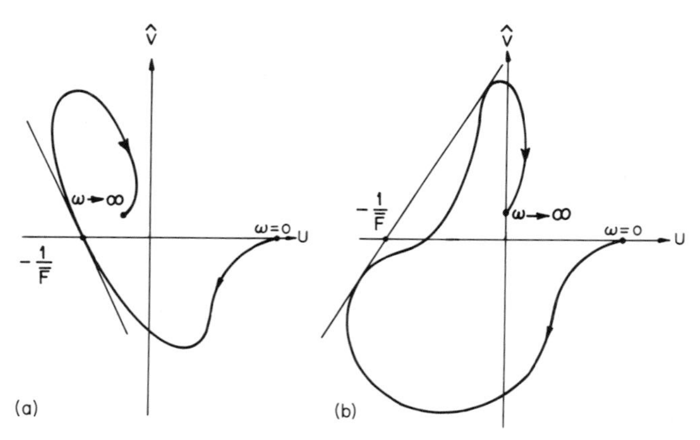

Fig. 7-6. Obtaining the largest range $[0, \bar{F})$ by Popov's criterion: (a) tangency point on the real axis; Aizerman's conjecture valid; (b) two tangency points, one above the real axis, one below it.

The necessity of the constraint $\rho \geqslant -\bar{F}^{-1}$ in addition to the graphical condition is due to the fact that the point $W_\infty = \lim_{\omega \to \infty} W(i\omega) = \rho$ must lie on the real axis of the U, V plane, that is, $U_\infty = \rho$, $V_\infty = 0$, while in the U, $\hat{V}$ plane we have $U_\infty = \rho$, $\hat{V}_\infty = -h^{\mathrm{T}}b = -h_n$, since for large frequency $W(i\omega) \approx \rho + h_n/i\omega$. Thus, although the Popov graphical condition prevents Γ_W or the Nyquist diagram of $W(i\omega)$ from contacting the real axis to the left of $-\bar{F}^{-1}$ for finite ω, it does not preclude W_∞ from satisfying $W_\infty < -\bar{F}^{-1}$, which thus must be explicitly avoided by this subsidiary condition. This

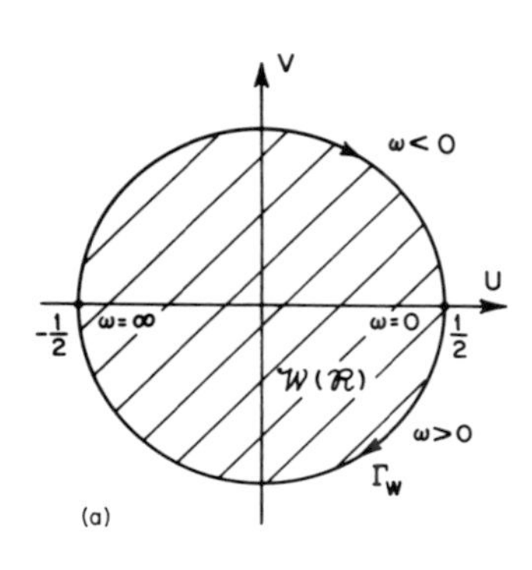

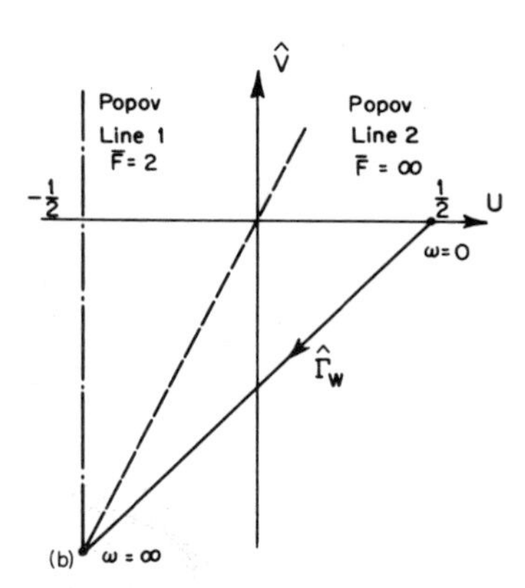

Fig. 7-7. The failure of the graphical Popov condition: (a) Nyquist range $-2 < \kappa < 2$; (b) absolute $\{F\}$ stability range $0 \leqslant f(\sigma_0)/\sigma_0 < 2$.

point is demonstrated in Fig. 7-7, where we treat the example $W(s) = -\frac{1}{2} + (s+1)^{-1}$ introduced in Chapter V, Section 1; the graphical condition alone [Popov Line 2] fails, that is, it incorrectly predicts absolute stability for $f(\sigma_0)/\sigma_0 \in [0, \infty)$.

There are some simple cases where the criterion can be applied directly to the Nyquist plot of $W(i\omega)$ rather than the modified frequency response curve specified in the above criterion. The modified Nyquist plot is obtained by multiplying the ordinate of each point on the Nyquist plot by the corresponding frequency ω. Hence, if all the points of the Nyquist plot are to the right of a vertical line through $U = -\bar{F}^{-1}$, then so are all the points of the Popov plot, and the absolute stability of the system can be concluded directly from the Nyquist plot. The circle criterion, however, guarantees that a system satisfying this condition strictly (the vertical line does not contact Γ_W) would be absolutely $\{G_i[F, K_0]\}$ stable for all nonlinear time-varying gains $g(\sigma_0, t)$ in the range $[0, \bar{F}]$, so that this observation is redundant.

A second situation where the Nyquist diagram alone suffices for a determination of absolute stability arises when Γ_W for $\omega \in [0, \infty)$ lies in the upper or lower half of the W-plane; we consider only the latter situation which is subdivided into two cases:

(i) $\rho \geqslant 0$ and $-\pi < \arg W(i\omega) \leqslant 0, \quad \omega \in [0, \infty)$
guarantees absolute stability for $f(\sigma_0)/\sigma_0 \in [0, \infty)$;

(ii) $\rho < 0$ and $-\pi < \arg [W(i\omega) - \rho] \leqslant 0, \quad \omega \in [0, \infty)$
guarantees absolute stability for $f(\sigma_0)/\sigma_0 \in [0, -\rho^{-1})$.

In treating these cases, the multiplier $Z(s) = (s + \varepsilon)$ may always be used in proving absolute stability, as shown in Fig. 7-8.

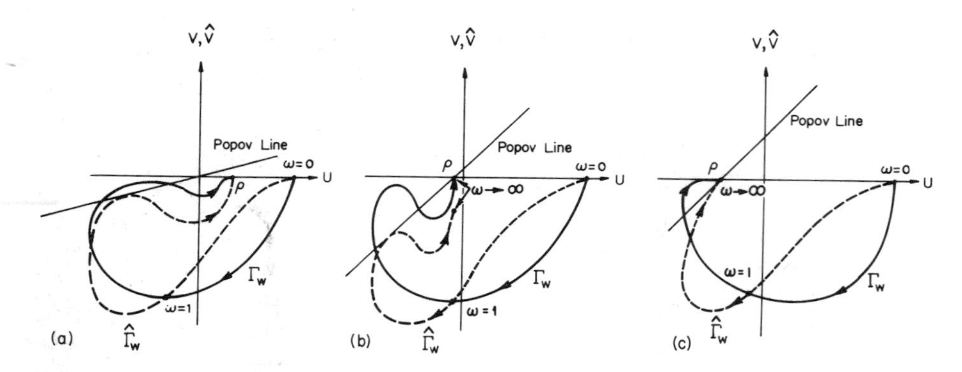

Fig. 7-8. The Nyquist plot of W lying in the lower half plane: (a) $\rho > 0$, with asymptotic behavior $\{W(i\omega) - \rho\} \approx (i\omega)^{-2}$; (b) $\rho < 0$, with asymptotic behavior $\{W(i\omega) - \rho\} \approx (i\omega)^{-1}$; (c) $\rho < 0$, with asymptotic behavior $\{W(i\omega) - \rho\} \approx (i\omega)^{-2}$.

In any situation where $\bar{F} = -\rho^{-1}$ (this includes the infinite sector case with $\rho = 0$), absolute stability is guaranteed only when a nonnegative parameter α exists such that

$$\operatorname{Re}\{[W(i\omega) + \bar{F}^{-1}](1 + \alpha i\omega)\} = \operatorname{Re}\{[h^{\mathrm{T}}(i\omega I - A)^{-1}b](1 + \alpha i\omega)\} \geqslant 0;$$

the asymptotic behavior of $h^{\mathrm{T}}(i\omega I - A)^{-1}b$ as $\omega \longrightarrow \infty$ precludes the use of $\alpha < 0$. Since the phase angle of $(1 + i\alpha\omega)$ lies between zero and $\pi/2$ for $\alpha > 0$, it is clear that such an α cannot be found if the Nyquist plot of $W(i\omega)$ enters the second quadrant. If $\bar{F} \neq -\rho^{-1}$, then it is always possible to use $\alpha < 0$ if necessary, in which case the Popov line has a negative slope.

Many of these observations, particularly the direct relation between the graphical condition for absolute stability and the verification of the Aizerman conjecture, and the validity of the use of a vertical line in the standard polar plot (Nyquist plot) of $W(i\omega)$, are due to Popov [3].

The geometric criterion stated above is merely an interpretation of one form of the Popov criterion, and therefore provides the same sufficient condition for absolute stability. It is thus no more restrictive than the frequency domain condition upon which it is based, Criterion 1a of Chapter V, Section 1. Since the interpretation of the other forms of the Popov condition for the principal and particular cases follows directly, we simply summarize these results (see also Fig. 7-9).

CRITERION 1b. We require in addition to the conditions of Criterion 1a that $A_{\bar{F}} \in \{A_1\}$, or that $1 + \bar{F}W(i\omega) \neq 0$, $\omega \in [-\infty, \infty]$. For finite ω, this condition is satisfied if $\hat{\Gamma}_W$ does not contact the Popov line on the real axis, and as $\omega \longrightarrow \infty$, by demanding that $\rho > -\bar{F}^{-1}$. □

CRITERION 1c. We add to the requirements of Criterion 1b the condition $H(s) \in \{\mathrm{SPR}_0\}$, or $J(\omega^2) > 0$ for finite ω. The Popov line can only be contacted as $\omega \longrightarrow \infty$ (provided $\hat{V}_\infty \neq 0$); again $\rho > -\bar{F}^{-1}$ is required. □

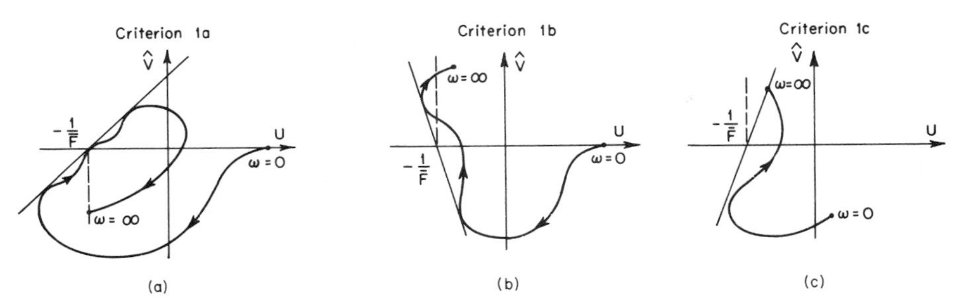

Fig. 7-9. Geometric interpretation of the Popov criteria: (a) $A \in \{A_1\}$, $f(\sigma_0)/\sigma_0 \in [0, \bar{F})$; (b) $A \in \{A_1\}$, $f(\sigma_0)/\sigma_0 \in [0, \bar{F}]$; (c) $A \in \{A_0\}$, $f(\sigma_0)/\sigma_0 \in (0, \bar{F}]$.

C. *The Parabola Criterion for Absolute $\{F\}$ Stability*

The finite sector stability problem considered in the previous section specifies a range of the NLTI gain having a lower bound of zero. For conditionally stable systems where several sectors of absolute stability can exist (see Fig. 7-10) or for systems having an unstable plant $W(s)$, it is not possible to obtain such sectors $[\underline{F}, \bar{F}]$ by the direct application of Criterion 1b, Chapter V, Section 1. In principle, it is possible to transform the problem into the form considered earlier so that the lower bound on the nonlinear gain of the transformed problem is zero as in Chapter V, Section 5, but this procedure is generally tedious. For every choice of $\underline{F}$, a corresponding value of $\bar{F}$ can be found so that the system is absolutely stable in the interval $[\underline{F}, \bar{F}]$. One of the main difficulties of the procedure lies in choosing $\underline{F}$ to maximize the sector $[\underline{F}, \bar{F}]$; this generally involves a process of trial and error, and the Popov plot of $W(i\omega)/(1 + \underline{F}W(i\omega))$ would have to be constructed for each value of $\underline{F}$.

In view of this difficulty, it would prove most helpful if the sectors for absolute stability could be determined without transformations. Such a method was proposed by Bergen and Sapiro [1] in 1967, and the corresponding result is called the parabola criterion. In deriving this criterion, the Popov condition is applied to the transformed system as in Chapter V, Section 5, and the resulting frequency domain condition for absolute stability is reinterpreted in terms of $(U, \hat{V})$ defined for $W(i\omega)$, as in the previous section.

Taking the general finite sector result of General Stability Criterion 1, absolute stability for $f(\sigma_0)/\sigma_0 \in [\underline{F}, \bar{F}]$ is guaranteed if $A_{\underline{F}} \in \{A_1\}$, $A_{\bar{F}} \in \{A_1\}$ [Eq. (4-2)] and if

$$\frac{1 + \bar{F}W(s)}{1 + \underline{F}W(s)}(1 + \hat{\alpha}s)^{\pm 1} \in \{\text{PR}\} \tag{7-16}$$

where $\hat{\alpha} \geqslant 0$. Provided that $A_\kappa \in \{A_1\}$, $\kappa \in [\underline{F}, \bar{F}]$, this is equivalent to

$$\underline{F}\bar{F}U^2 + (\underline{F} + \bar{F})U + 1 \geqslant \alpha\omega(\bar{F} - \underline{F})V - \underline{F}\bar{F}V^2, \qquad \alpha \in (-\infty, \infty). \tag{7-17}$$

This relation is useful only if it can be interpreted simply in the U, V or U,

$\hat{V}$ plane. If $\underline{F}\bar{F} > 0$, we may discard the last term of inequality (7-17) to obtain the stronger constraint

$$(\underline{F}U + 1)(\bar{F}U + 1) \geqslant \alpha(\bar{F} - \underline{F})\hat{V},$$

which is thus a sufficient condition for absolute stability.

GEOMETRIC CRITERION 3 (The parabola criterion). Given a system described by Eq. (7-1) with

$$g(\sigma_0, t) = f(\sigma_0) \in \{F\}, \qquad f(\sigma_0)/\sigma_0 \in [\underline{F}, \bar{F}] \tag{7-18}$$

and $A_\kappa \in \{A_1\}$, $\kappa \in [\underline{F}, \bar{F}]$, if in the $U, \hat{V}$ plane the parabola

$$(\underline{F}U + 1)(\bar{F}U + 1) = \alpha(\bar{F} - \underline{F})\hat{V} \tag{7-19}$$

does not intersect any part of the modified Nyquist plot of $W(i\omega)$, then the system is absolutely $\{F\}$ stable. □

In application, this criterion is most clear if both the normal and modified Nyquist plots of $W(i\omega)$ are made. In this way the range $[\underline{F}, \bar{F}]$ such that $A_\kappa \in \{A_1\}$ for $\kappa \in [\underline{F}, \bar{F}]$ may be easily ascertained from $\mathfrak{W}(\mathfrak{R})$, and then the parabola is drawn with respect to $\hat{\Gamma}_W$ or the Popov plot of $W(i\omega)$. The estimation of $\underline{F}$ and $\bar{F}$ such that a suitable parabola can be drawn is simplified by noting that the parabola specified in Eq. (7-19) has several useful properties:

(i) Real axis crossings ($\hat{V} = 0$) are at $U = -\underline{F}^{-1}$ and $U = -\bar{F}^{-1}$.
(ii) The parabola is tangent to straight lines drawn through these crossing points of slope $-1/\alpha$ and $1/\alpha$ respectively. Again, the Popov multiplier $(1 + \hat{\alpha}s)^{\pm 1}$ is explicitly available from the graphical analysis.
(iii) At $U = -\frac{1}{2}(\underline{F}^{-1} + \bar{F}^{-1})$, $d\hat{V}/dU = 0$ and $\hat{V} = -(\bar{F} - \underline{F})/(4\alpha\underline{F}\bar{F})$.
(iv) The intersection of the two straight lines of tangency occurs at $(U = -\frac{1}{2}(\underline{F}^{-1} + \bar{F}^{-1}), \hat{V} = -(\bar{F} - \underline{F})/(2\alpha\underline{F}\bar{F})$, which is just twice the value of $\hat{V}$ on the parabola at the same ordinate.

These relations are shown in Fig. 7-10 in one example of the application of the parabola criterion. For a given choice of $\underline{F}$ and $\bar{F}$, it is usually advantageous to choose $|\alpha|$ as large as possible, as estimated by drawing the pair of tangency lines. Taking $\alpha < 0$ merely reverses the sense of the parabola.

As in the case of the circle criterion, it is not necessary that $W(s)$ be stable if the range of $f(\sigma_0)$ does not include zero ($0 \notin [\underline{F}, \bar{F}]$). Again, consider the example $W(s) = (s + \beta)/(s(s - \alpha))$: We have

$$U(\omega^2) = -\frac{(\alpha + \beta)}{\omega^2 + \alpha^2}, \qquad \hat{V}(\omega^2) = \frac{\alpha\beta - \omega^2}{\omega^2 + \alpha^2} = -1 - \alpha U,$$

which demonstrates that the Popov plot $\hat{\Gamma}_W$ is a straight line segment with endpoints $(U = -(\alpha + \beta)/\alpha^2,\ \hat{V} = \beta/\alpha)$ corresponding to $\omega = 0$ and $(U = 0,\ \hat{V} = -1)$ as $\omega \rightarrow \pm\infty$. The result shown in Fig. 7-11 (in which

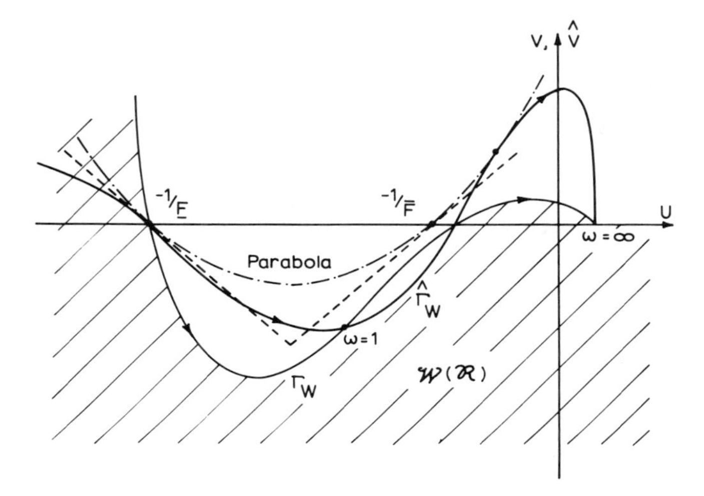

Fig. 7-10. An application of the parabola criterion.

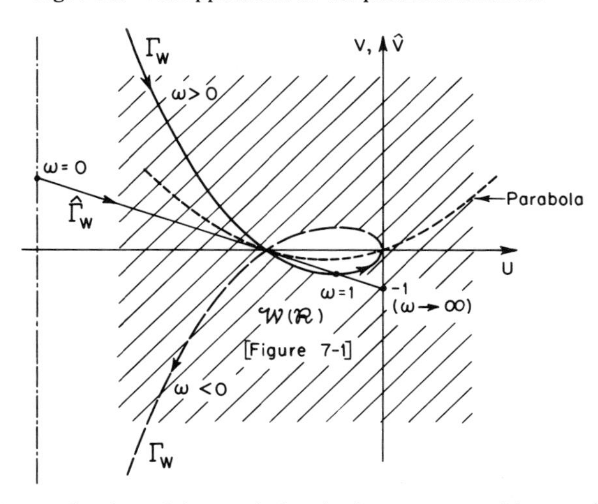

Fig. 7-11. An application of the parabola criterion to an unstable open-loop plant.

$\mathcal{W}(\mathcal{R})$ is an enlarged version of Fig. 7-1) is that the parabola criterion guarantees stability for all gains $f(\sigma_0) \in \{F\}$, $f(\sigma_0)/\sigma_0 \in [(\alpha + \varepsilon), \varepsilon^{-1}]$, where $\varepsilon > 0$. This system provides another example of a plant $W(s)$ for which a restricted type of Aizerman conjecture is valid.

Like the other geometric criteria, the principal advantage of this criterion is the ease with which it can be applied, compared with the direct application of Popov's criterion to the transformed system on a trial and error basis. However, it is considerably more difficult to apply than Geometric Criteria

1 and 2; the values of $\underline{F}$ and $\bar{F}$ must be estimated, then values of α must be chosen in the search for a parabola that does not intersect the modified Nyquist plot.

It must be noted that the criterion imposes stricter conditions than Popov's Criterion since it is a sufficient but not necessary condition for the existence of a Popov multiplier. While its advantages are most obvious for conditionally stable systems, it can be used in all situations where the Popov criterion is applicable. For the particular case where $\underline{F} = 0$, the parabola reduces to Geometric Criterion 2 (Popov) since the discarded term $-\underline{F}\bar{F}V$ in Eq. (7-17) is zero.

4. Monotonic Nonlinearities: An Off-Axis Circle Criterion

For a conventional feedback system having an LTI plant $W(s)$ in the forward path [Eq. (7-1)] and a single time-invariant monotonic nonlinearity in the return path,

$$g(\sigma_0, t) = f(\sigma_0) \in \{F_m\}, \qquad \Delta f(\sigma_0)/\Delta\sigma_0 \in [\underline{M}, \bar{M}], \tag{7-20}$$

the sufficient condition for absolute stability has been demonstrated to be that there must exist some $Z(s)$ such that $Z^{\pm 1}(s) \in \{Z_{F_m}(s)\}$ and

$$H(s) = \left[\frac{1 + \bar{M}W(s)}{1 + \underline{M}W(s)}\right] Z(s) \in \{\mathrm{PR}\}, \tag{7-21}$$

where $[1 + \underline{M}W(i\omega)] \neq 0$ and $[1 + \bar{M}W(i\omega)] \neq 0$, $\omega \in [-\infty, \infty]$. In treating the problem of absolute stability for monotonic gains (Chapter V, Sections 2 and 5), it has been determined that the class of multiplier functions $\{Z_{F_m}(s)\}$ contains the class of functions that can be realized as driving point impedances of RL networks; $\{Z_{F_m}\} = \{Z_{\mathrm{RL}}\}$.

When for a specific choice of $\underline{M}$, $\bar{M}$ and $W(s)$, it becomes necessary to extend the search for a suitable multiplier $Z(s)$ satisfying the above conditions to RL functions having several poles and zeros, say $0 < \lambda_1 < \mu_1 < \lambda_2$, for example, it becomes a very difficult task to determine whether values of these parameters exist such that $H(s)$ is positive real. For this reason, geometric criteria are especially desirable in cases when multiplier classes include complicated functions. An off-axis circle criterion has been derived by Cho and Narendra [2] to answer this need for monotonic gains. This result provides a sufficient condition to be satisfied by the Nyquist plot of $W(i\omega)$ that ensures the existence of some $Z(s)$ such that $Z^{\pm 1}(s) \in \{Z_{\mathrm{RL}}\}$ and condition (7-21) is satisfied.

The basis of this theorem can be most simply presented in terms of the phase characteristics (argument) of RL multiplier functions. The stability criterion follows from a basic lemma concerning the realizability of a useful class of functions $\{Z_\theta(s)\}$ that is a subset of the class $\{Z_{\text{RL}}(s)\}$.

LEMMA 7-1 (Z_{RL} multipliers). Let $[\omega_1, \omega_2]$ be any closed subinterval of $(0, \infty)$, θ any constant in $(0, \pi/2)$ and $\varepsilon > 0$ a constant that may be arbitrarily small. Then there exists a function $Z(s)$ such that $Z(s) \in \{Z_{\text{RL}}(s)\}$ and

$$|\arg Z(i\omega) - \theta| < \varepsilon, \qquad \omega \in [\omega_1, \omega_2],$$
$$0 < \arg Z(i\omega) \leqslant \theta + \varepsilon, \qquad \omega \notin [\omega_1, \omega_2].$$

A member of this class (depicted in Fig. 7-12) is denoted $Z(s) \in \{Z_\theta(s)\} \subset \{Z_{\text{RL}}(s)\}$. □

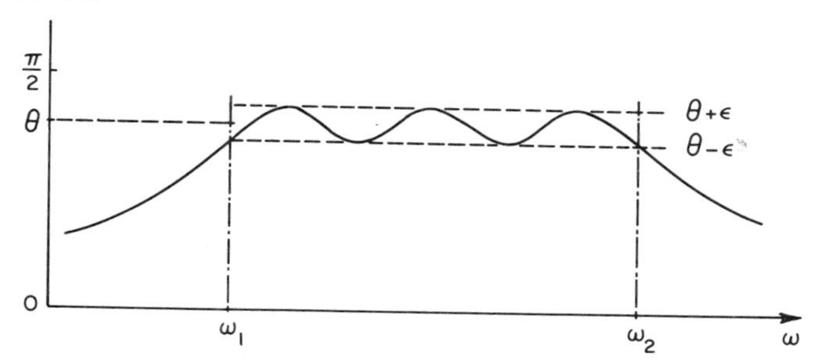

Fig. 7-12. The phase characteristic of $Z(s) \in \{Z_\theta(s)\}$.

A heuristic explanation of the method of proof may be outlined as follows. The multiplier $Z(s) \in \{Z_{\text{RL}}\}$ is defined as a function having alternating poles and zeros on the negative real axis; the first or smaller of these singularities is a zero. The argument of $Z(i\omega)$ can then be expressed as a convolution of a "step-function" and a "smoothing function" and this in turn may be used to obtain a bound on $|\arg Z(i\omega) - \theta|$ using Holder's inequality. For ε arbitrarily small, the first zero of $Z(s)$ is likewise arbitrarily small in magnitude. Since this proof is quite technical, the reader is referred to Cho and Narendra [2].

First consider the finite sector case, $\Delta f(\sigma_0)/\Delta\sigma_0 \in [0, \bar{M}]$ or $\Delta f(\sigma_0)/\Delta\sigma_0 \in (0, \bar{M}]$, depending on the stability properties of $W(s)$. Suppose on the Nyquist plot Γ_W of $W(i\omega)$ it is possible to draw a straight line of positive slope† through the point $(U = -\bar{M}^{-1}, V = 0)$ such that Γ_W lies strictly to the right of the line, $\omega \in [0, \infty]$. Denoting the angle between this line and the imaginary axis by θ, this means that

$$-\pi/2 - \theta < \phi_{\bar{M}}(\omega) \triangleq \arg[W(i\omega) + \bar{M}^{-1}] < \pi/2 - \theta, \qquad \theta \in (0, \pi/2).$$

† We consider only finite slopes here. If the slope is infinite, we have absolute $\{G_i[F, K_0]\}$ stability by the circle criterion, Section 2.

Since the points where Γ_W approaches this line must be in finite U, V-plane even if $W(s)$ has a pole at the origin and since $\phi_{\bar{M}}(\omega)$ cannot approach θ asymptotically as $\omega \longrightarrow \infty$ or as $\omega \longrightarrow 0$ ($\phi_{\bar{M}}(\omega)$ must approach 0 or $\pi/2$ in either case, that is, $[W(i\omega) + \bar{M}^{-1}]$ must approach a positive constant or it must behave like $(i\omega)^{-1}$ as $\omega \longrightarrow \infty$ or as $\omega \longrightarrow 0$, since $W(s)$ is rational), this assumed constraint is equivalent to

$$-\pi/2 - \theta + \varepsilon_1 \leqslant \phi_{\bar{M}}(\omega) \leqslant \pi/2 - \theta - \varepsilon_2, \tag{7-22}$$

where $\varepsilon_1 > 0$ and $\varepsilon_2 > 0$ are arbitrarily small. There are two cases that must be taken into account; the first is the only one to be considered if $A \in \{A_1\}$ (if $W(s)$ is asymptotically stable), whereas the second may occur if $A \in \{A_0\}$ (if $W(s)$ has a single pole at $s = 0$).

(i) Two frequencies ω_1 and $\omega_2 > \omega_1$ exist such that $-\pi/2 \leqslant \phi_{\bar{M}}(\omega) < \pi/2 - \theta$ for $\omega \in [0, \omega_1)$ and $\omega \in (\omega_2, \infty]$ [Fig. 7-13a]: Choose $Z(s) \in \{Z_\theta(s)\}$ to have its argument between $(\theta - \varepsilon)$ and $(\theta + \varepsilon)$ for $\omega \in [\omega_1, \omega_2]$ where $\varepsilon < \min(\varepsilon_1, \varepsilon_2)$. Then define $\hat{\varepsilon}_1 \triangleq \varepsilon_1 - \varepsilon > 0$, $\hat{\varepsilon}_2 \triangleq \varepsilon_2 - \varepsilon > 0$, and we have

$$-\pi/2 + \hat{\varepsilon}_1 \leqslant \arg\{[W(i\omega) + \bar{M}^{-1}]Z(i\omega)\} \leqslant \pi/2 - \hat{\varepsilon}_2, \tag{7-23}$$

$\omega \in [\omega_1, \omega_2]$. Since $0 < \arg Z(i\omega) \leqslant \theta + \varepsilon$ outside this closed range of ω, it is evident that this restriction on the argument of $[W(i\omega) + \bar{M}^{-1}]Z(i\omega)$ must hold for all $\omega \in [0, \infty]$. Thus there must exist some $Z(s) \in \{Z_\theta(s)\} \subset \{Z_{\text{RL}}(s)\}$ such that $[W(s) + \bar{M}^{-1}]Z(s) \in \{\text{PR}\}$, or $\in \{\text{SPR}_0\}$. Furthermore, the graphical restriction guarantees that $[1 + \bar{M}W(i\omega)] \neq 0$, $\omega \in [-\infty, \infty]$. These two conditions are sufficient to ensure absolute stability as demonstrated previously (Chapter V, Section 2).

(ii) A frequency ω_2 exists such that $\phi_{\bar{M}}(\omega) \in (-\pi/2, \pi/2 - \theta)$ for $\omega \in (\omega_2, \infty]$ but $\phi_{\bar{M}}(\omega) \in (-\pi/2 - \theta, -\pi/2)$ for $\omega \in (0, \omega_2)$ [Fig. 7-13b]: As $\omega \longrightarrow 0$, $\phi_{\bar{M}}(\omega)$ must approach $-\pi/2$ from below. In this case we choose

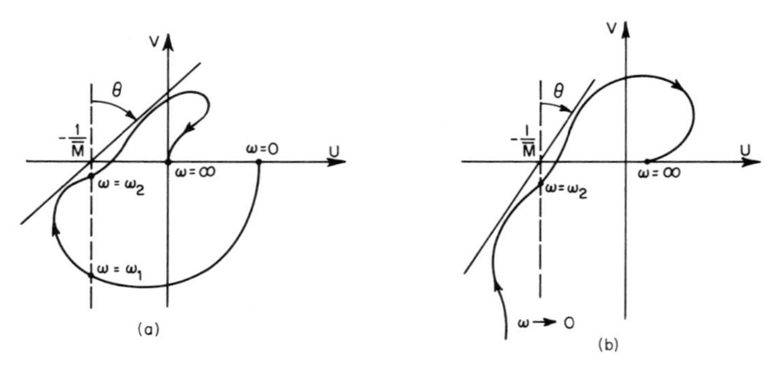

Fig. 7-13. Off-axis circle criterion, two cases: (a) $A \in \{A_1\}$, $\Delta f(\sigma_0)/\Delta\sigma_0 \in [0, \bar{M}]$; (b) $A \in \{A_0\}$, $\Delta f(\sigma_0)/\Delta\sigma_0 \in (0, \bar{M}]$.

$Z(s) \in \{Z_\theta(s)\}$ as in (i) except $\omega_1 > 0$ is taken to be arbitrarily small so that $\arg Z(i\omega)$ dominates $\phi_{\bar{M}}(\omega)$ for small ω, and thus we can ensure that $[W(s) + \bar{M}^{-1}]Z(s) \in \{\text{SPR}_0\}$.

Actually in all of these cases the fact that the straight line does not contact Γ_W at any point implies that $[W(s) + \bar{M}^{-1}]Z(s) \in \{\text{SPR}\}$ (if $A \in \{A_1\}$), or $[W(s) + \bar{M}^{-1}]Z(s) \in \{\text{SPR}_0\}$ (if $A \in \{A_0\}$), as seen in the inequality (7-23). Thus in either case we have a sufficient condition for absolute stability.

To this point we have considered $\theta \in (0, \pi/2)$ in which case $Z(s) \in \{Z_\theta(s)\} \subset \{Z_{\text{RL}}(s)\}$ exists to satisfy Eq. (7-21) with $\underline{M} = 0$. If $\theta \in (-\pi/2, 0)$ then as in previous developments $Z^{-1}(s) \in \{Z_\theta(s)\}$ exists such that $[W(s) + \bar{M}^{-1}]Z(s) \in \{\text{PR}\}$, that is, $Z(s)$ is a member of $\{Z_{\text{RC}}\}$. Thus Γ_W may lie to the right of a line of negative slope, and we proceed to prove absolute stability as above.

GEOMETRIC CRITERION 4a (Monotonic gains, finite sector). Given a system described by Eqs. (7-1) and (7-20), where $\Delta f(\sigma_0)/\Delta\sigma_0 \in [0, \bar{M}]$ if $A \in \{A_1\}$ or $\Delta f(\sigma_0)/\Delta\sigma_0 \in (0, \bar{M}]$ if $A \in \{A_0\}$. Then if Γ_W or the Nyquist plot of $W(i\omega)$ for $\omega \in [0, \infty]$ lies strictly to the right of a straight line through the point $(U = -\bar{M}^{-1}, V = 0)$ with nonzero slope, the system is absolutely $\{F_{\text{m}}\}$ stable. □

We should emphasize that the portion of Γ_W for $\omega < 0$ is not drawn in applying this theorem. In one respect, the conditions are actually more strict than those of the Popov criterion in that the monotonic line cannot be touched by Γ_W, whereas $\hat{\Gamma}_W$ can contact the Popov line. This is due to the argument of $Z(s) \in \{Z_\theta(s)\}$ lying in the range $(\theta - \varepsilon, \theta + \varepsilon)$ over the specified range of ω and not being strictly constant.

To prove the analogous result for the general finite sector case, $f(\sigma_0) \in [\underline{M}, \bar{M}]$, we again make use of the standard transformation of Chapter V, Section 5, to obtain a system $\{\tilde{W}, \tilde{f}\}$ that may be considered to lie within the ambit of Geometric Criterion 4a. In algebraic terms, we have $\tilde{U} \triangleq \text{Re}\{W(s)/(1 + \underline{M}W(s))\}$, $\tilde{V} \triangleq \text{Im}\{W(s)/(1 + \underline{M}W(s))\}$; thus the condition to be satisfied for absolute stability is that there must be some $\mu \neq 0$ (μ being the slope of the line specified in Geometric Criterion 4a) such that

$$\mu^{-1}\tilde{V} < (\tilde{U} + (\bar{M} - \underline{M})^{-1}), \qquad \omega \in [0, \infty].$$

Solving for $\tilde{U}$ and $\tilde{V}$ in terms of U and V of the original system, this condition is equivalent to

$$J_\mu(\omega^2) \triangleq [\underline{M}\bar{M}(U^2 + V^2) + (\underline{M} + \bar{M})U - ((\bar{M} - \underline{M})/\mu)V + 1] > 0. \quad (7\text{-}24)$$

To interpret this constraint in geometric terms, consider a circle passing through $(U_1 = -\underline{M}^{-1}, V_1 = 0)$ and $(U_2 = -\bar{M}^{-1}, V_2 = 0)$ with slope $-\mu$ and $+\mu$ at those points respectively. If $\mu \neq \infty$, then the center of the circle does not lie on the real axis; elementary geometric relationships reveal that

$$\text{center:}\quad v_0 = -(\underline{M} + \bar{M})/(2\underline{M}\bar{M}), \qquad \nu_0 = (\bar{M} - \underline{M})/(2\mu\underline{M}\bar{M})$$
$$\text{radius:}\quad \rho_0{}^2 = (\mu^2 + 1)\left[\frac{\bar{M} - \underline{M}}{2\mu\underline{M}\bar{M}}\right]^2. \tag{7-25}$$

Substituting these quantities into $(U - v_0)^2 + (V - \nu_0)^2 = \rho_0{}^2$ and simplifying, this circle corresponds to the condition $J_\mu(\omega^2) = 0$. The significance of the inequality in Eq. (7-24) depends upon the signs of $\underline{M}$ and $\bar{M}$: if $\underline{M}\bar{M} > 0$, that is, if both $\underline{M}$ and $\bar{M}$ have the same sign, then points (U, V) satisfying the condition (7-24) lie strictly outside the circle in the usual loose sense, while if $\underline{M}\bar{M} < 0$, then points that are consistent with this constraint are strictly inside. As in Geometric Criterion 1, we formally define the interior of the circle defined by $J_\mu(\omega^2) = 0$ to exclude the origin.

Definition. In the U, V plane, $C_\mu[\underline{M}, \bar{M}]$ denotes the generalized off-axis circle

$$C_\mu[\underline{M}, \bar{M}] \triangleq \begin{cases} \{(U, V): [\underline{M}\bar{M}(U^2 + V^2) + (\underline{M} + \bar{M})U \\ \qquad - ((\bar{M} - \underline{M})/\mu)V + 1] = 0\} \\ \{(U, V): V/\mu = (U + \bar{M}^{-1})\}, \qquad \underline{M} = 0 \\ \{(U, V): V/\mu = (U + \underline{M}^{-1})\}, \qquad \bar{M} = 0 \end{cases} \tag{7-26}$$

and the closed interior of $C_\mu[\underline{M}, \bar{M}]$ is the region in the U, V plane defined by

$$I_\mu[\underline{M}, \bar{M}] \triangleq \begin{cases} \{(U, V): [\underline{M}\bar{M}(U^2 + V^2) + (\underline{M} + \bar{M})U \\ \qquad - ((\bar{M} - \underline{M})/\mu)V + 1] \leqslant 0\} \\ \{(U, V): (\bar{M}U + 1) \leqslant \bar{M}V/\mu\}, \qquad \underline{M} = 0 \\ \{(U, V): (\underline{M}U + 1) \leqslant \underline{M}V/\mu\}, \qquad \bar{M} = 0. \ \square \end{cases} \tag{7-27}$$

As stated previously in considering the finite sector case, only the Nyquist plot for $\omega \geqslant 0$ must avoid the line or circle specified by this criterion. Thus rather than being interested in the mapping of the entire right half plane, that is, in $\mathcal{W}(\mathcal{R}) = W(s \in \mathcal{R})$ as defined in Section 1A, we must ascertain the mapping of the infinite first quadrant ($\text{Re}\, s \geqslant 0$, $\text{Im}\, s \geqslant 0$) alone. If we denote the finite first quadrant of the s-plane by

$$\mathfrak{F}_R \triangleq \{s = \rho e^{i\theta};\quad \rho \in [0, R],\ \theta \in [0, \pi/2]\},$$

then the comparable region in the U, V plane is $\mathcal{W}(\mathfrak{F}) \triangleq W(s \in \mathfrak{F})$ where again, R is taken to be arbitrarily large if $\lim_{\omega\to\infty} |W(i\omega)| = \infty$, or we formally take the limit as $R \to \infty$ if $\lim_{\omega\to\infty} |W(i\omega)| < \infty$. If $W(s)$ has any poles on the boundary of $\mathfrak{F}$, that is, on the positive real or imaginary axes, then the contour must be extended to include these points.

Definition of Region $\mathfrak{F}$. For any $W(s)$ having poles at $s = \lambda_i$ satisfying

$$[\text{Re}\, \lambda_i \geqslant 0, \qquad \text{Im}\, \lambda_i \geqslant 0], \qquad i = 1, 2, \ldots, m;$$
$$\measuredangle \lambda_i \in (\pi/2, 2\pi), \qquad i = (m + 1), \ldots, n,$$

the region $\mathfrak{F}$ in the s-plane consists of the union of all points satisfying:

(i) $|s| \in [0, R]$ and $\measuredangle s \in [0, \pi/2]$ where R is arbitrarily large (or $R \longrightarrow \infty$ if $W(\infty) < \infty$), and

(ii) $|s - \lambda_i| \leqslant \rho_0$, $i = 1, 2, \ldots, m$ where ρ_0 may be arbitrarily small†. □

GEOMETRIC CRITERION 4b (The off-axis circle criterion). A system described by Eqs. (7-1) and (7-20) is absolutely $\{F_m\}$ stable if $\mathcal{W}(\mathfrak{F}) \cap I_\mu[\underline{M}, \bar{M}] = \phi$ where $\mu \in [-\infty, \infty]$ is not zero. □

EXAMPLE 1. Again we consider $W_1(s) = (s + \beta)/(s(s - \alpha))$. Taking $\mathfrak{F}$ or the first quadrant of the s-plane extended to enclose both of the poles, we plot the mapping $\mathcal{W}(\mathfrak{F}) \triangleq W(s \in \mathfrak{F})$ using the conventional techniques; see Fig. 7-14. We note that an off-axis circle may be drawn touching Γ_W at $(U = -\alpha^{-1}, V = 0)$ and at the origin. Thus any monotonic gain $f(\sigma_0) \in \{F_m\}$, $\Delta f(\sigma_0)/\Delta\sigma_0 \in [\alpha + \varepsilon, \varepsilon^{-1}]$ may be used in conjunction with $W_1(s)$ to result in a closed loop system that is absolutely $\{F_m\}$ stable since $I_\mu[\alpha + \varepsilon, \varepsilon^{-1}] \cap \mathcal{W}(\mathfrak{F}) = \phi$ where $\varepsilon > 0$ may be arbitrarily small. This is exactly the same range obtained by use of the parabola criterion (Fig. 7-11), so this result is only included for demonstrative purposes.

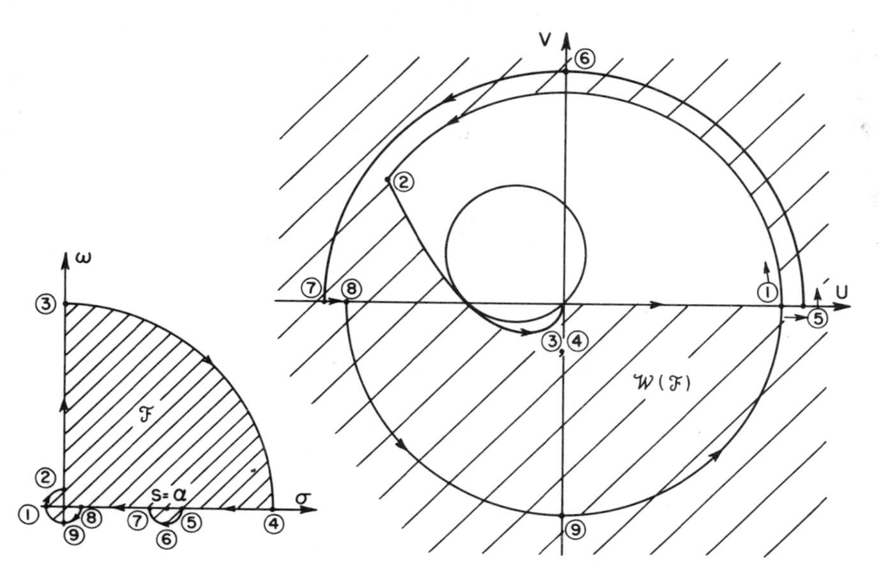

Fig. 7-14. Example 1: the off-axis circle criterion for an unstable plant.

† None of the remaining poles at $s = \lambda_i$, $i = (m + 1), \ldots, n$ may lie in $\mathfrak{F}$.

EXAMPLE 2. Given a feedback system described by Eqs. (7-1) and (7-20) with the linear plant having a transfer function

$$W(s) = \frac{s + 1}{s(s + 0.1)(s^2 + 0.5s + 9)}.$$

(a) The Nyquist plot of $W(i\omega)$ and the region $\mathcal{W}(\mathcal{R})$ are shown in Fig. 7-15, which indicates that the Hurwitz range for the stability of an LTI feedback system incorporating this plant is (0, 4.28).

(b) Since the Nyquist plot is asymptotic to a vertical line through the point $(U = -10.06,\ V = 0)$ as $\omega \to 0$ [analytically, $U(\omega) \geqslant U(0) = -815/81$], the system is stable for all nonlinear time-varying gains $g(\sigma_0, t) \in \{G_i[F, K_0]\}$ in the sector $[\varepsilon, 0.0993]$ by the circle criterion (Geometric Criterion 1).

(c) The modified Nyquist plot is also indicated in Fig. 7-15. The straight line that is almost tangential to the modified Nyquist plot at two points intersects the negative real axis at $(-2.85, 0)$, yielding a Popov gain upper bound

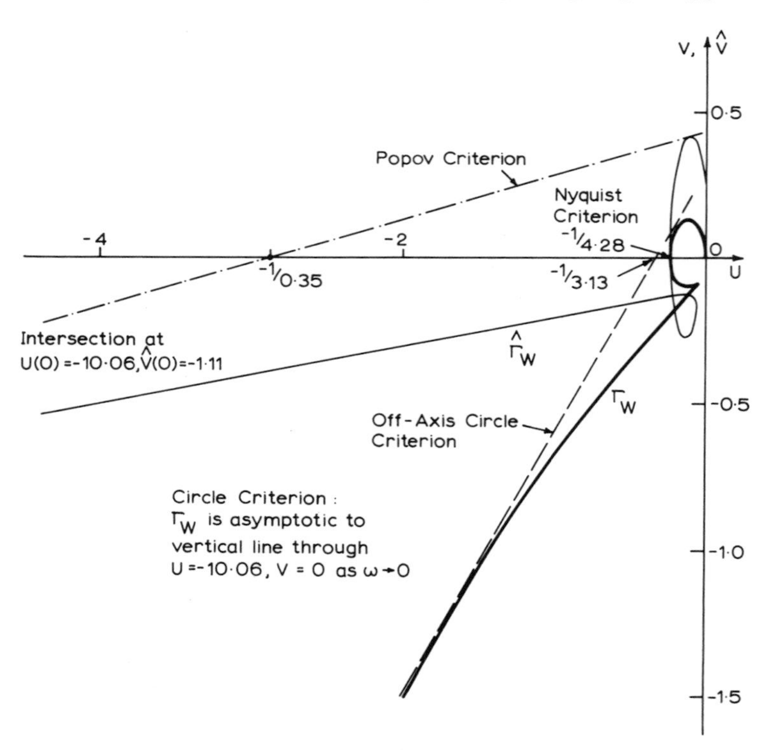

Fig. 7-15. Example 2: Comparison of stability criteria; (—) $\hat{V} = \omega \operatorname{Im} W(i\omega)$; (——) $V = \operatorname{Im} W(i\omega)$.

of $\bar{F} = 0.35$. The system is thus stable for all nonlinear functions $f(\sigma_0) \in \{F\}$ in the sector $(0, 0.35]$ by Criterion 1c (see Fig. 7-9c).

(d) A straight line nearly tangential to the Nyquist plot at two points intersects the negative real axis at $(-\bar{M}^{-1}, 0)$ where $\bar{M} = 3.13$. The system is consequently stable for all monotonic nonlinearities in the sector $(0, 3.13]$ by Geometric Criterion 4a.

EXAMPLE 3. Consider $W(s) = 3(s+1)/(s^2(s^2+s+25))$, with the lower bound on the nonlinearity being specified to be unity.

(a) The Nyquist criterion applied to the system indicates that the system is stable for all linear gains in the range $\kappa \in (0, 8)$ [Fig. 7-16].

(b) A circle $C[1, \bar{G}]$ with its center on the negative real axis is drawn passing through the point $(-1, 0)$ to be nearly tangential to the Nyquist plot. Since $C[1, \bar{G}]$ intersects the negative real axis again at $(-2.22^{-1}, 0)$, by the circle criterion the system is stable for all nonlinear and time-varying gains $g(\sigma_0, t) \in \{G_i[F, K_0]\}$ in the sector $[1, 2.22]$.

(c) A parabola satisfying the Geometric Criterion 3 passing through $(-1, 0)$ and tangential to the Popov plot $\hat{\Gamma}_W$ intersects the negative real axis at $U = -0.37$. Hence, by the parabola criterion, the system is stable for all nonlinear gains $f(\sigma_0) \in \{F\}$ in the sector $[1, 2.70]$.

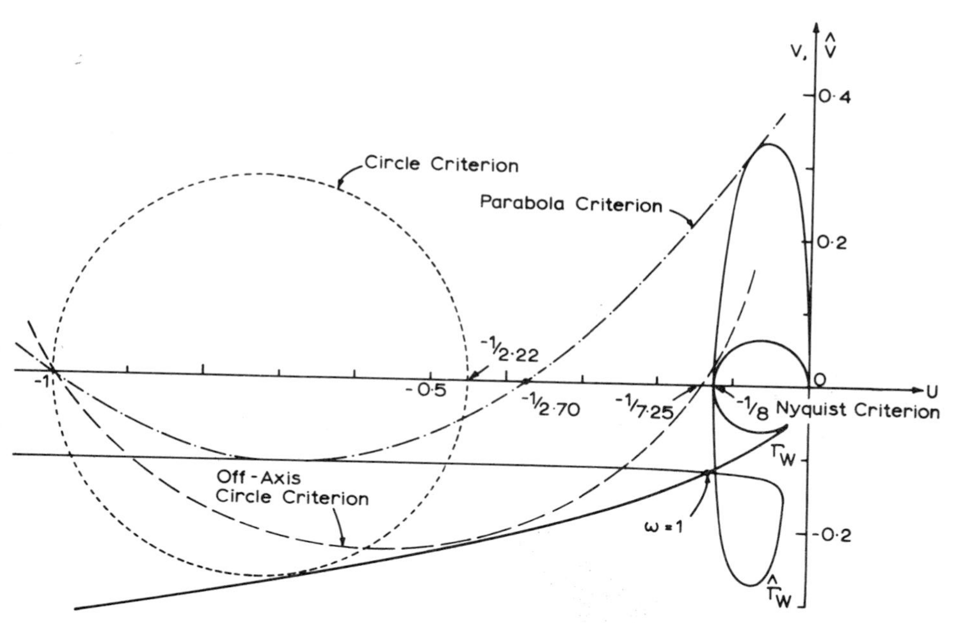

Fig. 7-16. Example 3: Further comparison of stability criteria.

(d) A circle $C_\mu[1, \bar{M}]$ passing through $(-1, 0)$ and nearly touching the Nyquist curve Γ_W at two points intersects the negative real axis at $(-7.25^{-1}, 0)$. By the off-axis circle criterion, the system is stable for all monotonic gains in the range $[1, 7.25]$.

Comments on the Off-Axis Circle Criterion

1. The circle criterion (Section 2) for nonlinear and time-varying systems calls for a circle with its center on the negative real axis. The circle criterion for monotonic nonlinearities stated in Geometric Criterion 4b calls for any circle intersecting the negative real axis at the two points $(-\underline{M}^{-1}, 0)$ and $(-\bar{M}^{-1}, 0)$. In the very rare cases where $W(i\omega)$ is symmetrical about the real axis for $\omega \geqslant 0$ (not including Γ_W for $\omega < 0$), the two criteria become identical.
2. The off-axis circle criterion may be applied in many ways in various practical situations. In particular, if the lower limit $\underline{M}$ is specified, the maximum range of stability given by the criterion can be determined by drawing a circle through $(-\underline{M}^{-1}, 0)$ and touching the Nyquist plot of $W(i\omega)$ at two points. The second intersection of the circle with the negative real axis yields $(-\bar{M}^{-1}\ 0)$, see Example 3; and similarly the best (least restrictive) lower limit $\underline{M}$ may be found if $\bar{M}$ is specified.
3. Since this criterion only assures the existence of a multiplier function $Z(s)$ such that $Z^{\pm 1}(s) \in \{Z_\theta(s)\} \subset \{Z_{\mathrm{RL}}(s)\}$, it is only sufficient and not necessary for the existence of an absolute $\{F_m\}$ Lyapunov function. In this respect, the off-axis circle criterion is analogous to the parabola criterion for the absolute $\{F\}$ stability problem. This implies that even when the Nyquist plot of $W(i\omega)$ does not satisfy Geometric Criterion 4, it may be possible to find a multiplier $Z(s)$ such that $Z^{\pm 1}(s) \in \{Z_{\mathrm{RL}}\}$ and condition (7-21) is satisfied.
4. If $W(s)$ is asymptotically stable and the lower limit $\underline{G} = 0$, the sector $\kappa \in [0, \bar{K}]$ given by the Nyquist criterion, the range $[0, \bar{M}]$ given by the monotonic criterion, the Popov sector $[0, \bar{F}]$ and the range $[0, \bar{G}]$ provided by the circle criterion can all be determined given the Nyquist plot and Popov plot of $W(i\omega)$. The bounds $\bar{K}$, $\bar{F}$, and $\bar{G}$ necessarily satisfy the inequalities

$$\bar{G} \leqslant \bar{F} \leqslant \bar{K}.$$

In general, the monotonic gain upper bound $\bar{M}$ lies between the Nyquist gain limit $\bar{K}$ and the Popov gain bound $\bar{F}$, that is, $\bar{F} \leqslant \bar{M} \leqslant \bar{K}$. There may be cases, however, where the application of the off-axis circle criterion yields a sector that is more restrictive than the Popov sector. Since $\bar{F}$ is obtained using a modified Nyquist plot and $\bar{M}$ directly from the Nyquist plot, the relative magnitude of $\bar{F}$ and $\bar{M}$ depend on the behavior of the frequency response diagrams in the vicinity of the negative real axis. It may be possible for the line specified by the off-axis circle criterion to cut the negative real axis at a point $-1/\bar{M}$ which is to the left of the Popov line intersection at $-1/E$, that is, that $\bar{M} < E$ is obtained as a result of the application of the off-axis circle criterion. In such a case the criterion is a failure since absolute $\{F\}$ stability in the sector $[0, \bar{F}]$ is also applicable to monotonic functions.

5. Further Geometric Interpretations for Time-Varying Systems

The various point or instantaneous constraints (Chapter VI, Sections 2–4) on the rate of time variation of $k(t)$ in the linear and nonlinear feedback systems treated previously all permit simple graphical interpretations if $k(t)$ is represented in the phase plane or the plot of dk/dt versus k. This plot is in wide use in the analysis of oscillations and other nonlinear phenomena by graphical and analytic techniques (see Cunningham [1]). Thus this condition may even be applied to time-varying gains that are generated as nonlinear oscillations which cannot readily be expressed in analytic form (see Fig. 7-17).

We consider only the most general criterion of this category, General Stability Criterion 2 (Chapter VI, Section 4), which deals with the general finite sector problem for nonlinear time-varying gains. We have the restriction

$$\begin{aligned} dk/dt &\leqslant \alpha_0[k(t) - \underline{K}][1 - k(t)/\bar{K}]\bar{G}_N/(\bar{G}_N - \underline{G}_N), \qquad Z(s) \in \{Z_N\} \\ dk/dt &\geqslant -\alpha_0[k(t) - \underline{K}][1 - k(t)/\bar{K}]\bar{G}_N/(\bar{G}_N - \underline{G}_N), \quad Z^{-1}(s) \in \{Z_N\} \end{aligned} \tag{6-34$^\pm$}$$

where

$$\alpha_0 \triangleq \bar{\Lambda}\underline{\Phi}. \tag{7-28}$$

Assuming that all of the parameters involved are known or can be simply estimated (some useful comments pertinent to the determination of $\bar{\Lambda}$ and $\underline{\Phi}$ are given in this section), this constraint is easily visualized in terms of k and $\dot{k}$. Two special cases occur when the ranges $[0, \bar{G}_N]$ and $[0, \infty)$ are considered. All three types of restriction are shown in Fig. 7-17. In the first and third case, we assume that $Z^{-1}(s) \in \{Z_N(s)\}$, so the condition correspond-

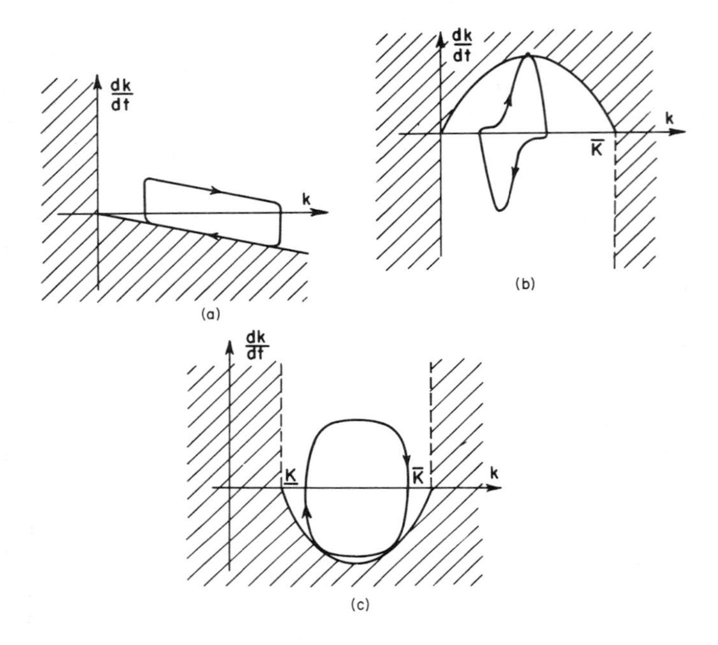

Fig. 7-17. Typical phase plane restrictions of $k(t)$: (a) infinite sector restriction $Z^{-1}(s) \in \{Z_N\}$; (b) finite sector restriction $Z(s) \in \{Z_N\}$; (c) general finite sector restriction $Z^{-1}(s) \in \{Z_N\}$.

ing to Eq. (6-34$^-$) is depicted, whereas in the second example it has been assumed that $Z(s) \in \{Z_N(s)\}$. The oscillations shown in these examples are (a) a solution of the Raleigh equation, (b) a solution of the van der Pol equation, and (c) the solution to a differential equation representing the dynamic behavior of a mass on a hard spring (see Cunningham [1]).

A. Comments on Constraint (6-34$^\pm$)

(1) It is a rare occurrence that $k(t)$ actually occupies the entire range $[\underline{K}, \bar{K}]$. In general, the shape of the phase plane portrait is not compatible with the parabolic constraint, so usually $k(t) \in [K_1, K_2]$ where $\underline{K} < K_1 < K_2 < \bar{K}$. This is particularly true when dealing with periodic functions: The phase plane trajectory of a periodic function must cross the k-axis with infinite slope, which is not permitted by the parabolic restriction on dk/dt at

$\underline{K}$ and $\bar{K}$. Thus in all of the examples indicated in Fig. 7-17, $k(t)$ occupies a range that is smaller than $[\underline{K}, \bar{K}]$.

(2) If $k(t)$ is periodic the choice of the extremes $\underline{K}$ and $\bar{K}$ in terms of the mean value of $k(t)$,

$$\kappa_0 \triangleq T^{-1} \int_0^T k(t)\, dt \tag{7-29}$$

is generally not straightforward. If the phase trajectory is not symmetric with respect to a vertical line passing through $k = \kappa_0$, it is usually easiest to do this graphically. If it is symmetric, however, we can simply choose $\underline{K}$ and $\bar{K}$ so that $\bar{K} - \kappa_0 = \kappa_0 - \underline{K}$. Sinusoidal gains give rise to elliptical phase plane plots, for example, so in Chapter VIII, Section 4B, we take this course of action. These two cases are depicted in Fig. 7-18.

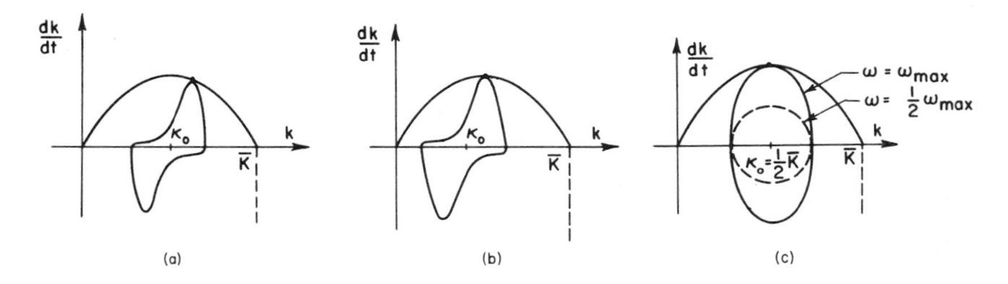

Fig. 7-18. The choice of $\kappa_0 = T^{-1} \int_0^T k(t)\, dt$ to maximize the range or frequency: (a) unsymmetric phase plane portrait: $\kappa_0 = \bar{K}/2$ is suboptimal; (b) optimal κ_0 to yield maximum range of $\kappa(t)$; (c) $\bar{K} - \kappa_0 = \kappa_0 - \underline{K}$ for symmetric phase plane plots.

(3) The constraint on dk/dt can be applied in several ways. The shape of the phase plane plot of k is determined by the wave form in the time domain; considering periodic gains, sinusoidal gains correspond to ellipses in the phase plane, while nonlinear oscillations are generally irregular (see Fig. 7-17). The horizontal dimension of the trajectory corresponds to the range $[K_1, K_2]$ of $k(t)$, while the vertical dimension is directly proportional to the range and frequency. Thus if the range is prespecified, we extend the trajectory only in the vertical direction until it touches the parabola to determine the maximum frequency that is allowed. If the frequency is fixed, then the entire portrait is enlarged until it touches the parabola which determines the maximum range permitted by the constraint. The choice of κ_0 and determination of the maximum permissible range of a gain is shown in Fig. 7-18a and b and ascertaining the maximum frequency is demonstrated in Fig. 7-18c.

(4) Finally, if $k(t)$ is of prespecified frequency ω_0, it is often advantageous to choose $\bar{\Lambda}$ first before applying the frequency domain constraint on $W(s)$. This is important, because ω_0 determines the proportions of the phase plane trajectory of $k(t)$ and it is desirable to match the shape of the parabola to that of the portrait of the gain $k(t)$. We see in Fig. 7-19 that if $\bar{\Lambda}$ is too small, then we must make poor use of the nominal range, that is, $\underline{K} \ll K_1 < K_2 \ll \bar{K}$, and if $\bar{\Lambda}$ is too large (or if $\underline{K}$ and $\bar{K}$ are too close to K_1 and K_2 in value) then the frequency domain condition which gives this large value of $\bar{\Lambda}$ is probably too restrictive. A reasonable choice of $\bar{\Lambda}$ is shown in Fig. 7-19 for the range $\underline{K}_2$, $\bar{K}_2$.

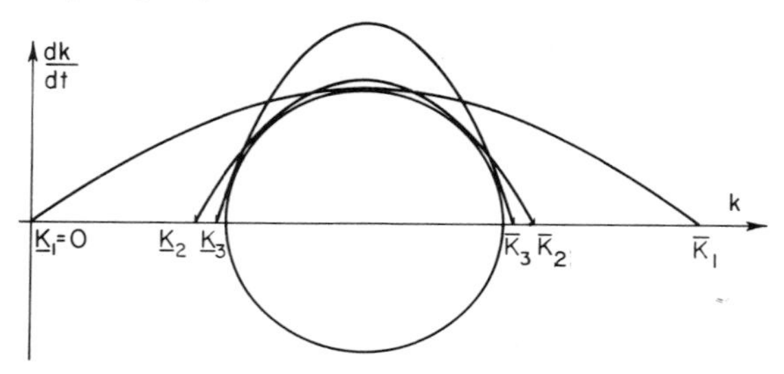

Fig. 7-19. Three choices of $\bar{\Lambda}$ for a specified gain.

If the system under consideration is completely determined, then it is a relatively simple matter to ascertain whether or not it satisfies a given absolute stability criterion. If, however, certain parameters are free (generally these are range and/or frequency) and it is desired to find the maximum values they can take while still ensuring absolute stability, then several or all of the above points must be taken into consideration and the analysis becomes quite involved. Many of these same points are made in Chapter VIII in applying this constraint analytically to the damped Mathieu equation in its linear and nonlinear forms.

With these comments and guidelines as a basis, the parameters $\bar{\Lambda}$ and $\underline{\Phi}$ for the infinite sector and finite sector cases have to be determined so that α_0 [Eq. (7-28)] may be evaluated. In the general finite sector case [Eq. (6-34$^{\pm}$)], it is also useful to know the range $[\underline{G}_N, \bar{G}_N]$ of the nonlinearity as well. The determination of $\bar{\Lambda}$ and $\underline{\Phi}$ depends upon $W(s)$ or the behavior of the linear plant, on the nonlinear behavior of the gain (on $\{N\}$), and on the corresponding class of multipliers $\{Z_N(s)\}$.

B. Nonlinear Time-Varying Gains in the First and Third Quadrants

The first step in the determination of α_0 in Eq. (7-28) is the evaluation of $\underline{\Phi}$. For a nonlinear function lying in the first and third quadrants, this parameter can be determined by inspection in some simple cases, or estimated quite accurately in other situations. A detailed discussion of this matter is provided in Chapter II, Section 1B.

Once $\underline{\Phi}$ is known, the absolute stability criterion is completely graphical in nature. The value of $\bar{\Lambda}$ may be found directly from the geometric form of the frequency domain condition. For the infinite sector and finite sector cases, the Popov line (see Geometric Criterion 2) has a slope of absolute value $|\alpha^{-1}| = \gamma_0/\beta_0 \triangleq \bar{\Lambda}$ (the slope is positive if $Z(s) \in \{Z_F(s)\}$ and negative if $Z^{-1}(s) \in \{Z_F(s)\}$). When treating the general finite sector case, the parabola criterion may be used to determine $\bar{\Lambda}$; if the parabola cuts the real axis with slope $\pm\gamma_0/\beta_0$, then $|\gamma_0/\beta_0| = \bar{\Lambda}$. For the NLTV case, the only change that has to be made in applying these graphical frequency domain conditions is that the Popov plot of $W(i\omega)$ must not touch the constraining line or parabola for all $\omega \in [-\infty, \infty]$ if $k(t)$ is not periodic (refer to Chapter VI, Section 5).

GEOMETRIC CRITERION 5. The system described by Eq. (7-1) with

$$\tau = -k(t) f(\sigma_0),$$

$$\left.\begin{array}{ll} k(t) \in \{K_1\}, & k(t) \in [\underline{K}, \bar{K}] \\ f(\sigma_0) \in \{F\}, & f(\sigma_0)/\sigma_0 \in [\underline{F}, \bar{F}] \end{array}\right\} \begin{array}{l} \underline{G}_F = \underline{K}\underline{F} \\ \bar{G}_F = \bar{K}\bar{F} \end{array} \tag{7-30}$$

is absolutely $\{G_1[F, K_1]\}$ stable if

(1) the Popov plot of $W(i\omega)$ lies outside [strictly outside if $k(t)$ is aperiodic] a parabola passing through the points $(U_1 = -(\underline{G}_F)^{-1}, \hat{V} = 0)$ and $(U_2 = -(\bar{G}_F)^{-1}, \hat{V} = 0)$, $\omega \in [-\infty, \infty]$, with slope $-\nu$ at $U = U_1$, $+\nu$ at $U = U_2$ where $\nu \neq 0$;

(2) the phase plane portrait of k lies in the semi-infinite strip bounded by $k = \underline{K}$, $k = \bar{K}$ and the concave parabola

$$\dot{k} = \nu\underline{\Phi}(k - \underline{K})(1 - k/\bar{K})\bar{G}_F/(\bar{G}_F - \underline{G}_F), \tag{7-31}$$

where $\underline{\Phi}$ is determined by $f(\sigma_0)$ [Eq. (2-11)]. □

We note that the slope of the frequency domain parabola (or Popov line if $\underline{G}_F = 0$) at $U_2 = -(\bar{G}_F)^{-1}$ determines not only $\bar{\Lambda}$ but whether $Z(s) \in \{Z_F\}$ or $Z^{-1}(s) \in \{Z_F\}$in the original analytic frequency domain restriction; thus in constraining $k(t)$, we need to specify only one parabola, Eq. (7-31), as a bound of the permitted region of $\dot{k}$ versus k in the phase plane. The sense in which this parabola is said to be concave is evident in Fig. 7-17.

For the case when $k(t)$ is completely specified, it is most effective to choose $\bar{\Lambda}$ (or ν in Geometric Criterion 5) and $[\underline{K}, \bar{K}]$ first as mentioned in Comment (4). The interrelation between $\bar{\Lambda}$ and $\bar{G}_F$ for the periodic finite sector case is shown in Fig. 7-20.

Some of these relationships are given by Brockett and Forys [1] in their discussion of a stability criterion for LTV gains. In that result (see Chapter VI, Section 2), $Z(s) = (1 + \alpha s)/(1 + \beta s)$ is used as a multiplier and $\bar{\Lambda} = \min(\alpha^{-1}, \beta^{-1})$. They noted that a Popov line of positive slope μ corresponded

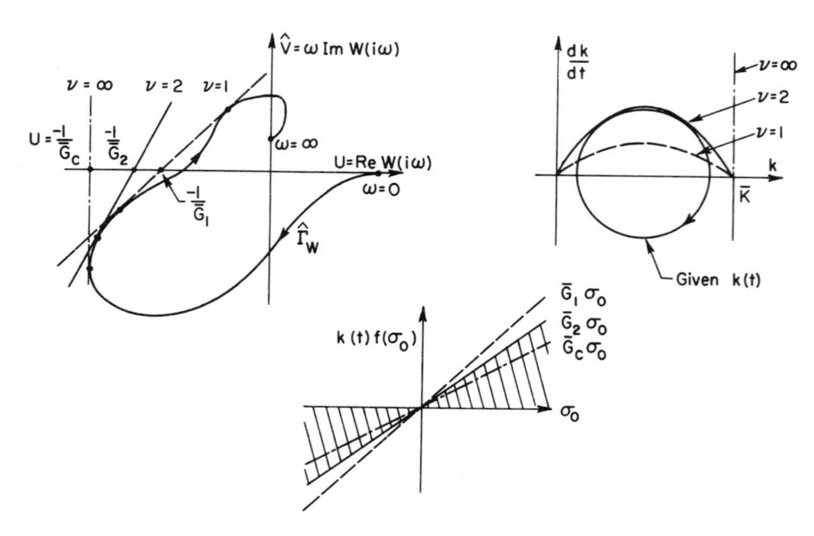

Fig. 7-20. The extended Popov criterion for periodic NLTV systems.

to $\alpha = \mu^{-1}$, $\beta = 0$, $\bar{\Lambda} = \mu$, and one of negative slope $-\hat{\mu}$ to the conditions $\alpha = 0$, $\beta = \hat{\mu}^{-1}$, $\bar{\Lambda} = \hat{\mu}$.

C. *Nonlinear Time-Varying Gains with Monotonic Nonlinear Behavior*

In this case, $f(\sigma_0) \in \{F_m\}$ (Chapter II, Section 1) and a reasonable estimate of $\underline{\Phi} = 1$ may be used. If $f(\sigma_0)$ is specified completely, however, a better estimate of $\underline{\Phi}$ may be obtained, as in the previous case, either analytically or with the aid of a computer.

The determination of $\bar{\Lambda}$ is a more difficult problem in this case as the corresponding frequency domain condition is considerably more complex. If a multiplier $Z(s)$ exists such that $Z(s)^{\pm 1} \in \{Z_{F_m}(s)\}$ satisfying the condition $H(s) \in \{\text{SPR}\}$ (Eq. (7-2)) if $k(t)$ is aperiodic or $H(s) \in \{\text{PR}\}$ if $k(t) = k(t + T)$, then $\bar{\Lambda}$ is the distance from the origin to the nearest singularity of $Z(s)$. Choosing a multiplier $Z(s)$ that maximizes $\bar{\Lambda}$ in any specific situation may be difficult to accomplish and trial and error methods may have to be resorted to.

The off-axis circle criterion (Geometric Criterion 4, Section 4) provides a method of obtaining an estimate of $\bar{\Lambda}$ (Taylor [2]). We cannot apply Geometric Criterion 4 directly, since although the criterion guarantees the existence of a multiplier $Z(s) \in \{Z_\theta(s)\} \subset \{Z_{\text{RL}}(s)\}$ satisfying the required frequency domain condition, the exact form of $Z(s)$ (and thus the multiplier margin $\bar{\Lambda}$) cannot be determined. This problem can be circumvented, if $W(s)$ is known analytically, by using a frequency domain shift: If the Nyquist plot of $W(i\omega - \lambda)$ satisfies the off-axis circle criterion (Geometric Criterion 4b) for some $\lambda > 0$, then λ is a conservative estimate of $\bar{\Lambda}$ since if $[W(s-\lambda) + \bar{M}^{-1}]Z(s) \in \{PR\}$ by Geometric Criterion 4, then $[W(s) + \bar{M}^{-1}]Z(s+\lambda) \in \{SPR\}$ (Chapter III, Section 5) and the multiplier margin of $Z(s+\lambda)$ must be at least λ. In applying this condition, we formally define $\mathfrak{F}_\lambda$ to be the s-plane region $\operatorname{Re} s \geqslant -\lambda$, $\operatorname{Im} s \geqslant 0$ suitably extended to enclose any poles on the boundary (as in Fig. 7-14). In the special case when $\underline{G}_M = 0$ and the off-axis circle reduces to a straight line, the condition is considerably easier to apply.

GEOMETRIC CRITERION 6 (Taylor [2]). The system described by Eq. (7-1) with

$$\tau = -k(t)f(\sigma_0)$$

$$\begin{array}{lll} k(t) \in \{K_1\}, & k(t) \in [\underline{K}, \bar{K}] & \underline{G}_M = \underline{K}\underline{M}, \\ f(\sigma_0) \in \{F_m\}, & \Delta f(\sigma_0)/\Delta\sigma_0 \in [\underline{M}, \bar{M}] & \bar{G}_M = \bar{K}\bar{M}, \end{array}$$

is absolutely $\{G_1\ [F_m, K_1]\}$ stable if for some $\lambda > 0$

(1) $\mathcal{W}(\mathfrak{F}_\lambda) \cap I_\mu[\underline{G}_M, \bar{G}_M] = \phi$, where $\mu \in [-\infty, \infty]$ is not zero;

(2) condition (2) of Geometric Criterion 5 is satisfied with $\nu = +\lambda$ if $C_\mu[\underline{G}_M, \bar{G}_M]$ is centered above the real axis, or $\nu = -\lambda$ if $C_\mu[\underline{G}_M, \bar{G}_M]$ is centered below the real axis, and $\bar{G}_F/(\bar{G}_F - \underline{G}_F)$ is replaced with $\bar{G}_M/(\bar{G}_M - \underline{G}_M)$. □

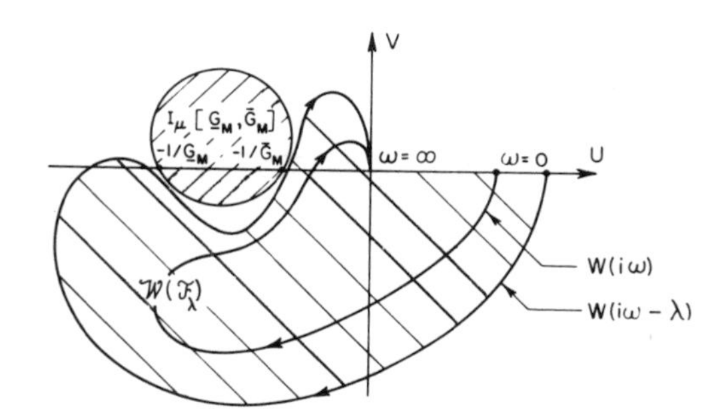

Fig. 7-21. Geometric Criterion 6: applying the shifted off-axis circle criterion to determine $\bar{\Lambda}$.

Note that even if $k(t)$ is periodic, the off-axis circle cannot be allowed to contact $W(i\omega - \lambda)$ for $\omega \geqslant 0$; refer to Geometric Criterion 4.

D. $g(\sigma_0, t) = k(t)\sigma_0$: *Three Methods for Estimating* $\bar{\Lambda}$

In the case of systems with a single time-varying gain $k(t)$ in the feedback path $\underline{\Phi} = 2$. Three methods are outlined for estimating $\bar{\Lambda}$ given the transfer function $W(s)$ of the linear part of the system in analytic form.

(i) Define $\mathcal{R}_\lambda$ to be the half plane $\operatorname{Re} s \geqslant -\lambda$, extended where necessary to enclose any poles of $W(s)$ on the line $\operatorname{Re} s = -\lambda$ as before. Locate the region $\mathcal{W}(\mathcal{R}_\lambda)$ by plotting the Nyquist diagram of $W(i\omega - \lambda)$, denoted $\Gamma_{W,\lambda}$, and locating its interior in the usual manner. Choose any $\underline{K}$ and $\bar{K}$ such that the line segment ($U = -\kappa^{-1}$, $\kappa \in [\underline{K}, \bar{K}]$, $V = 0$) is touched by $\Gamma_{W,\lambda}$ but none of the points of this segment lie in the interior of $\mathcal{W}(\mathcal{R}_\lambda)$. The system $\{W(s - \lambda), \kappa\}$ for $\kappa \in [\underline{K}, \bar{K}]$ is thus at least marginally stable, so from the argument presented in Chapter IV, Section 3, there exists some $Z(s) \in \{Z_{LC}(s)\}$ such that

$$\frac{1 + \bar{K}W(s - \lambda)}{1 + \underline{K}W(s - \lambda)} Z(s) \in \{\text{PR}\},$$

or, according to Definition PR2, Chapter III, Section 5,

$$\frac{1 + \bar{K}W(s)}{1 + \underline{K}W(s)} Z(s + \lambda) \in \{\text{SPR}\}. \tag{7-32}$$

This condition immediately yields λ as an estimate of $\bar{\Lambda}$.

(ii) A second equivalent method for obtaining the estimate of $\bar{\Lambda}$ is based on the root-locus technique. Suppose that for $\kappa \in [\underline{K}, \bar{K}]$, each of the n root loci lies on or to the left of the line defined by $\operatorname{Re} s = -\lambda$. This implies that the closed loop system

$$W_\kappa(s - \lambda) = \frac{W(s - \lambda)}{1 + \kappa W(s - \lambda)}$$

is at least marginally stable for all $\kappa \in [K, \bar{K}]$. Since this is precisely the assumption made in the previous case, λ again provides a conservative estimate of $\bar{\Lambda}$.

(iii) A third equivalent method for the determination of $\bar{\Lambda}$ uses the Hurwitz criterion (Chapter IV, Section 2). The Hurwitz criterion can be used to determine whether or not a characteristic equation has its roots in the open left half plane. If $W(s) \triangleq n(s)/d(s)$, we consider the shifted characteristic polynomial

$$\begin{aligned} p_\lambda(s) &\triangleq p(s - \lambda) = d(s - \lambda) + \kappa n(s - \lambda) \\ &= s^n + \hat{a}_n s^{n-1} + \cdots + \hat{a}_2 s + \hat{a}_1 \end{aligned} \tag{7-33}$$

and set up the array indicated in Eq. (4-33). If the n Hurwitz determinants are positive for all $\kappa \in [\underline{K}, \bar{K}]$ then the feedback system is asymptotically stable, so we find $\bar{\Lambda}$ such that $p_\lambda(s)$ has its roots in the open left half plane for $0 \leqslant \lambda < \bar{\Lambda}$.

In all three cases there is a shifted LC multiplier $Z(s) \in \{Z_\lambda(s)\}$ that satisfies condition (7-32). Since this class contains its own inverse, we may use either of the constraints ($6\text{-}34^\pm$). If the phase plane portrait of $k(t)$ is not symmetrical about the k-axis, then this choice is useful.

Geometric Criterion 7 ($g(\sigma_0, t) = k(t)\sigma_0$, general finite sector case). The system described by Eq. (7-1) with

$$\tau = -k(t)\sigma_0; \qquad k(t) \in \{K_1\}, \quad k(t) \in [\underline{K}, \bar{K}],$$

is absolutely $\{K_1\}$ stable if

(1) (a) the segment of the real axis corresponding to $U = -\kappa^{-1}$, $\kappa \in [\underline{K}, \bar{K}]$ is only touched by $\Gamma_{W,\bar{\Lambda}}$ (no point is interior to $\mathcal{W}(\mathcal{R}_{\bar{\Lambda}})$), or

(b) the root locus of $W(s)$ for $\kappa \in [\underline{K}, \bar{K}]$ lies on or to the left of the vertical line Re $s = -\bar{\Lambda}$, or

(c) the characteristic polynomial $p_\lambda(s)$ [Eq. (7-33)] satisfies the Hurwitz condition for $\kappa \in [\underline{K}, \bar{K}]$ and $0 \leqslant \lambda < \bar{\Lambda}$;

(2) condition (2) of Geometric Criterion 5 is satisfied with $\nu = \pm\bar{\Lambda}$, $\underline{\Phi} = 2$ and $\bar{G}_F/(\bar{G}_F - \underline{G}_F)$ replaced by $\bar{K}/(\bar{K} - \underline{K})$. □

The three completely equivalent criteria presented provide sufficient conditions for absolute $\{K_1\}$ stability. All three require us to know $W(s)$ analytically; empirical information as to the frequency response of $W(i\omega)$ is not useful. Which of the three parts of condition (1) is used is in part a matter of personal preference. The main difference is that the root locus technique requires a knowledge of the poles and zeros of $W(s)$, so it is necessary to factor $n(s)$ and $d(s)$ for this form and not for the others. This method has the advantage that only one root locus diagram needs to be made for $W(s)$, whereas the other two criteria may require several applications of the Nyquist or Hurwitz criterion to $W(s - \lambda)$ for various values of λ on a trial and error basis. This point again depends on the type of stability problem being considered.

VIII

THE MATHIEU EQUATION: AN EXAMPLE

The theorems developed in the foregoing chapters giving sufficient conditions for the stability of nonlinear and time-varying systems may appear to be deceptively simple. It is only when they are applied to specific problems in an attempt to derive the least restrictive results that the real difficulties involved in their utilization become apparent. Further, it is also desirable to know how these results compare with conditions which are both necessary and sufficient for the stability of the system under investigation. In most cases, deriving necessary and sufficient stability conditions is an almost impossible task, and consequently no comparisons can be made to evaluate how stringent the absolute stability conditions (which are merely sufficient) really are. One significant exception is the Mathieu equation, which has been studied extensively in the past and for which stability boundaries in parameter space have been derived.

In this chapter, some of the theorems developed in this book are applied to a damped version of the Mathieu equation as well as to a nonlinear equation derived from it. The various steps involved in applying the theorems to determine the least conservative conditions for stability are carried out in detail to acquaint the reader with the magnitude of this task. In addition, since the actual stability boundaries can be calculated for the linear problem, the conditions derived can also be compared with those given by the known stability boundaries.

Many authors have used various techniques and assumptions in deriving estimates of stability regions for the linear damped case. Since our intent is only the comparison of the stability criteria of Chapter VI with the necessary and sufficient conditions for stability, a comparison with other results is not attempted here.

The Damped Mathieu Equation

The Mathieu equation is a special case of the Hill's equation

$$d^2x/dt^2 + p(t)\,dx/dt + r(t)x = 0 \tag{8-1}$$

(where $p(t)$ and $r(t)$ are periodic time functions). This equation is central to the analysis of the infinitesimal stability of periodic solutions in a nonlinear system (see Stoker [1]). If $y_0(t)$ is a periodic solution $[y_0(t+T) = y_0(t)]$ of a second-order nonlinear differential equation, Hill's equation is obtained as the variational equation about the nominal trajectory $y_0(t)$. The differential equation considered by Mathieu is of the form†

$$\ddot{x} + [a - 2q\cos(2t)]x = 0, \tag{8-2}$$

where a and q are constant parameters. It first arose in the study of vibrational modes of a stretched membrane having an elliptical boundary. The Mathieu equation also comes up naturally in a variety of problems including wave propagation in elliptical wave guides, study of parametric amplifiers, and modulation theory. The solutions of Eq. (8-2) have been studied by many mathematicians (see McLachlan [1] for details) and the form of the stability boundaries has been established in some regions of the (a, q) plane.

The stability of two forms of the Mathieu equation are investigated in this chapter. The first is linear and contains a damping term (one involving $\dot{x}$):

$$\ddot{x} + 2\zeta\dot{x} + [a - 2q\cos(2t)]x = 0. \tag{8-3}$$

This is referred to as the linear damped Mathieu equation in the following sections. The second has a nonlinear term in x:

$$\begin{gathered}(\ddot{x} + 2\zeta\dot{x} + \zeta^2 x) + [a - \zeta^2 - 2q\cos(2t)]f(x) = 0,\\ f(x) \in \{F\}; \qquad 0 \leqslant xf(x),\end{gathered} \tag{8-4}$$

and is referred to as the nonlinear damped Mathieu equation.

Since our primary aim is to compare the sectors of stability given by the different stability criteria with those obtained using traditional analytic methods, the first section deals with some of the solution properties of the linear forms of the Mathieu equation, and the next two sections will be devoted to determining stability boundaries in the (a, q) plane by classical means.

† There are many different formulations in terms of parameter nomenclature (a, q), which makes comparison of results difficult at times. This form seems to be dominant.

1. Solutions of the Mathieu Equation

The behavior of the system described by Eq. (8-2) as determined by the parameters a and q is considered in detail by McLachlan [1], Cunningham [1], and Stoker [1], among others. We confine our attention to a few salient points that are relevant to this study.

We are interested in the stability characteristics of Eq. (8-2) only for positive values of a and q. Since the theorems to be applied require that the time-varying gain in the system be positive, negative values of a cannot be considered. The behavior of solutions for $q = q_0 < 0$ is identical to that for $\hat{q} = |q_0| > 0$, that is, the stability boundaries in the (a, q) plane are symmetric about the a-axis, and negative q thus need not be considered.

Our principal concern is with the stability boundary or the locus of transition points in the space of the parameters a and q. We first consider the undamped system [Eq. (8-2)] and the corresponding stability boundaries designated by the curves $\{\hat{q}(a, 0)\}$. Subsequently, we consider the damped Mathieu equation (8-3) and once again determine the stability boundaries, denoted by $\{\hat{q}(a, \zeta)\}$ for any specified value of damping ζ, in the same parameter space.

The boundary $\{\hat{q}(a, 0)\}$ divides the parameter space so that the values of the parameters (a, q) on one side of the curve render Eq. (8-2) unstable with exponentially growing solutions, while any value of (a, q) corresponding to a point on the other side leads to a stable system. Typical curves are depicted in Fig. 8-1; $P_1 \in \{U\}$ corresponds to solutions that grow exponentially, $P_3 \in \{S\}$ to stable solutions and $P_2 \in \{\hat{q}(a, 0)\}$ to a transition point.

Floquet theory may be used to show that the following property is enjoyed by at least one solution of Eq. (8-3) for $P \in \{\hat{q}(a, \zeta)\}$:

> For all $P \in \{\hat{q}(a, \zeta)\}$, there must exist at least one periodic solution of Eq. (8-3) with period π or 2π. Except at $(q = 0;\ a = 1, 2^2, 3^2, \ldots, n^2)$ where the Mathieu equation degenerates into the ordinary harmonic oscillator, other solutions become unbounded.

This result is basic to the analysis of the damped Mathieu equation that follows (Sections 2 and 3).

The points $P \in \{U\}$ may be associated with specific rates of exponential growth. Curves have been established by McLachlan [1] and more recently and extensively by Smirnov [1] that correspond to given growth rates in the $\{U\}$ regions. An iso-μ curve in the (a, q) plane is defined as the curve on which all points correspond to the same exponential growth rate. The point P_0 on the iso-μ curve for μ_0 then corresponds to values of a and q that lead

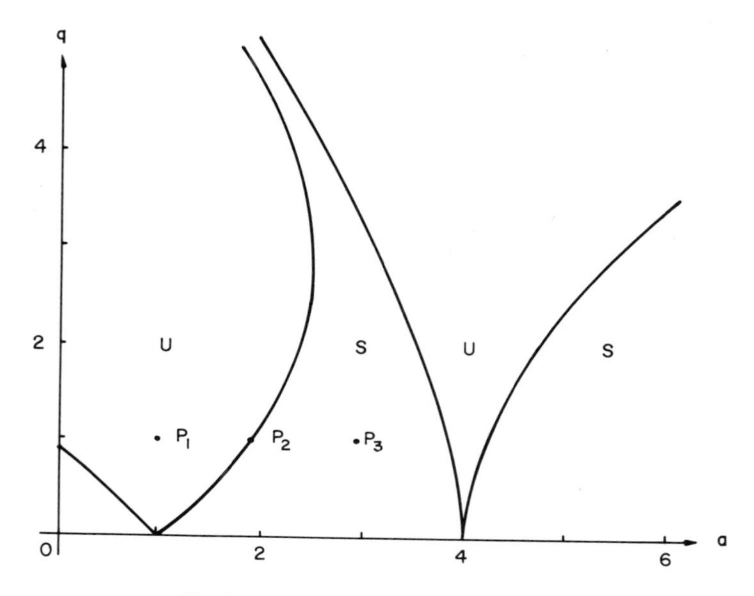

Fig. 8-1. The stability boundary $\{\hat{q}(a, 0)\}$.

to solutions of the form

$$x(t) = \exp(\mu_0 t)\phi(t),$$

where $\phi(t)$ is periodic. Iso-μ curves in their most comprehensive form are reproduced in Fig. 8-2.

If we consider the damped Mathieu equation [Eq. (8-3)] under the transformation $y(t) = e^{\zeta t}x(t)$, we obtain

$$e^{-\zeta t}[\ddot{y} + (a - \zeta^2 - 2q \cos 2t)y] = 0. \tag{8-5}$$

Equation (8-5) implies that iso-μ curves corresponding to $\mu = \zeta$ for the undamped Mathieu equation $\ddot{y} + (a - \zeta^2 - 2q \cos 2t)y = 0$ are the stability boundaries of the damped Mathieu equation (8-3). It is evident that the stability of the linear damped Mathieu equation (8-3) may be ascertained by locating the point $(a - \zeta^2, q)$ on Fig. 8-2; if this point lies below the iso-μ curve for $\mu = \zeta$, then solutions to Eq. (8-3) are stable.

2. Linear Case ($a \approx 1$): A Perturbation Analysis

In the undamped case, the solutions of the Mathieu equation are unstable for any $q \neq 0$ if $a = 1$. Introducing damping as in Eq. (8-3) produces stability for small q, as is evident from the iso-μ curves of Fig. 8-2. Our objective in

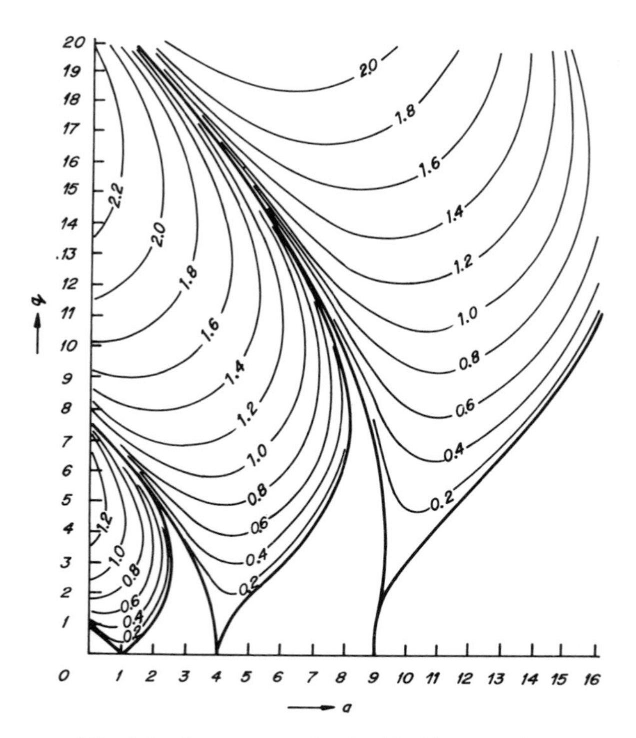

Fig. 8-2. Iso-μ curves for the Mathieu equation.

this analysis is to obtain a first-order approximation to the exact stability boundary.

The damping ζ ($\zeta \ll 1$) will be used as the perturbation parameter in this analysis. It is subsequently seen that on the boundary $\hat{q}_0 = 0(\zeta)$, so we define $q \triangleq \rho\zeta$. Substituting $x = x_0 + \zeta x_1 + \cdots$, $a = 1 + \zeta a_1 + \cdots$ into the differential equation and equating terms of equal powers of ζ, we obtain

(i) ζ^0: $\ddot{x}_0 + x_0 = 0$;

(ii) ζ^1: $\ddot{x}_1 + x_1 = [2\rho\cos(2t) - a_1]x_0 - 2\dot{x}_0$.

The solution of (i) [the generating solution] is

$$x_0 = B_0 \sin t + C_0 \cos t;$$

substituting into (ii) and using standard trigonometric identities,

$$\ddot{x}_1 + x_1 = [-(a_1 + \rho)B_0 + 2C_0]\sin t + [-2B_0 - (a_1 - \rho)C_0]\cos t + \rho[C_0\cos(3t) + B_0\sin(3t)].$$

The driving terms proportional to $\sin(t)$ and $\cos(t)$ would produce unstable solutions, as they are at the resonant frequency of the undriven system $\ddot{x}_1$

$+ x_1 = 0$. Stability to the first approximation is thus ensured if these secular driving terms may be eliminated for some (B_0, C_0) other than the trivial solution (0, 0), that is, if

$$\begin{bmatrix} -(a_1 + \rho) & +2 \\ -2 & -(a_1 - \rho) \end{bmatrix} \begin{bmatrix} B_0 \\ C_0 \end{bmatrix} = \begin{bmatrix} 0 \\ 0 \end{bmatrix}$$

for $(B_0, C_0) \neq (0, 0)$. This can only occur if

$$\begin{vmatrix} -(a_1 + \rho) & +2 \\ -2 & -(a_1 - \rho) \end{vmatrix} = {a_1}^2 - \rho^2 + 4$$
$$= ((a-1)/\zeta)^2 - (q/\zeta)^2 + 4 = 0$$

Hence for $a \approx 1$, the stability boundary is defined by the hyperbola

$$(\hat{q}_0)^2 = (a-1)^2 + (2\zeta)^2. \tag{8-6}$$

This hyperbola is asymptotic to the lines $q = \pm(a-1)$, which are the stability boundaries for the undamped case to the first approximation. It should be stressed that this boundary is only valid in the small region where $|a - 1| = 0(\zeta)$. At $a = 1$, $\hat{q}_0 = 2\zeta$. This value of $\hat{q}_0$ is important for comparison purposes in the succeeding sections.

3. Linear Case ($a \approx 4$): A Floquet Analysis

As previously noted, for values of a, q on the stability boundary, Eq. (8-3) must admit to one solution that has period π or 2π. The hyperbola for $a \approx 1$ (Section 2) may be found by an analysis like that to follow taking $T = 2\pi$ and three terms of the Fourier expansion;

$$x_0 = c_0 + d_1 \sin t + c_1 \cos t.$$

However, the boundary at $a \approx 4$ corresponds to $T = \pi$, and must be found by taking a second order approximation:

$$x_0 = a_0 + a_1 \cos(2t) + b_1 \sin(2t) + a_2 \cos(4t) + b_2 \sin(4t). \tag{8-7}$$

The solution x_0 (8-7) must be an approximate solution to the differential equation (8-3); to facilitate substitution, we inspect the last term of (8-3), namely, $[a - 2q\cos(2t)]x_0$. Using standard trigonometric identities, this can be expanded into

$$\begin{aligned} x_0[a - 2q\cos(2t)] \\ = (aa_0 - qa_1) + [aa_1 - 2qa_0 - qa_2]\cos(2t) \\ + [ab_1 - qb_2]\sin(2t) + [aa_2 - qa_1]\cos(4t) \\ + [ab_2 - qb_1]\sin(4t) - q[a_2\cos(6t) + b_2\sin(6t)]. \end{aligned} \tag{8-8}$$

To this relation, $\ddot{x}_0 + 2\zeta\dot{x}_0$ is added and the coefficients of the constant term and the $\sin(2t)$, $\cos(2t)$, $\sin(4t)$ and $\cos(4t)$ terms are set to zero. In order for the resultant fifth-order system of simultaneous equations to have solutions other than the trivial one, it is required that the following determinant be zero:

$$\begin{vmatrix} a & -q & 0 & 0 & 0 \\ -2q & (a-4) & 4\zeta & -q & 0 \\ 0 & -4\zeta & (a-4) & 0 & -q \\ 0 & -q & 0 & (a-16) & 8\zeta \\ 0 & 0 & -q & -8\zeta & (a-16) \end{vmatrix} = 0. \tag{8-9}$$

Expanding the determinant results in a quadratic equation in q^2:

$$\begin{aligned}(32 - 3a)q^4 &+ 4(a - 8)[(a - 4)(a - 16) + (4\zeta)^2]q^2 \\ &-a[(a - 4)^2 + (4\zeta)^2][(a - 16)^2 + (8\zeta)^2] = 0.\end{aligned} \tag{8-10}$$

It is no longer a simple matter to interpret this relation in the a, q parameter plane. It is not difficult to calculate a single point on the curve $\{\hat{q}(a, \zeta)\}$ for a specified value of ζ, but to obtain a segment of such a curve would be tedious without the use of computer calculations. The complexity of this result (and it is only an approximation valid for $a \approx 4$) should enforce our motivation in practical situations to try the stability criteria of Chapter VI rather than resorting directly to classical methods where iso-μ curves (Fig. 8-2) are not precomputed.

4. Application of Stability Criteria to the Linear Case

To apply the stability criteria developed in the preceding chapters, Eq. (8-3) must be expressed in the standard form,

$$\begin{aligned} W(s) &= (s^2 + 2\zeta s + \delta)^{-1}, \\ \tau(t) &= -k(t)\sigma_0(t) \triangleq -[a - \delta - 2q\cos(2t)]\sigma_0. \end{aligned} \tag{8-11}$$

Since $W(s)$ must be asymptotically stable and $k(t) \geqslant 0$ for all t, δ must satisfy the condition

$$0 < \delta \leqslant (a - 2q).$$

Further, as before, we also assume that $\zeta \ll 1$ [$\zeta = 0(0.1)$] and $a \geqslant 0(1)$. The problem, then, is to choose δ so that q in Eq. (8-11) is maximized under the constraints of the stability criteria.

A. *The Circle Criterion*

Of the three criteria we consider in this section, the circle criterion is the easiest to apply to any specific problem. In view of the simplicity of the system (8-11), we do not use the geometric criterion but instead calculate the stability sector analytically. According to the theorem given in Chapter VI, Section 5, we have:

THEOREM 1. The system governed by Eq. (8-11) is absolutely $\{K_0\}$ stable if $k(t) \in [0, \bar{K}_c]$ and $W(s)$ satisfies $[W(s) + (\bar{K}_c)^{-1}] \in \{\mathrm{PR}\}$. $\square$

If we define $\hat{h}_1 \triangleq \min_\omega h_1(\omega^2) \triangleq \min_\omega\{\mathrm{Re}\, W(i\omega)\}$, then by this theorem, the solution of the damped Mathieu equation is stable if

$$k(t) \leqslant \bar{K}_c = -(\hat{h}_1)^{-1}. \tag{8-12}$$

By inspection,

$$h_1(\omega^2) = \frac{\delta - \omega^2}{(\delta - \omega^2)^2 + (2\zeta\omega)^2}.$$

Since

$$\frac{dh_1}{d\omega} = \frac{2\omega[(\delta - \omega^2)^2 - (2\zeta)^2\delta]}{[(\delta - \omega^2)^2 + (2\zeta\omega)^2]^2},$$

the extrema of $h_1(\omega^2)$ are located at $\omega^2 = 0, \sqrt{\delta}[\sqrt{\delta} - 2\zeta]$ and $\sqrt{\delta}[\sqrt{\delta} + 2\zeta]$. The minimum is given by $\omega^2 = \sqrt{\delta}[\sqrt{\delta} + 2\zeta]$, and the corresponding value of $h_1(\omega^2)$ is

$$\hat{h}_1 = -(4\zeta[\sqrt{\delta} + \zeta])^{-1}.$$

By inequality (8-12), we require that $k(t) \leq 4\zeta[\sqrt{\delta} + \zeta]$; defining $\hat{q}_1$ to correspond to the supremum of $k(t)$, we have

$$\sup_t k(t) = a - \delta + 2\hat{q}_1 = 4\zeta[\sqrt{\delta} + \zeta].$$

The circle criterion then guarantees that the system is uniformly asymptotically stable for any $q \leq \hat{q}_1$.

The parameter δ is still unrestricted in the range $(0, a - 2q]$, so it can be chosen to maximize $\hat{q}_1$. This yields

$$\hat{q}_1 = \zeta(a - \zeta^2)^{1/2} \approx \zeta\sqrt{a} \tag{8-13}$$

which defines a stability sector in the parameter space (a, q). At $a = 1$, $\hat{q}_1 \approx \zeta$, which is just one half of the limit obtained using the perturbation method. This result was obtained earlier by Graham [1] and Narendra and Goldwyn [3].

It must be noted that the circle criterion assures absolute $\{K_0\}$ stability for all nonlinear and time-varying gains in any sector $[0, \bar{K}_c]$ where $\bar{K}_c \leq \hat{q}_1$

and not merely for the sinusoidal variations that arise in the Mathieu equation. It is therefore not surprising that the range of stability predicted by the criterion is less than that given by the perturbation analysis which deals specifically with a time-varying gain of the form $[a - \delta - 2q \cos 2t]$. Further, for values of $a \neq n^2$ $(n = 1, 2, \ldots)$ the range of q given by Eq. (8-13) is found to be quite conservative as compared to even the actual stability ranges of the undamped Mathieu equation, which are clearly within the stability regions of Eq. (8-3). This again is to be expected since the nature of the criterion is such that the best results can be expected precisely at those values of a at which the system has the smallest range of stability; the sector $\hat{q}_1 = \zeta\sqrt{a}$ must lie below the curves at their lowest points, that is, at $a \approx 1, 4, \ldots, n^2$.

B. The Extended Popov Criterion; Point Conditions

The circle criterion assures stability for a range of $k(t)$ independent of the rate at which $k(t)$ changes with time. The time-varying function under consideration is sinusoidal, however, and has a well-behaved derivative with respect to time. In such a case, the application of the extended Popov criterion (Chapter VI, Section 5) may be expected to yield a larger range of q for stability, since the criterion takes this factor into account.

THEOREM 2. The system governed by Eq. (8-11) is absolutely $\{K_1\}$ stable for $k(t) \in [0, \bar{K}_p]$ if some $\lambda > 0$ exists such that

$$H_2(s) \triangleq [W(s) + (\bar{K}_p)^{-1}](s + \lambda) \in \{\text{PR}\} \tag{8-14a}$$

and

$$dk/dt \leqslant 2\lambda k(1 - k(t)/\bar{K}_p). \quad \square \tag{8-14b}$$

Before proceeding to apply Theorem 2 to the stability problem of the damped Mathieu equation it is instructive to outline the various steps involved and interpret the restriction on $\dot{k}(t)$ in terms of the parameter q to be estimated. For a specific choice of $\bar{K}_p$, Eq. (8-14a) may be used to determine a maximum value for $\lambda \triangleq \hat{\lambda}(\bar{K}_p)$. Since the form of $k(t)$ is known, inequality (8-14b) yields a relation between q, ζ, δ, and $\bar{K}_p$ that guarantees uniform asymptotic stability; the upper bound on q can then be maximized by appropriate choices of δ and $\bar{K}_p$.

In the phase plane $(k, \dot{k})$ the trajectory of $k(t)$ must lie in the range $[0, \bar{K}_p]$; by the inequality (8-14b), this curve must lie below a parabola with a vertical axis passing through the point $(\frac{1}{2}\bar{K}_p, 0)$. Since the phase plane plot of $[a - \delta - 2q \cos(2t)]$ is an ellipse as shown in Fig. 8-3, with its major axis the vertical line through the point $(a - \delta, 0)$, to obtain the maximum value of q under the constraints (8-14b) we choose

$$\bar{K}_p = 2(a - \delta)$$

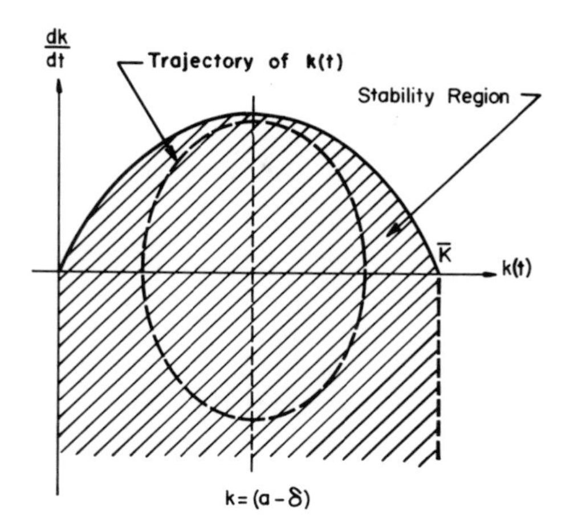

Fig. 8-3. The phase plane representation of inequality (8-14b).

in accordance with the discussion given in Chapter VII, Section 5. Defining the parameter α by $(a - \delta) \triangleq 2\alpha q$ yields

$$W(s) + (\bar{K}_p)^{-1} = (4\alpha q)^{-1} \cdot \frac{s^2 + 2\zeta s + (a + 2\alpha q)}{s^2 + 2\zeta s + (a - 2\alpha q)},$$

where by inspection $1 \leqslant \alpha < a/2q$ from the constraint on δ. If we define $h_2(\omega^2)$ to be the numerator of $\operatorname{Re}\{H_2(i\omega)\}$ (Eq. (8-14a)), we have

$$\lambda^{-1} h_2(\omega^2) \triangleq \omega^4 - 2\omega^2[a - 2\zeta^2 - 4\alpha q\zeta/\lambda] + [a^2 - (2\alpha q)^2].$$

For $H_2(s) \in \{\text{PR}\}$, it is necessary that $h(\omega^2) \geq 0$ for all $\omega \in (-\infty, \infty)$; this condition is satisfied if λ satisfies

$$\lambda \leq \hat{\lambda} \triangleq \frac{4\alpha q\zeta}{(a - 2\zeta^2) - (a^2 - (2\alpha q)^2)^{1/2}}. \tag{8-15}$$

Substituting $\hat{\lambda}$ into Eq. (8-14b) yields the requirement that $q \leq \hat{q}_2$, where $\hat{q}_2$ satisfies

$$f(\theta) \triangleq \frac{\sin\theta}{\alpha^2 - \cos^2\theta} \leqslant \frac{2\zeta\hat{q}_2}{(a - 2\zeta^2) - (a^2 - (2\alpha\hat{q}_2)^2)^{1/2}}, \tag{8-16}$$

where $\theta \triangleq 2t$. The maximum $\hat{f}$ of the left hand side of inequality (8-16) occurs at $\theta_1 = \pi/2$ if α lies in the range $1 \leqslant \alpha < \sqrt{2}$, or at two points $\theta_2 \in (0, \pi/2)$ and $\theta_3 \in (\pi/2, \pi)$ for $\alpha \geqslant \sqrt{2}$ and has the values

$$\hat{f} = \begin{cases} [2(\alpha^2 - 1)^{1/2}]^{-1}, & 1 \leqslant \alpha < \sqrt{2}, \\ (\alpha^2)^{-1}, & \sqrt{2} \leqslant \alpha. \end{cases} \tag{8-17}$$

To solve for $\hat{q}_2$ using the relation (8-16), it is useful to make the approximation

$$(a^2 - (2\alpha\hat{q}_2)^2)^{1/2} \approx a[1 - \tfrac{1}{2}(2\alpha\hat{q}_2/a)^2],$$

which is valid, as it turns out that $(2\alpha\hat{q}_2/a)^2 \ll 1$. This yields $\hat{q}_2$ (the optimum upper bound on q) defined by

$$\hat{f} = a\zeta\hat{q}_2/((\alpha\hat{q}_2)^2 - a\zeta^2). \tag{8-18}$$

It may be seen that the maximum value of $\hat{q}_2$ occurs at $\alpha_0 \triangleq [1 + a/(1 + a)]^{1/2} \in [1, \sqrt{2})$ for $a \in [0, \infty)$, and this is

$$\hat{q}_2 = \zeta(a(1 + a))^{1/2}. \tag{8-19}$$

For large values of a we have $\hat{q}_2 \approx a\zeta$, so that since $\alpha < \sqrt{2}$ and $a \geqslant 0(1)$, the assumption $(2\alpha q/a)^2 \ll 1$ is valid.

For large values of a, the stability range given by Eq. (8-19) is found to be significantly larger than that given by the circle criterion (Eq. (8-13)). Further comparisons are made subsequently.

C. *The Extended Popov Criterion; Integral Conditions*

Since this criterion involves a time-averaged condition, it imposes even less stringent requirements on $k(t)$ than the point criterion of Theorem 2 [Eq. (8-14b)]. In the problem under consideration, this leads to a larger value of $\hat{q}$, the upper bound on q that ensures absolute stability.

Since General Stability Criterion 4 is more complex to apply than Theorem 2, we consider only the infinite sector problem; $0 < k(t) < \bar{K} < \infty$, where $\bar{K}$ is arbitrarily large.

THEOREM 3. The system governed by Eq. (8-11) is absolutely $\{K_2\}$ stable for $k(t) \in (0, \bar{K} < \infty)$ if $W(s - \mu)$ is asymptotically stable for some $\mu > 0$, if some $\lambda > 0$ exists such that $H_3(s) \triangleq \{W(s - \mu)[s + \lambda - \mu]\} \in \{\text{PR}\}$ and if

$$p(t) \triangleq \sup\{-2\mu; -[2\lambda - k^{-1}(dk/dt)]\} \tag{8-20}$$

satisfies

$$I \triangleq \int_0^T p(t)\,dt < 0. \ \square \tag{8-21}$$

The inequality (8-21) is used since $k(t)$ (and thus $p(t)$) is periodic.

If $\hat{\lambda}$ is the largest value of λ such that $H_3(s) \in \{\text{PR}\}$ and $\hat{q}_3$ is defined as that value of q such that

$$\int_0^T \hat{p}(t)\,dt = 0 \qquad \text{where} \quad \hat{p}(t) \triangleq \sup\{-2\mu; -[2\hat{\lambda} - k^{-1}(dk/dt)]\},$$

then Theorem 3 ensures the uniform asymptotic stability of the solutions of the Mathiew equation for all $q < \hat{q}_3$.

From Eq. (8-11), $W(s - \mu)$ is asymptotically stable if (i) $\mu < \zeta$, and (ii) $\zeta^2 \leqslant \delta$, and $H_3(s) \in \{\text{PR}\}$ for any $\lambda \leqslant \hat{\lambda}$ given by (iii) $\hat{\lambda} = 2\zeta - \mu$. The requirements (i) and (iii) are simultaneously satisfied for any $\alpha \in (0, 1)$ if we define

$$\mu \triangleq (1-\alpha)\zeta, \qquad \hat{\lambda} \triangleq (1+\alpha)\zeta. \tag{8-22}$$

The function $p(t)$ can now be expressed in terms of α, ζ, and $k(t)$:

$$\hat{p}(t) = \sup\left\{-2(1-\alpha)\zeta; \quad -\left[2(1+\alpha)\zeta - \frac{4q\sin(2t)}{a-\delta-2q\cos(2t)}\right]\right\} \tag{8-23}$$

The main difficulty now lies in choosing α and δ such that q is maximized while satisfying inequality (8-21). If δ is chosen as $\delta \approx \zeta^2 \ll a$, however, then $p(t)$ may be simplified considerably by setting $a - \delta - 2q\cos 2t \approx a$, which is valid since $q = 0(a\zeta) \ll a$; hence consider

$$\tilde{p}(t) \triangleq \sup\{-2(1-\alpha)\zeta; \quad -[2(1+\alpha)\zeta - (4q/a)\sin 2t]\}. \tag{8-24}$$

The form of $\tilde{p}(t)$ is shown in Fig. 8-4. Inequality (8-21) may be expressed as

$$\begin{aligned}\tfrac{1}{2}\tilde{I} &\triangleq \tfrac{1}{2}\int_0^T \tilde{p}(t)\,dt \\ &= \int_0^{x_1} \{-2(1+\alpha)\zeta + (4q/a)\cos x\}\,dx - 2(1-\alpha)\zeta(\pi - x_1) < 0\end{aligned}$$

by the change of variable indicated in Fig. 8-4, where x_1 lies in the range $0 \leqslant x_1 \leqslant \pi/2$ and is defined by

$$2(1-\alpha)\zeta = 2(1+\alpha)\zeta - (4q/a)\cos x_1.$$

The value of α that minimizes the integral of $\tilde{p}(t)$ is $\alpha = 0$, and the corresponding value of $\tilde{I}$ is $\frac{1}{2}\tilde{I} = 4q/a - 2\pi\zeta$. This implies that for all $q < \hat{q}_3$ where

$$\hat{q}_3 = \frac{\pi}{2}a\zeta \tag{8-25}$$

the system (8-3) is uniformly asymptotically stable.

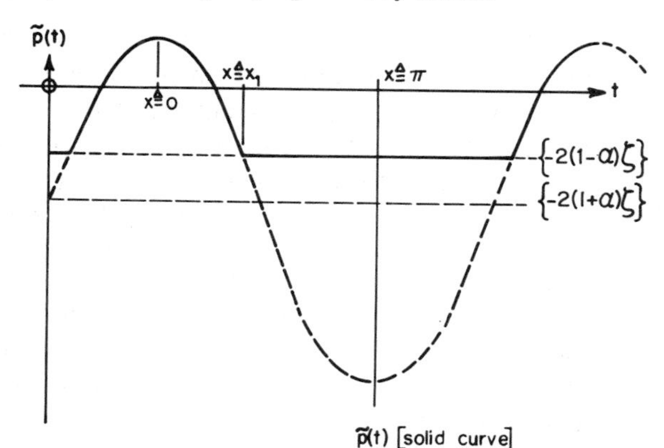

Fig. 8-4. The determination of $\tilde{p}(t)$.

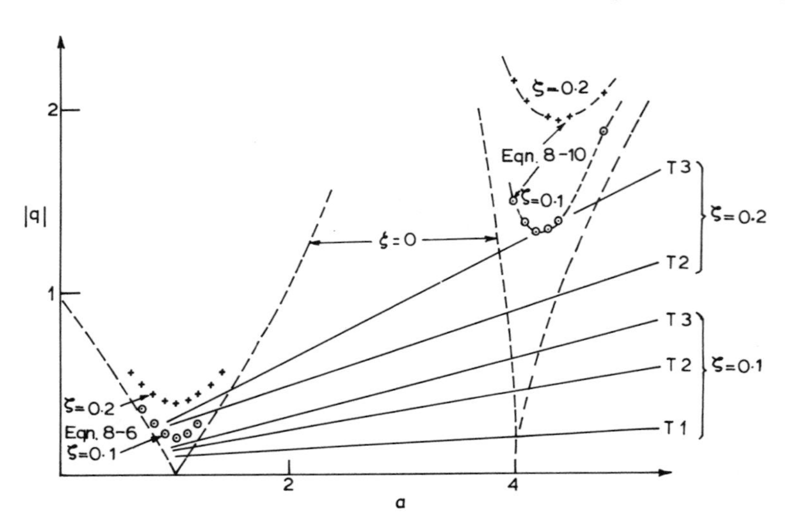

Fig. 8-5. Application of three absolute stability criteria to the linear damped Mathieu equation.

For $a < a_0 \triangleq .681$, Theorem 3 seems to yield a smaller value of $\hat{q}$ than Theorem 2 (a_0 corresponds to the value of a when $(a(a+1))^{1/2} = (\pi/2)a$). This occurs because $\bar{K}$ was assumed to be arbitrarily large, which is very restrictive for $a \leqslant 0(1)$. It can be seen, however, that as $\mu \longrightarrow 0$, the stability conditions of Theorem 3 are the same as those of Theorem 2, so it is never more strict if applied in the same manner. The three stability boundaries are shown in Fig. 8-5.

D. Comments on the Stability Theorems

The preceding analyses indicate how the different stability theorems developed in Chapter VI for general nonlinear time-varying problems can be applied to a specific problem, that is, the damped Mathieu equation. It is clear that Theorem 1 (the circle criterion) is the most direct in application, even without recourse to the simple frequency domain geometric interpretation of Chapter VII. Condition (8-13) can be used for all values of a and ζ, since no simplifications requiring $\zeta \ll a$ were necessary.

Theorem 2 is more complex to apply than Theorem 1 and simplifying assumptions which are valid for $\zeta \ll 1$ had to be made to render the problem analytically tractable. Theorem 3 requires further simplifying assumptions (that is, $\bar{K} = \infty$) if an analytic expression of the maximum value $\hat{q}$ of q in terms of a and ζ is to be obtained. It must be noted, however, that if values of a and ζ are specified and a numerical estimate of $\hat{q}$ is desired, Theorems 2 and 3 are still applicable. A digital computer may have to be used in such a case.

As mentioned earlier, these three stability theorems can be applied to any time-varying system, resulting in sufficient conditions for absolute stability. Such an application generally leads to the determination of a large region in the appropriate parameter space that ensures absolute stability, while classical techniques often only give small segments of stability boundaries. This point is one of the strongest arguments that can be made in favor of the use of absolute stability criteria for the analysis of simple systems. (If a system is of higher order or of greater complexity, the use of most classical methods is virtually out of the question.) Although the results of Section 4 are presented mainly for illustrative purposes, they have theoretical interest as well: The iso-μ curves discussed earlier have not been obtained (to the best of the authors' knowledge) for $a > 16$, whereas the use of Theorem 3 has resulted in the definition of a sector of absolute stability that is of infinite extent [$a \in (0(1), \infty)$].

In judging the efficacy of these criteria, it is of particular interest to compare the result given by their application with the necessary and sufficient condition for stability obtained by a classical analytic technique. In considering the damped Mathieu equation, the points of greatest importance correspond to $a \approx 1, 4, \ldots, n^2$, since the known stability boundaries approach the parameter-space absolute stability boundaries most closely at those points. In making such a comparison, we use $a = 1$, $\zeta = 0.1$, $q = 0.2$ as obtained in Section 2, and $a = 4.4$, $\zeta = 0.2$, $q = 1.94$ as obtained from Eq. (8-10) (refer to Fig. 8-5). The pertinent information is given in Tables 8-1 and 8-2.

The result of the analysis at $a = 1$ (Table 8-1) shows that the circle criterion is the least effective, but not decisively so. The difference between the per-

TABLE 8-1 COMPARISON OF STABILITY BOUNDARIES, $a = 1, \zeta = 0.1$

Stability method	$\hat{q}$	NASC (%)
Theorem 1	0.1000	50.0
Theorem 2	0.1414	70.7
Theorem 3	0.1571	78.5
Perturbation analysis	0.2000	—

TABLE 8-2 COMPARISON OF STABILITY BOUNDARIES, $a = 4.4, \zeta = 0.2$

Stability method	$\hat{q}$	NASC (%)
Theorem 1	0.4195	21.6
Theorem 2	0.9749	50.2
Theorem 3	1.3823	71.3
Floquet analysis	1.94	—

centages 70.7% (Theorem 2) and 78.5% (Theorem 3) obtained by Theorems 2 and 3 is not significant. A much more compelling comparison may be made at $a = 4.4$; in Table 8-2 we see that the circle criterion is extremely conservative for large values of a since $\hat{q}_1 = \zeta\sqrt{a}$, whereas $\hat{q}_2 \approx a\zeta$ and $\hat{q}_3 = (\pi/2)a\zeta$ for $a \gg 1$. Theorem 3 gives an upper bound on q that surpasses that obtained using Theorem 2 by the factor $\pi/2$ (57%) for large a.

The primary aim of this analysis is to indicate in a rather detailed fashion how the stability theorems can be applied in a specific situation. Comparison of the results obtained with actual stability boundaries indicates how effective the various theorems are. For more complex problems where the actual stability regions may be impossible to derive theoretically, the results provide a qualitative idea of what might be expected from an application of these criteria.

5. Application of Stability Criteria to the Nonlinear Case

We consider in this section the stability of Eq. (8-4) where $f(x)$ is any nonlinear function in the first and third quadrants of the $x, f(x)$ plane, that is, $0 \leqslant f(x)/x$. The advantages of the stability theorems of the form used here become particularly evident while dealing with problems of this type where stability boundaries are not known. The application of classical techniques as in Section 3 is significantly more complicated than in the linear case, if not impossible. Also, it is not evident that these classical stability regions coincide with the actual stability regions of the problem, that is, that they are both necessary and sufficient. The stability theorems 2 and 3 of Section 4, on the other hand, can be applied directly to the problem with but slight modifications. The circle criterion, however, cannot be used since the nonlinear function lies in the infinite sector and Re $W(i\omega)$ cannot be made nonnegative for any choice of δ.

THEOREM $\hat{2}$. The system governed by Eq. (8-4) is absolutely $\{G_1[F, K_1]\}$ stable for $k(t) \in [0, \bar{K} < \infty]$ if some $\lambda > 0$ exists such that

$$\hat{H}_2(s) \triangleq W(s) \cdot (s + \lambda) \in \{\text{PR}\} \tag{8-26}$$

and

$$dk/dt \leqslant \lambda \underline{\Phi} k(t) \tag{8-27}$$

where

$$\underline{\Phi} \triangleq \min_x \left\{ x f(x) \Big/ \int_0^x f(\xi)\, d\xi \right\}. \ \square$$

THEOREM $\hat{3}$. The system governed by Eq. (8-4) is absolutely $\{G_1[F, K_2]\}$ stable for $k(t) \in (0, \bar{K} < \infty)$ if $W(s - \mu)$ is asymptotically stable for some

$\mu > 0$, if some $\lambda > 0$ exists such that

$$\hat{H}_3(s) \triangleq W(s-\mu)\cdot(s+\lambda-\mu) \in \{\text{PR}\},$$

and if

$$p(t) \triangleq \sup\{-2\mu; -[\lambda\underline{\Phi} - k^{-1}\,dk/dt]\} \tag{8-28a}$$

satisfies

$$\hat{I} \triangleq \int_0^T p(t)\,dt < 0. \ \square \tag{8-28b}$$

Since Theorem $\hat{2}$ is essentially a special case of Theorem $\hat{3}$, we treat the application of Theorem $\hat{3}$ in detail and derive the corresponding result for Theorem $\hat{2}$ in passing.

The preliminary development is identical to the linear case (Section 4C) except that $\underline{\Phi}$ replaces 2 where it is appropriate. From Eq. (8-22), $W(s-\mu)$ is asymptotically stable and $\hat{H}_3(s)$ is positive real if for $\alpha \in (0, 1)$ μ and λ satisfy

$$\mu \triangleq (1-\alpha)\zeta, \qquad \lambda \triangleq (1+\alpha)\zeta.$$

Using these definitions for the parameters λ and μ in terms of the auxiliary parameter α, $\hat{I}$ may be expressed as

$$\tfrac{1}{2}\hat{I} = \int_0^{x_1} \{-(1+\alpha)\underline{\Phi}\zeta + (4q/a)\cos x\}\,dx - 2(1-\alpha)\zeta(\pi - x_1). \tag{8-29}$$

Defining $\beta \triangleq 4q/(a\zeta)$, x_1 in Eq. (8-29) is given by

$$\beta \cos x_1 = (\underline{\Phi} - 2) + \alpha(\underline{\Phi} + 2), \tag{8-30}$$

as depicted in Fig. 8-4. If β_3 represents the value of β for which $\hat{I}$ in Eq. (8-29) becomes zero, by the application of Theorem $\hat{3}$, absolute stability is ensured for all $q < \hat{q}_3 \triangleq a\beta_3\zeta/4$. Corresponding to every value of α, $x_1(\alpha)$ and $\beta(\alpha)$ can be determined by solving the simultaneous equations that arise from setting $\hat{I} = 0$ and from the definition of x_1:

$$\begin{aligned} \beta_3 \cos x_1 &= (\underline{\Phi} - 2) + \alpha(\underline{\Phi} + 2), \\ \beta_3(\sin x_1 - x_1 \cos x_1) &= 2\pi(1-\alpha). \end{aligned} \tag{8-31}$$

The final step is the maximization of β_3 with respect to α. It can be shown that the only zero of $d\beta_3/d\alpha$ occurs for $\hat{x}_1 = 2\pi/(\underline{\Phi} + 2)$ and corresponds to a maximum of $\beta_3(\alpha)$. For $\underline{\Phi}$ in the interval $(0, \infty)$, $\hat{x}_1 \in (0, \pi)$. As long as $\hat{x}_1 \in (0, \pi/2)$, it is possible to show using Eq. (8-31) that $\alpha \in (0, 1)$. However, if $\hat{x}_1 \in [\pi/2, \pi)$, α is zero or negative; hence, α must be chosen to be greater than zero, since α by definition has to lie in the range (0, 1). Thus there are seen to be two cases, that is, $2 < \underline{\Phi} < \infty$ and $0 \leqslant \underline{\Phi} \leqslant 2$, which must be considered separately.

(a) $2 < \underline{\Phi} < \infty \qquad [\hat{x}_1 \in (0, \pi/2)]$:

In this case,

$$\hat{\beta}_3 = \frac{4\pi\underline{\Phi}}{(\underline{\Phi}+2)\sin(2\pi/(\underline{\Phi}+2))}. \tag{8-32}$$

(b) $0 \leqslant \underline{\Phi} \leqslant 2$ $\quad [\hat{x}_1 \in [\pi/2, \pi]]$:

The parameter α must be equal to 0^+; hence $\hat{\beta}_3$ may be obtained by solving the simultaneous equations

$$\hat{\beta}_3 \cos x = (\underline{\Phi} - 2) < 0$$
$$\hat{\beta}_3(\sin x - x\cos x) = 2\pi.$$

Substituting $y = (\pi - x)$ yields the solution

$$\tan y = y + \pi\underline{\Phi}/(2 - \underline{\Phi}).$$

In Fig. 8-6, values of y are found for $\underline{\Phi} = 0.5$, 1.0, and 1.5 by a graphical method and the corresponding values of $\hat{\beta}_3$ are calculated. Using Eq. (8-32) values of $\hat{\beta}_3$ are also calculated for $\underline{\Phi} \in (2, \infty)$. Table 8-3 indicates how $\hat{\beta}_3$ varies with $\underline{\Phi}$.

TABLE 8-3

$\underline{\Phi}$	y	$\hat{\beta}_3$
0.	0.	2.0
0.5	1.1425	3.61
1.0	1.352	4.61
1.5	1.479	5.45
2.0	—	6.28
2.5	—	7.09
3.0	—	7.93
4.0	—	9.67
5.0	—	11.50

It must be noted that as $\mu \longrightarrow 0$ (or $\alpha \longrightarrow 1$), the stability requirements of Theorem $\hat{3}$ become those of Theorem $\hat{2}$. In particular the integral criterion (8-28b) becomes $\{-2\underline{\Phi} + \beta\cos x\} < 0$, which is satisfied for all x if

$$\beta < \hat{\beta}_2 \triangleq 2\underline{\Phi}.$$

The values of $\hat{q}/a\zeta$, which according to Theorem $\hat{2}$ and Theorem $\hat{3}$ guarantee absolute stability, are plotted as functions of $\underline{\Phi}$ in Fig. 8-7. Theorem $\hat{3}$ is found to give a considerably larger upper bound for q ($q = \beta a\zeta/4$) for $\underline{\Phi}$ in the neighborhood of 2. The improvement is particularly significant for $0 \leqslant \underline{\Phi} \leqslant 2$.

The primary goal of the preceding analysis is to demonstrate the various steps involved in the application of the absolute stability theorems to a specific nonlinear time-varying system. The analysis indicates that the upper

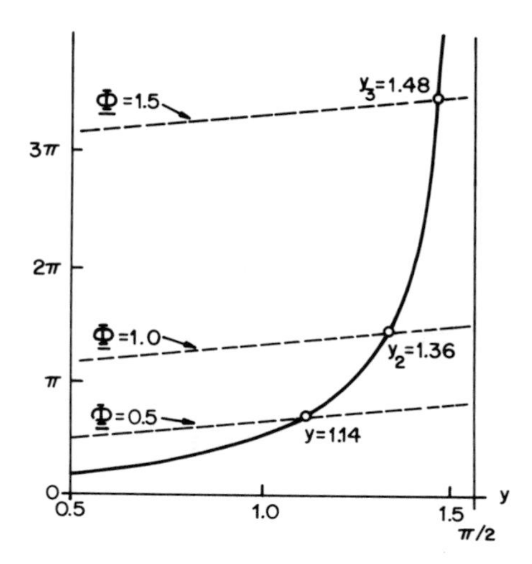

Fig. 8-6. The graphical solution of $\tan y = \pi\underline{\Phi}/(2 - \underline{\Phi}) + y$.

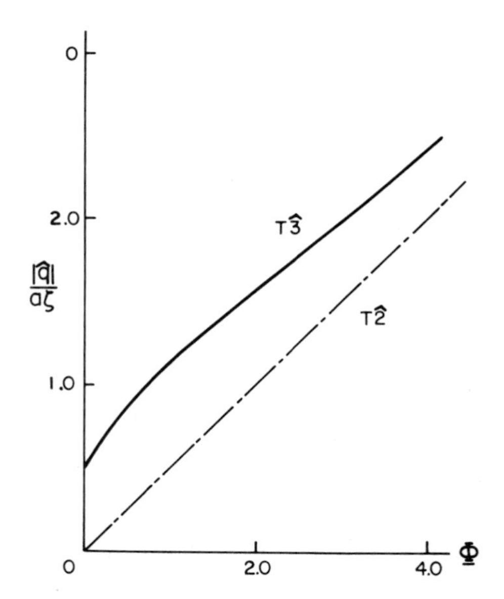

Fig. 8-7. The application of point and time-averaged constraints on $k(t)$.

bound on the time-varying part of the gain (that is, on q) can be derived as a function of $\underline{\Phi}$ and that the increased complexity of the analysis required by Theorem $\hat{3}$ as compared to Theorem $\hat{2}$ may be justified by the considerably less stringent stability condition obtained.

IX

ABSOLUTE STABILITY OF SYSTEMS WITH MULTIPLE NONLINEAR TIME-VARYING GAINS

1. Introduction

An obvious and significant question that arises in considering the introductory comments and results of the previous chapters is the possibility of generalizing the stability criteria derived therein to render them applicable to systems with more than one gain that is nonlinear, time-varying, or both. The development of stability criteria for such systems has lagged behind the generation of comparable results for the scalar case considered heretofore; this is due in part to the difficulty encountered in extending the Kalman–Yakubovich Lemma. Despite the problem of establishing appropriate matrix analogs to Lemmas 1–5 of Chapter III, Section 4, however, preliminary efforts make it seem reasonable to expect that most of the criteria of the previous chapters may be generalized in a quite straightforward manner.

The developments presented here are necessarily quite skeletal. The basic definitions and tools are followed by the derivation of stability criteria; the organization of this material represents an abbreviated version of Chapters II–VI.

2. Problem Statement

In this section, a precise description of systems to be considered subsequently is established. The definitions of stability (Chapter II, Section 2) and the formal problem statement (Chapter II, Section 3) remain essentially unchanged except for the direct extension involved in the concept of absolute $\{\underline{G}_i\}$ stability; these are not repeated here.

A. System Definition

The primary working system description is again given by the state vector differential equation†

$$\begin{aligned}
&\dot{x} = Ax + Bp; \qquad (A, B) \text{ completely controllable}, \quad A \in \{A_1\} \\
&r = Cx + Rp; \\
&p = -\underline{g}(r, t) \triangleq -\underline{G}(r, t)r; \\
&g_j(r_j, t) \in \{N_j, T_j\}, \qquad \underline{G} \in [\underline{0}, \bar{\underline{G}}_N].
\end{aligned} \tag{9-1}$$

A, B, C, and R are real matrices of dimension $(n \times n)$, $(n \times m)$, $(m \times n)$ and $(m \times m)$, respectively. The properties of these matrices are further discussed in Section B, and the gain classes and bounds are defined in Section C below.

By direct application of the Laplace transform we obtain

$$\begin{aligned}
L(r) = [R + C(sI - A)^{-1}B]L(p) &\triangleq \underline{W}(s)L(p), \\
p &= -\underline{g}(r, t),
\end{aligned} \tag{9-2}$$

which provides an equivalent representation of the system described by Eq. (9-1). As before, we refer to $\underline{W}(s)$ as the LTI plant and $\underline{g}(r, t)$ as the nonlinear time-varying controller for convenience; this nomenclature is depicted in Fig. 9-1.

A special case of the general form given in Eq. (9-1) is discussed in Lemma 9-1. The input matrix B in this case has rank one. As shown by Gilbert [1], this implies that there is effectively only one input to the LTI plant, as is evident in Fig. 9-2, which shows the structure of a system in this class. We

† The underline is used to indicate a change in dimensionality notation from chapters I–VIII. ($\underline{W}(s)$ denotes an $(m \times m)$ matrix of transfer functions; $\underline{f}(r)$ or $\underline{g}(r, t)$ denotes a vector function of the vector r, namely $\underline{g}(r, t) = \text{col}\,[g_1(r_1, t), \ldots, g_m(r_m, t)]$; $\bar{\underline{G}}_N$ represents the diagonal matrix of upperbounds of $G(r, t)$ in the appropriate sense; $\underline{0}$ is the null matrix.) $G(r, t) \in [\underline{0}, \bar{\underline{G}}_N]$ implies that $\underline{0} \leqslant G(r, t) \leqslant \bar{\underline{G}}_N$ or that $G(r, t)$ and $(\bar{\underline{G}}_N - G(r, t))$ are positive semidefinite matrices. As in the scalar case, this inequality is understood to be more restrictive for any element $g_j(r_j, t)$ which is monotonic.

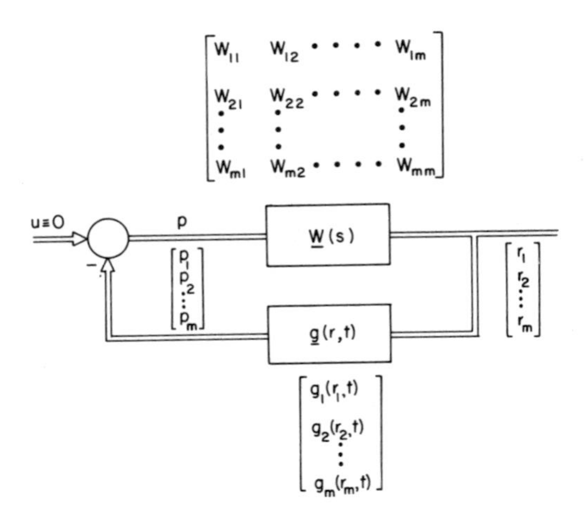

Fig. 9-1. A general system in transfer function form.

henceforth refer to this special form of the system as SF1; it is a useful particular case since many systems can be represented by a single nth order differential equation with m nonlinear time-varying terms and because the frequency domain conditions for absolute stability are notably simplified (Lemma 9-2).

LEMMA 9-1. Any system that may be described by the NLTV differential equation

$$[D^n + \sum_{j=1}^{n} a_j D^{j-1}]\xi + \sum_{i=1}^{m} g_i[\{\rho_i D^n + \sum_{j=1}^{n} d_{ij} D^{j-1}\}\xi, t] = 0 \tag{9-3}$$

may be equivalently represented by the first-order state vector differential equation (9-1), where $\hat{C} \triangleq [c_{ij} \triangleq d_{ij} - \rho_i a_j]$,

$$\hat{R} \triangleq \begin{bmatrix} \rho_1 & \rho_1 & \cdots & \rho_1 \\ \rho_2 & \rho_2 & \cdots & \rho_2 \\ \cdot & & & \cdot \\ \cdot & & & \cdot \\ \cdot & & & \cdot \\ \rho_m & \rho_m & \cdots & \rho_m \end{bmatrix}, \qquad \hat{A} \triangleq \left[\begin{array}{c|ccc} 0 & & & \\ 0 & & & \\ \cdot & & I & \\ \cdot & & & \\ \cdot & & & \\ 0 & & & \\ \hline -a_1 & -a_2 & \cdots & -a_n \end{array}\right],$$

$$\hat{B} \triangleq \begin{bmatrix} 0 & 0 & \cdots & 0 \\ 0 & 0 & \cdots & 0 \\ \cdot & \cdot & & \cdot \\ \cdot & \cdot & & \cdot \\ \cdot & \cdot & & \cdot \\ 0 & 0 & \cdots & 0 \\ 1 & 1 & \cdots & 1 \end{bmatrix}; \tag{9-4}$$

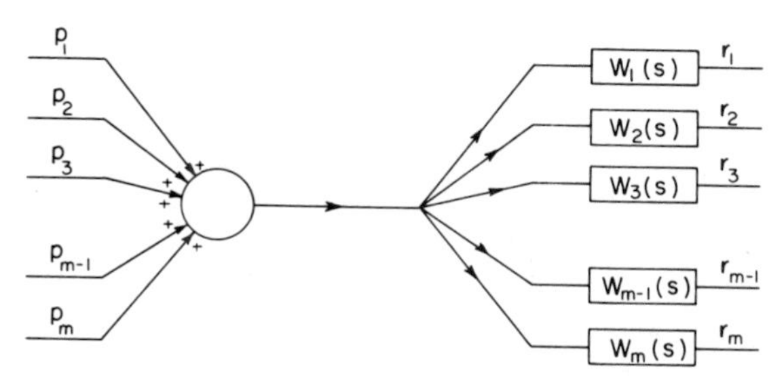

Fig. 9-2. The structure of the LTI plant in SF1.

$(\hat{A}, \hat{B})$ is completely controllable, and $\underline{W}(s)$ [Eq. (9-2)] has common row elements ($W_{i1} = W_{i2} = \cdots = W_{im} \triangleq W_i(s)$) in this representation. We denote all matrices with common row elements by indicating only the first column:

$$\hat{R} = \begin{bmatrix} \rho_1 & \cdots \\ \rho_2 & \\ \cdot & \\ \cdot & \\ \cdot & \\ \rho_m & \end{bmatrix}, \quad \hat{B} = \begin{bmatrix} 0 & \cdots \\ 0 & \\ \cdot & \\ \cdot & \\ \cdot & \\ 0 & \\ 1 & \end{bmatrix}, \quad \underline{W}(s) = \begin{bmatrix} W_1(s) & \cdots \\ W_2(s) & \\ \cdot & \\ \cdot & \\ \cdot & \\ W_m(s) & \end{bmatrix}. \quad \square$$

Proof: This result may be demonstrated by construction: choose $\hat{C} \triangleq [c_{ij} \triangleq d_{ij} - \rho_i a_j]$, and define the variables of x, p, and r by

$$x \triangleq \begin{bmatrix} \xi \\ D\xi \\ \cdot \\ \cdot \\ \cdot \\ D^{n-1}\xi \end{bmatrix}, \quad p \triangleq -\begin{bmatrix} g_1(r_1, t) \\ g_2(r_2, t) \\ \cdot \\ \cdot \\ \cdot \\ g_m(r_m, t) \end{bmatrix}, \quad r \triangleq \begin{bmatrix} \rho_1 \sum_{k=1}^{m} p_k + \sum_{j=1}^{n} c_{1j} x_j \\ \rho_2 \sum_{k=1}^{m} p_k + \sum_{j=1}^{n} c_{2j} x_j \\ \cdot \\ \cdot \\ \cdot \\ \rho_m \sum_{k=1}^{m} p_k + \sum_{j=1}^{n} c_{mj} x_j \end{bmatrix},$$

and the first part of the lemma follows directly. The complete controllability of $(\hat{A}, \hat{B})$ is an obvious result obtained as in considering the phase variable canonical form in the scalar case, Chapter II, Section 1. $\square$

B. *Properties of the Linear Time-Invariant Plant*

Given an LTI plant in transfer function form, Eq. (9-2), where elements $W_{ij}(s)$ of $\underline{W}(s)$ are rational, having no more zeros than poles, there are infinitely many quadruples $\{C, A, B, R\}$ such that Eqs. (9-1) and (9-2) are equivalent. Each of these sets of matrices is called a realization of $\underline{W}$ (Kalman [3]). As in the scalar case $C(i\omega I - A)^{-1}B$ approaches $\underline{0}$ as $\omega \longrightarrow \infty$, so $W_\infty \triangleq \lim_{\omega\to\infty} \underline{W}(i\omega) = R$; $\{C, A, B\}$ is thus a realization of $\underline{W}(s) - W_\infty$. A realization of $\underline{W}(s)$ that incorporates an A matrix of minimal dimension n_0 is said to be a minimal realization, and if $n > n_0$ then the realization is reducible. We have seen in the scalar case (Chapter II, Section 1; Chapter V, Section 2) that it is possible to increase the dimensionality of A by introducing poles and zeros of $W(s)$ that cancel and that if the numerator and denominator of $h^T(sI - A)^{-1}b$ are relatively prime, complete controllability and observability are guaranteed and vice versa. In the matrix case the first result holds but the second does not:

(i) The triple $\{C, A, B\}$ is a minimal realization of $\underline{W}(s) - W_\infty$ if and only if the triple is completely controllable and observable (Kalman [3]; see below).

(ii) If $\{C, A, B\}$ does constitute a minimal realization of some $\underline{W}(s)$ where $W_\infty = \underline{0}$ it is not implied that the elements $W_{ij}(s)$ are made up of ratios of relatively prime polynomials, nor that $|\underline{W}(s)|$ displays all of the eigenvalues of A (see the example given below).

The necessary and sufficient condition that (A, B) is completely controllable is that the $(n \times nm)$ matrix

$$[A^{n-1}B, A^{n-2}B, \ldots, AB, B] \tag{9-5}$$

must have rank n (Kalman, Ho, and Narendra [1]).

The necessary and sufficient condition that (C, A) is completely observable is that the $(n \times nm)$ matrix

$$[(CA^{n-1})^T, (CA^{n-2})^T, \ldots, (CA)^T, C^T] \tag{9-6}$$

must have rank n (Kalman [3]). (C, A) is completely observable if and only if (A^T, C^T) is completely controllable.

These concepts are clarified by an elementary example:

$$\underline{W}(s) = \begin{bmatrix} \alpha & 1 \\ \beta & 1 \end{bmatrix} \begin{bmatrix} s & -1 \\ \alpha\beta & (s+\alpha+\beta) \end{bmatrix}^{-1} \begin{bmatrix} 0 & 0 \\ 1 & 1 \end{bmatrix} = \begin{bmatrix} (s+\beta)^{-1} & (s+\beta)^{-1} \\ (s+\alpha)^{-1} & (s+\alpha)^{-1} \end{bmatrix};$$

$$[AB, \quad B] = \begin{bmatrix} 1 & 1 & 0 & 0 \\ -(\alpha+\beta) & -(\alpha+\beta) & 1 & 1 \end{bmatrix}; \qquad \text{rank} = 2;$$

$$[(CA)^T, \quad C^T] = \begin{bmatrix} -\alpha\beta & -\alpha\beta & \alpha & \beta \\ -\beta & -\alpha & 1 & 1 \end{bmatrix}; \qquad \text{rank} = 2 \quad \text{if} \quad \alpha \neq \beta.$$

Thus, if $\alpha \neq \beta$, then $\{C, A, B\}$ is completely controllable, completely observable and thus a minimal realization of $\underline{W}(s)$, even though $|\underline{W}| \equiv 0$ and each element $W_{ij}(s)$ has a pole-zero cancellation. If $\alpha = \beta$, the controllability is unaffected but observability is lost; A may be reduced to the (1×1) matrix (scalar) $-\alpha$; $\underline{W}(s) = (s + \alpha)^{-1} CB$.

If $\{C, A, B\}$ and $\{\hat{C}, \hat{A}, \hat{B}\}$ are two minimal realizations of $\underline{W}(s) - W_\infty$, then a nonsingular matrix T exists such that $\hat{C} = CT^{-1}$, $\hat{A} = TAT^{-1}$, $\hat{B} = TB$; conversely, if $\{C, A, B\}$ is a minimal realization of $\underline{W}(s) - W_\infty$ and T is a nonsingular matrix, then $\{CT^{-1}, TAT^{-1}, TB\}$ is also minimal.

C. *Properties of the Nonlinear Time-Varying Controller*

Each output r_j of the LTI plant $\underline{W}$ is the input of a nonlinear device whose output $g_j(r_j, t)$, $j = 1, 2, \ldots, m$, in turn, is used in a negative feedback configuration as the jth input p_j (Fig. 9-1). The matrix $\underline{W}(s)$ allows a wide latitude in the interconnection of subsystems, as demonstrated in Lemma 9-1. To cite another example given by Anderson [1], any single-loop system with m nonlinear and/or time-varying gains separated by LTI subsystems may be expressed in this form (Fig. 9-3).

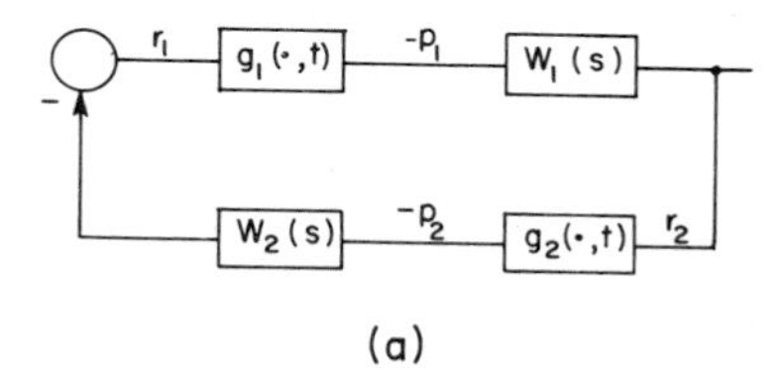

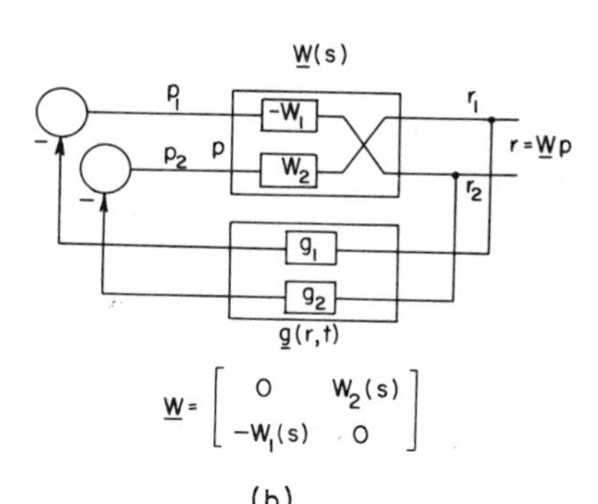

Fig. 9-3. (a) A single-loop system with multiple nonlinearities; (b) the equivalent formulation.

In extending the results of Chapter VI for separable NLTV gains (inseparable gains are not considered due to the difficulty encountered in applying the resulting criteria), we have

$$\underline{g}(r\ \ t) = K(t)\underline{f}(r)$$

where $K(t) \triangleq \text{diag}[k_i(t), i = 1, 2, \ldots, m]$ and $K(t) \in [\underline{0}, \underline{\bar{K}}]$. Since separability is always assumed, we simply specify the behavior of $g_j(r_j, t)$ by $g_j(r_j, t) \in \{N_j, T_j\}$.

The previous definition of range is directly generalized: we take

$$\bar{G}_{N_j} \triangleq \begin{cases} \max\limits_{r_j, t} \left[\dfrac{g_j(r_j, t)}{r_j}\right], & N_j = F \\ \max\limits_{r_j, \hat{r}_j \neq r_j, t} \left[\dfrac{g_j(r_j, t) - g_j(\hat{r}_j, t)}{r_j - \hat{r}_j}\right], & N_j = F_m \end{cases} \tag{9-7}$$

and define the matrix $\underline{\bar{G}}_N \triangleq \operatorname{diag}[\bar{G}_{N_1}, \bar{G}_{N_2}, \ldots, \bar{G}_{N_m}]$; the lower bound is invariably $\underline{0}$.

It is also necessary to consider $\underline{\Phi}_j$ or the index of each nonlinearity in treating any NLTV situation:

$$\underline{\underline{\Phi}} \triangleq \operatorname{diag}[\underline{\Phi}_j] \triangleq \operatorname{diag}\left[\min_{\sigma}\left\{\frac{\sigma f_j(\sigma)}{\int_0^\sigma f_j(\xi)\, d\xi}\right\}\right]. \tag{9-8}$$

The ranges of $\underline{\Phi}_j$ that correspond to each class $\{N_j\}$ of nonlinear behavior are unchanged (Table 2-1).

D. *The Origin as the Sole Equilibrium Point*

The system differential equation (9-1) is alternatively given by

$$\begin{aligned} \dot{x} &= Ax - BG(r, t)r \\ r &= Cx - RG(r, t)r. \end{aligned} \tag{9-9}$$

It is a necessary precondition to the generation of absolute stability criteria that the differential equation (9-9) must be asymptotically stable for all constant matrices $G(r, t) \equiv \underline{\kappa}$ of the form $\underline{\kappa} \triangleq \operatorname{diag}[\kappa_j]$, $\kappa_j \in [0, \bar{G}_{N_j}]$. If this is so, $I + R\underline{\kappa}$ must be nonsingular for all such matrices $\underline{\kappa}$ so that r is well defined, and consequently $[I + RG(r, t)]^{-1}$ must exist for all r and t. Equation (9-9) then reduces to

$$\dot{x} = [A - BG(r, t)[I + RG(r, t)]^{-1}C]x. \tag{9-10}$$

Furthermore, the asymptotic stability of

$$\dot{x} = [A - B\underline{\kappa}(I + R\underline{\kappa})^{-1}C]x \triangleq A_{\kappa}x \tag{9-11}$$

guarantees that $|A_\kappa| \neq 0$, which in turn ensures that the only equilibrium (solution to $A_\kappa x \equiv 0$) is $x \equiv 0$. Similarly, the matrix $[A - BG(r, t)\{I + RG(r, t)\}^{-1}C]$ must be nonsingular for all r and t if A_κ is nonsingular for all $\underline{\kappa} \in [\underline{0}, \underline{\bar{G}}_N]$, so no equilibrium of Eq. (9-10) exists other than $x \equiv 0$, provided that $A_\kappa \in \{A_1\}$, $\underline{\kappa} \in [\underline{0}, \underline{\bar{G}}_N]$.

3. Mathematical Preliminaries

At the heart of the development of stability criteria for systems with multiple nonlinearities lie the matrix version of the Kalman–Yakubovich Lemma and the concept of positive real transfer function matrices. As in the scalar case, this lemma is found to be crucial in relating the frequency domain criteria to the existence of Lyapunov functions of the quadratic plus integral type discussed in earlier chapters. While different forms of the matrix analog of the Kalman–Yakubovich Lemma have been stated in the literature (Anderson [2], Popov [4], Yakubovich [3]), in this section we follow closely the results reported in the excellent paper by Anderson. The lemma as stated by Anderson can be directly applied to the stability problem stated in Section 2 using an approach identical to that for the scalar case.

The stability criteria developed in Chapters V and VI for systems with a single memoryless nonlinearity are expressed in terms of the positive realness of a function of a complex variable. For the case when the system contains many nonlinearities, the corresponding frequency domain criteria can be expressed in terms of the positive realness of a matrix $\underline{H}(s)$ of transfer functions. Further, since all the systems considered are finite dimensional, the matrix $\underline{H}(s)$ is generally a matrix of real rational functions. In view of this, the properties of positive real matrices of real rational functions are first discussed prior to considering the details of the matrix version of the Kalman–Yakubovich Lemma.

DEFINITION PR4. (Newcomb [1]) The $(m \times m)$ matrix $\underline{H}(s)$ is positive real ($\underline{H}(s) \in \{\underline{\text{PR}}\}$) if:

(i) The elements of $\underline{H}(s)$ are analytic for $\operatorname{Re} s > 0$;
(ii) $\underline{H}^*(s) = \underline{H}^{\mathrm{T}}(s^*)$;
(iii) The Hermitian part of $\underline{H}(s)$,

$$\underline{J}(s) \triangleq \operatorname{He}\{\underline{H}(s)\} \triangleq \tfrac{1}{2}[\underline{H}(s) + \underline{H}^*(s)], \tag{9-12}$$

is positive semidefinite for $\operatorname{Re} s > 0$. □

In all the cases that we consider conditions (i) and (ii) are automatically satisfied so that only (iii) has to be checked to establish the positive realness of $\underline{H}(s)$.

The Hermitian part $\underline{J}(s) \triangleq \operatorname{He}\{\underline{H}(s)\}$ of an $(m \times m)$ matrix $\underline{H}(s)$ of real rational functions is real only if $\underline{H}(s)$ is symmetric; for a general matrix $\underline{H}(s)$, the real part is symmetric and the imaginary part is antisymmetric. The condition that $\underline{J}(s) \geqslant \underline{0}$ means that $b^*\underline{J}(s)b \geqslant 0$ for all complex vectors b. If we define b and $\underline{J}$ in terms of real vectors and matrices, say $b = b_1 + ib_2$,

$\underset{\sim}{J} = \underset{\sim}{J}_1 + i\underset{\sim}{J}_2$, then

$$b^*\underset{\sim}{J}(s)b = [b_1^{\mathrm{T}} \mid b_2^{\mathrm{T}}]\left[\begin{array}{c|c} \underset{\sim}{J}_1 & \underset{\sim}{J}_2^{\mathrm{T}} \\ \hline \underset{\sim}{J}_2 & \underset{\sim}{J}_1 \end{array}\right]\left[\begin{array}{c} b_1 \\ \hline b_2 \end{array}\right] \triangleq \hat{b}^{\mathrm{T}}\hat{\underset{\sim}{J}}\hat{b}.$$

Since b_1 and b_2 may be specified independently, it follows that $b^*\underset{\sim}{J}(s)b \geqslant 0$ implies that the $(2m \times 2m)$ real symmetric matrix $\hat{\underset{\sim}{J}}$ must be positive semidefinite (Chapter III, Section 2). It is not necessary to test a $(2m \times 2m)$ matrix, however, as conditions for positive (semi) definiteness may be stated directly in terms of the principal minors of $\underset{\sim}{J}(s)$.

DEFINITION. A Hermitian matrix $\underset{\sim}{J}$ is positive definite (positive semidefinite) if and only if all the leading principal minors of $\underset{\sim}{J}$ are positive (all the principal minors of $\underset{\sim}{J}$ are nonnegative). □

As indicated in Chapter III for real symmetric matrices, if all the leading principal minors of a Hermitian matrix are positive, then all principal minors are positive. However, this is not true for semidefinite matrices: nonnegativeness of the leading principal minors does not imply that all principal minors are nonnegative.

Lemmas 1–5 of Chapter III, Section 4 are concerned with the necessary and sufficient conditions for a scalar function of a complex variable to be positive real. The three lemmas presented in this section are due to Anderson and are concerned with similar conditions for a matrix of rational functions $\underset{\sim}{H}(s)$ to be positive real. Lemma 9 includes both Lemmas 7 and 8 as special cases, but following Anderson we present all three lemmas separately for completeness and clarity. Lemma 7 considers matrices whose elements have no poles on the imaginary axis and which are zero at $s = \infty$, Lemma 8 deals with matrices with poles only on the imaginary axis, and Lemma 9 with general positive real matrices.

LEMMA 7. Let a matrix $\underset{\sim}{H}(s)$ of rational functions be such that $H_\infty = \underset{\sim}{0}$ and $\underset{\sim}{H}$ has poles only in $\operatorname{Re} s < 0$. If $\{C_1, A_1, B_1\}$ is a minimal realization of $\underset{\sim}{H}(s)$, then $\underset{\sim}{H}(s)$ is positive real if and only if there exist a positive definite matrix P and a matrix Q such that

$$PA_1 + A_1^{\mathrm{T}}P = -QQ^{\mathrm{T}}; \qquad PB_1 = C_1^{\mathrm{T}}. \quad \square \tag{9-13}$$

LEMMA 8. Let a positive real matrix $\underset{\sim}{H}(s)$ have all imaginary poles with $H_\infty = \underset{\sim}{0}$ and let $\{C_1, A_1, B_1\}$ be a minimal realization for $\underset{\sim}{H}$. Then there exists a positive definite matrix P such that

$$PA_1 + A_1^{\mathrm{T}}P = 0; \qquad PB_1 = C_1^{\mathrm{T}}. \quad \square \tag{9-14}$$

LEMMA 9 (Anderson [2]). Let $\underset{\sim}{H}(s)$ be a matrix of rational transfer functions such that H_∞ is finite, $\underset{\sim}{H}(s)$ has poles which lie in $\operatorname{Re} s < 0$ or are simple for $\operatorname{Re} s = 0$. If $\{C_1, A_1, B_1, H_\infty\}$ is a minimal realization of $\underset{\sim}{H}$, then $\underset{\sim}{H}(s)$ is positive real if and only if a positive definite matrix P and matrices D and

Q exist such that

$$PA_1 + A_1^{\mathrm{T}}P = -QQ^{\mathrm{T}}; \qquad PB_1 = C_1^{\mathrm{T}} - QD; \qquad D^{\mathrm{T}}D = H_\infty + H_\infty^{\mathrm{T}}. \quad \square \tag{9-15}$$

While proofs of these lemmas are presented in Appendix I, we briefly consider here some of the significant features of the lemmas, in particular the assumption of minimality of the realization of $\underline{H}(s)$.

The proofs of Lemmas 7–9 depend on an important lemma on spectral factorization due to Youla [1]. If $\underline{Y}(s) \triangleq \underline{H}(s) + \underline{H}^{\mathrm{T}}(-s)$, $\underline{Y}(s)$ is termed *para-hermitian* and $\underline{Y}(i\omega)$ (for all real ω) is nonnegative definite if $\underline{H}(i\omega)$ is positive real. According to Youla, if $\underline{H}(s)$ is positive real and $\underline{Y}(s)$ has rank r almost everywhere, then there exists an $(r \times m)$ matrix $\underline{G}(s)$ such that

$$\underline{Y}(s) = \underline{H}(s) + \underline{H}^{\mathrm{T}}(-s) = \underline{G}^{\mathrm{T}}(-s)\underline{G}(s). \tag{9-16}$$

If $\underline{H}(s)$ has a minimal realization $\{C_1, A_1, B_1, H_\infty\}$ it was shown by Anderson that $\underline{G}(s)$ has a realization $\{Q^{\mathrm{T}}, A_1, B_1, D\}$. The matrix Q determines the positive definite matrix P by the relation

$$A_1^{\mathrm{T}}P + PA_1 = -QQ^{\mathrm{T}}; \tag{9-17}$$

P, Q, and D satisfy the relations of Eq. (9-15) when $\underline{H}(s)$ is positive real.

In Lemma 9 the case of simple imaginary axis poles is included; in that eventuality, the complete observability of (Q^{T}, A) is clearly not obtained: If (Q^{T}, A) is completely observable, then $P = P^{\mathrm{T}} > \underline{0}$ if and only if $A \in \{A_1\}$ (refer to Corollary F2, Chapter III, Section 6). Since this property of Q is required in our developments, we use the following result, also due to Anderson [2].

Corollary to Lemma 9. If $A \in \{A_1\}$ in Lemma 9, then (Q^{T}, A_1) in Eq. (9-15) is completely observable. $\square$

In the stability criteria discussed in this chapter, the matrix P occurs in the Lyapunov function candidate, while $A_1 = A$ and $B_1 = B$ are specified by the LTI part of the system. The frequency domain criteria are expressed in terms of the matrix $\underline{H}(s)$ which is generally the product of $\underline{Z}(s)$, a diagonal matrix of multiplier functions, and $[\underline{W}(s) + \underline{G}_N^{-1}]$ or the transfer matrix of the LTI plant plus the inverse of the NLTV gain upper bound matrix [Eq. (9-7)].

When $\underline{H}(s)$ is positive real, $\underline{H}(s)$ and $\underline{G}(s)$ are related by Eq. (9-16); since $\underline{G}(s)$ has been shown to have a minimal realization $\{Q^{\mathrm{T}}, A_1, B_1, D\}$ when $\underline{H}(s)$ has a minimal realization $\{C_1, A_1, B_1, H_\infty\}$, then Q and P satisfying relation (9-15) can be directly determined. This result provides the needed link between $\underline{H}(s)$ and the matrices P and Q and hence the Lyapunov function.

This result is more restrictive than the corresponding lemma used in the scalar case (Lemma 3, Meyer) in requiring that $\{C_1, A_1, B_1\}$ be a minimal realization of $\underline{H}(s) - H_\infty$. However, there has been no rigorously derived analog to the Meyer lemma, so this constraint must be retained. To simplify the statement of the absolute stability criteria, we incorporate this restriction and the positive real condition into one definition. Actually, with the Popov multiplier $\underline{Z}(s) = (\beta_0 s + \gamma_0)$ the final matrix $\underline{H}(s)$ may have more zeros than poles, that is, $\underline{H}(s) = C_1(sI - A_1)^{-1}B_1 + H_\infty + H_1 s$, so this case is included also.

DEFINITION PR 5. The matrix $\underline{H}(s)$ is minimal positive real ($\underline{H}(s) \in \{\underline{\text{MPR}}\}$) if $\underline{H}(s) \in \{\underline{\text{PR}}\}$ and if the realization $\{C_1, A_1, B_1\}$ of $[\underline{H}(s) - H_\infty - H_1 s]$ is minimal. □

The significance of the condition of minimality on $\underline{H}(s)$ is clear when the system is given as a state vector differential equation. If, however, the system is specified by $\underline{W}(s)$, the procedure to be followed is to inspect $\underline{H}(s) \triangleq \underline{Z}(s)[\underline{W}(s) + \bar{G}_N^{-1}] \in \{\underline{\text{PR}}\}$, and then to determine whether a minimal realization of $\underline{H}(s)$ of the form $\{C_1, A, B, H_\infty, H_1\}$ exists such that $\underline{W}(s)$ can be realized using the same matrices A and B; the realization of $\underline{W}(s)$ need not be minimal. These points are demonstrated in the discussion following MSC 2, Section 5B.

The matrix $\underline{H}(s)$ arising from the use of the special form SF1 (Lemma 9-1) is often found to have the structure

$$\underline{H}(s) = \begin{bmatrix} H_1(s) & \\ H_2(s) & \cdots \\ \vdots & \\ H_m(s) & \end{bmatrix}. \tag{9-18}$$

In this case, the following lemma provides a direct means of testing for the condition $\underline{H}(s) \in \{\underline{\text{MPR}}\}$.

LEMMA 9-2. The matrix $\underline{H}(s)$ of Eq. (9-18) satisfies $\underline{H}(s) \in \{\underline{\text{MPR}}\}$ if and only if $H_1(s) = H_2(s) \cdots = H_m(s) \triangleq H_0(s)$ and $H_0(s)$ is a positive real transfer function which is the ratio of relatively prime polynominals. □

Proof: Sufficiency follows by inspection. For necessity we assume that for some value of s, $H_j(s) \triangleq r_j + ip_j$; then

$$\text{He}\{\underline{H}(s)\} = \frac{1}{2}\begin{bmatrix} 2r_1 & \{(r_1 + r_2) + i(p_1 - p_2)\} & \cdots \\ \{(r_1 + r_2) - i(p_1 - p_2)\} & 2r_2 & \cdots \\ \vdots & \vdots & \end{bmatrix}.$$

A necessary condition for semidefiniteness is that the m one-element principal minors be nonnegative, that is, $r_i \geqslant 0$. Considering the $m(m-1)/2$

four-element principal minors leads to the requirements $-[(r_i - r_j)^2 + (p_i - p_j)^2] \geqslant 0$. These conditions together demonstrate that $\underline{J}(s)$ cannot be positive definite, and that positive semidefiniteness requires $r_i = r_j$ and $p_i = p_j$. Thus $\underline{H}(s)$ must reduce to

$$\underline{H}(s) = H_0(s)\begin{bmatrix} 1 & & & \\ & 1 & & \\ & & \ddots & \\ & & & 1 \end{bmatrix}, \tag{9-19}$$

which completes the proof: all higher order principal minors are zero. For minimality, we see from the structure of SF1 that a reduced matrix $\hat{A}$ can be used to realize $\underline{H}(s)$ if and only if there are one or more pole-zero cancellations in $H_0(s)$. □

In treating time-varying systems, it is necessary to show that $\dot{v}$ is negative definite rather than negative semidefinite as provided by the application of this lemma (refer to Section 6). In the scalar case this condition is guaranteed by use of Lemma 2 (Lefschetz) which again has no direct analog in the matrix case. We can obtain the desired result by use of the frequency domain shift technique given in Lemma 4.

LEMMA 10. Let $\underline{H}(s)$ be a matrix of real rational transfer functions such that H_∞ is finite and $\underline{H}(s)$ has poles which lie in Re $s < -\mu$. $\underline{H}(s - \mu) \in \{\underline{\text{MPR}}\}$ if and only if a matrix P, $P = P^{\mathrm{T}} > 0$, and matrices D and Q exist such that

$$PA_1 + A_1{}^{\mathrm{T}}P = -QQ^{\mathrm{T}} - 2\mu P; \quad PB_1 = C_1{}^{\mathrm{T}} - QD; \quad D^{\mathrm{T}}D = H_\infty + H_\infty{}^{\mathrm{T}}. \;\square$$

DEFINITION PR6. The matrix $\underline{H}(s)$ is minimal strictly positive real ($\underline{H}(s) \in \{\underline{\text{MSPR}}\}$) if for some $\varepsilon > 0$ $\underline{H}(s - \varepsilon) \in \{\underline{\text{MPR}}\}$. □

Point constraints on $d/dt\, K(t)$ are obtained using the direct method of Lyapunov (specifically, if $\underline{H}(s) \in \{\underline{\text{MSPR}}\}$, then $\dot{v} \leq -\varepsilon x^{\mathrm{T}}Px$, $P = P^{\mathrm{T}} > 0$, by Lemma 10 and Definition PR6), while time-averaged conditions for absolute stability are derived utilizing Lemma 10 in the usual manner (Chapter VI, Section 6) in conjunction with the theorem of Corduneanu.

4. Linear Time-Invariant Systems and Absolute Stability

The general LTI case of the differential equation (9-1) is given by specifying

$$p = -\underline{\kappa} r; \qquad \underline{\kappa} \triangleq \text{diag}[\kappa_i] \in [\underline{0}, \underline{\bar{K}}), \tag{9-20}$$

which may be reduced to the form indicated in Eq. (9-11), since $I + R\underline{\kappa}$ must be nonsingular for all values of $\underline{\kappa}$ under consideration.

A. *The Existence of* $x^T Px$ *and the Hurwitz Conditions*

In Chapter IV, it is demonstrated that it is possible to prove the equivalence of the Hurwitz conditions for asymptotic stability of LTI systems and the result that the existence of $v(x) = x^T Px$ as a Lyapunov function for $\dot{x} = A_0 x$ [Eq. (4-32)] is a necessary and sufficient condition for asymptotic stability. Only the differential equation $\dot{x} = A_0 x$ is considered; hence, the result of Chapter IV, Section 2 is equally valid in the case involving multiple LTI feedback gains.

B. *The Existence of* $x^T Px + \sum_{i=1}^{m} \kappa_i x^T M_i x$ *and the Nyquist Criterion*

It is reasonable to make the conjecture that the indicated $\underset{\sim}{\kappa}$-dependent Lyapunov function must exist if the LTI system defined by Eq. (9-11) with $R = \underset{\sim}{0}$ is asymptotically stable for all diagonal feedback gain matrices $\underset{\sim}{\kappa}$ in the range $[\underset{\sim}{0}, \underset{\sim}{\bar{K}})$ where $\underset{\sim}{\bar{K}} \triangleq \mathrm{diag}[\bar{K}_i]$ has been obtained by an application of the extended Nyquist criterion.† In contrast to the scalar gain case, the frequency domain multiplier conditions for the absolute stability of systems with LTI gains (Section 5) do not presently appear to be sufficiently general to prove this result. A discussion of what is currently known that is pertinent to this question is provided.

(1) *The Extension of the Nyquist Criterion*

Many efforts have been made in seeking a generalization of the criterion of Nyquist that is applicable to the stability analysis of LTI systems with multiple feedback gains κ_i. As in the scalar case, the most general work has been done by Desoer and Wu [2]. For our purposes, however, the research reported by Chen [1] and Hsu and Chen [1] is of sufficient sophistication.

As in the scalar case, the condition that the characteristic equation of the system (9-11), namely,

$$p(\lambda) \triangleq |\lambda I - A + B\underset{\sim}{\kappa}(I + R\underset{\sim}{\kappa})^{-1}C| = 0$$

is Hurwitz is a necessary and sufficient condition for the asymptotic stability of the differential equation. The characteristic polynomial may be recast as

$$p(\lambda) = |\lambda I - A||I + (\lambda I - A)^{-1}B\underset{\sim}{\kappa}(I + R\underset{\sim}{\kappa})^{-1}C|;$$

then the result of Sandberg [1] that for any two matrices $F(p \times q)$ and $G(q \times p)$, we have $|I + FG| = |I + GF|$ where I is of appropriate dimension (p or q) yields

$$p(\lambda) = |\lambda I - A||I + (I + R\underset{\sim}{\kappa})^{-1}C(\lambda I - A)^{-1}B\underset{\sim}{\kappa}|,$$

† If $R \neq \underset{\sim}{0}$, then $v(x)$ must have a term that is quadratic in p as in the single gain case; see Eq. (9-27).

which in turn may be reduced to

$$p(\lambda) = \frac{|\lambda I - A|}{|I + R\underline{\kappa}|} |I + \underline{\kappa}\underline{W}(\lambda)|.$$

The determinant $|I + \underline{\kappa}\underline{W}(\lambda)|$ is rational, so the denominator of this factor of $p(\lambda)$ must cancel some or all of the zeros of $|I\lambda - A| = 0$, as $p(\lambda)$ is a polynomial in λ. The denominator of $|I + \underline{\kappa}\underline{W}|$ does not necessarily cancel all of the zeros of $|\lambda I - A|$, so the necessary and sufficient conditions for asymptotic stability must take into account not only the zeros of $|I + \underline{\kappa}\underline{W}|$ but the uncancelled zeros of $|\lambda I - A|$ as well. If we assume that $A \in \{A_1\}$, then any right half plane zeros of $|\lambda I - A|$ are precluded.

THEOREM 9-1 (Hsu and Chen [1]). If $A \in \{A_1\}$ in the system represented by Eq. (9-11), then a necessary and sufficient condition for the asymptotic stability of this system is that the critical point ($U_T = -1$, $V_T = 0$) does not lie in $\mathcal{W}_T(\mathcal{R})$ where $\mathcal{W}_T(\mathcal{R})$ is the mapping of the s-plane region $\mathcal{R}$ (Chapter VII, Section 1) given by $W_T(s \in \mathcal{R})$ and

$$W_T(s) \triangleq |I + \underline{\kappa}\underline{W}(s)| - 1. \quad \square \tag{9-21}$$

By using SF1 as introduced in Lemma 9-1, this result may be significantly simplified. By direct expansion,

$$\hat{W}_T(s) = \sum_{i=1}^{m} \kappa_i W_i(s). \tag{9-22}$$

If all of the gains κ_i are permitted to take on values in the range $[0, \infty)$, then clearly the necessary and sufficient condition for asymptotic stability is that $|\measuredangle W_i(i\omega)| < \pi$, $i = 1, 2, \ldots, m$.

(2) *Stability Results from Lyapunov's Direct Method*

From the developments outlined in Section 5, the condition that is sufficient to guarantee the absolute stability of an LTI system with $\underline{\kappa} \in [\underline{0}, \underline{\bar{K}})$ is that there must exist some multiplier $\underline{Z}(s) = \mathrm{diag}[Z_i(s) \in \{Z_\alpha(s)\}]$ where $\{Z_\alpha(s)\}$ is the class of shifted LC functions [Eq. (5-44)] such that $\underline{H}(s) \triangleq \underline{Z}(s)[\underline{W}(s) + \underline{\bar{K}}^{-1}]$ is positive real and $\underline{Z}(s)\underline{W}(s)$ is proper.† It does not appear to be possible to relate this condition to the necessary and sufficient condition for asymptotic stability as it was in the scalar case.

(3) *Necessary and Sufficient Conditions for Stability*

For the LTI case it has been assumed so far that the system is stable for all $\underline{\kappa} \in [\underline{0}, \underline{\bar{K}})$ which defines a hyperparallelopiped in the m dimensional parameter space. In general the region of stability defined by the Hurwitz conditions

† Elements of $\underline{Z}(s)\underline{W}(s)$ must have no more zeros than poles.

can have any arbitrary form, and hence $\underline{\kappa} \in [\underline{0}, \underline{\bar{K}})$ would represent only sufficient conditions for stability. It is only in special cases such as systems described by Eq. (9-22) with $\underline{\kappa} \in [\underline{0}, \underline{\infty})$ that such a condition may also be necessary.

5. Stability of Nonlinear Systems

A. The Matrix Popov Stability Criterion

The system whose stability is to be investigated is defined by Eq. (9-1) with

$$p = -\underline{f}(r) \triangleq -\underline{F}(r)r; \qquad f_i(r_i) \in \{F\}, \qquad \underline{F}(r) \in [\underline{0}, \underline{\bar{F}}) \tag{9-23}$$

As a consequence of $\underline{F}(r) \in [\underline{0}, \underline{\bar{F}})$ we have that $f_i^2(r_i)/\bar{F}_i < r_i f_i(r_i)$, or, for any $\underline{\gamma}_0 \triangleq \mathrm{diag}[\gamma_{0i}] > \underline{0}$, that

$$\underline{f}^{\mathrm{T}}\underline{\bar{F}}^{-1}\underline{\gamma}_0\underline{f} < r^T\underline{\gamma}_0\underline{f}, \qquad \underline{f} \neq 0. \tag{9-24}$$

We assume initially that the differential equation specified by Eq. (9-23) with $\underline{f}(r) = \underline{\kappa}r$ where $\underline{\kappa} = \mathrm{diag}[\kappa_i]$ and $\underline{\kappa} \in [\underline{0}, \underline{\bar{F}})$ is asymptotically stable, that is, $A_{\underline{\kappa}}$ [Eq. (9-11)] satisfies $A_{\underline{\kappa}} \in \{A_1\}$, $\underline{\kappa} \in [\underline{0}, \underline{\bar{F}})$.

In order to express the Lyapunov function candidate concisely in matrix notation, we define

$$\Theta^{\mathrm{T}} \triangleq \left[\int_0^{r_1} f_1(\xi)\,d\xi, \ldots, \int_0^{r_m} f_m(\xi)\,d\xi\right] \triangleq \left(\int_0^{r} \underline{f}(z)\,dz\right)^{\mathrm{T}} \tag{9-25}$$

and

$$l^{\mathrm{T}} \triangleq [1, 1, \ldots, 1]. \tag{9-26}$$

Using these vectors, the generalized Lur'e–Postnikov form is

$$v(x) = x^{\mathrm{T}}Px + \Theta^{\mathrm{T}}\underline{\hat{\beta}}_0 l + l^{\mathrm{T}}\underline{\hat{\beta}}_0\Theta + \tfrac{1}{2}p^{\mathrm{T}}(\underline{\hat{\beta}}_0 R + R^{\mathrm{T}}\underline{\hat{\beta}}_0)p \tag{9-27}$$

where $P = P^{\mathrm{T}} > \underline{0}$ is unspecified, and we assume, at least for preliminary analysis, that $\underline{\hat{\beta}}_0 = \mathrm{diag}[\hat{\beta}_{0i}] \geqslant \underline{0}$ and $(\underline{\hat{\beta}}_0 R + R^{\mathrm{T}}\underline{\hat{\beta}}_0) \geqslant \underline{0}$ to assure the validity of $v(x)$ according to Lemma V2 (Chapter III, Section 2).

The total time derivative of this candidate along the trajectories of the system is

$$\begin{aligned}\dot{v} = {} & x^{\mathrm{T}}(A^{\mathrm{T}}P + PA)x - x^{\mathrm{T}}[PB - A^{\mathrm{T}}C^{\mathrm{T}}\underline{\hat{\beta}}_0 - C^{\mathrm{T}}\underline{\gamma}_0]\underline{f} - \underline{f}^{\mathrm{T}}[B^{\mathrm{T}}P - \underline{\hat{\beta}}_0 CA \\ & - \underline{\gamma}_0 C]x - \underline{f}^{\mathrm{T}}[\underline{\hat{\beta}}_0 CB + B^{\mathrm{T}}C^{\mathrm{T}}\underline{\hat{\beta}}_0 + \underline{\gamma}_0 R + R^{\mathrm{T}}\underline{\gamma}_0 + 2\underline{\gamma}_0\underline{\bar{F}}^{-1}]\underline{f} \\ & - [r^{\mathrm{T}}\underline{\gamma}_0\underline{f} - \underline{f}^{\mathrm{T}}\underline{\bar{F}}^{-1}\underline{\gamma}_0\underline{f}] - [\underline{f}^{\mathrm{T}}\underline{\gamma}_0 r - \underline{f}^{\mathrm{T}}\underline{\gamma}_0\underline{\bar{F}}^{-1}\underline{f}].\end{aligned} \tag{9-28}$$

As in the scalar case the last term of $v(x)$ is chosen so that the terms in dp/dt cancel. Since the nonlinearities lie in the finite sector $[\underline{0}, \underline{\bar{F}})$ and $\underline{\gamma}_0 > \underline{0}$ is assumed, the last two terms (which are equal) are negative semidefinite by Eq. (9-24). To make the first four terms into a perfect square we identify

$$\begin{aligned} A_1 &= A; \qquad B_1 = B; \qquad C_1 \triangleq \hat{\underline{\beta}}_0 CA + \underline{\gamma}_0 C; \\ D^T D &\triangleq \hat{\underline{\beta}}_0 CB + B^T C^T \hat{\underline{\beta}}_0 + \underline{\gamma}_0 R + R^T \underline{\gamma}_0 + 2\underline{\gamma}_0 \underline{\bar{F}}^{-1} \triangleq H_\infty + H_\infty{}^T, \end{aligned} \tag{9-29}$$

and by Lemma 9 obtain

$$\dot{v} = -[Q^T x + D\underline{f}]^T [Q^T x + D\underline{f}] - 2[r^T \underline{\gamma}_0 \underline{f} - \underline{f}^T \underline{\bar{F}}^{-1} \underline{\gamma}_0 \underline{f}] \leqslant 0 \tag{9-30}$$

if and only if

$$\underline{\hat{H}}(s) \triangleq \underline{\gamma}_0 [C(sI - A)^{-1} B + R + \underline{\bar{F}}^{-1}] + s\hat{\underline{\beta}}_0 C(sI - A)^{-1} B \in \{\underline{\mathrm{MPR}}\}, \tag{9-31}$$

where the relation $CB + CA(sI - A)^{-1}B = sC(sI - A)^{-1}B$ has been used.

Some of the final arguments of Chapter V, Section 1 carry through by inspection. First, $\dot{v} \equiv 0$ only if $\underline{f}(r) \equiv 0$ (or $\dot{x} \equiv Ax$) and $x^T Q Q^T x \equiv 0$. By the Corollary to Lemma 9, (Q^T, A) is completely observable, so $x^T Q \equiv 0$ where x satisfies $\dot{x} = Ax$ only for $x \equiv 0$, and thus absolute stability is guaranteed. The initial requirements that $\hat{\underline{\beta}}_0 \geqslant \underline{0}$ and $(\hat{\underline{\beta}}_0 R + R^T \hat{\underline{\beta}}_0) \geqslant 0$ may be removed since the sign of $\dot{v}$ is unaffected; as in Chapter V, Section 1, asymptotic stability for $p = -\underline{\kappa} r$, $\underline{\kappa} \in [\underline{0}, \underline{\bar{F}})$ and the condition that $\dot{v} \leqslant 0$ where $\dot{v} \equiv 0$ only if $x \equiv 0$ ensures that $v(x)$ is positive definite.

Thus, without dwelling on these aspects of the development in great detail, we state a preliminary generalization of the theorem of Popov:

Multiple-Gain Stability Criterion 1a (MSC 1a). The system defined by Eqs. (9-1) and (9-23) with $A_{\underline{\kappa}} \in \{A_1\}$, $\underline{\kappa} \in [\underline{0}, \underline{\bar{F}})$ is absolutely $\{\underline{F}\}$ stable if matrices

$$\begin{aligned} \hat{\underline{\beta}}_0 &\triangleq \mathrm{diag}[\hat{\beta}_{0i}, i = 1, 2, \ldots, m], \qquad \hat{\beta}_{0i} \in (-\infty, \infty) \\ \underline{\gamma}_0 &\triangleq \mathrm{diag}[\gamma_{0i}, i = 1, 2, \ldots, m] > \underline{0} \end{aligned}$$

exist such that $\underline{\hat{H}}(s)$ in Eq. (9-31) satisfies $\underline{\hat{H}}(s) \in \{\underline{\mathrm{MPR}}\}$. □

This result, while of broad generality, is not in a satisfactory final form when compared with the results obtained in the scalar case. One difficulty arises in the assumption that $W_\infty = R \neq \underline{0}$; if we consider the Hermitian part of $\underline{\hat{H}}(i\omega)$, then we can add $i\omega\hat{\underline{\beta}}_0 \underline{\bar{F}}^{-1}$ and $\frac{1}{2}i\omega(\hat{\underline{\beta}}_0 R + R^T \hat{\underline{\beta}}_0)$, which are imaginary and symmetric and thus do not contribute to $\mathrm{He}\{\underline{\hat{H}}(i\omega)\}$, to show that

$$\begin{aligned} \mathrm{He}\{\underline{\hat{H}}(i\omega)\} = \mathrm{He}\{\underline{H}_1(i\omega)\} &\triangleq \mathrm{He}\{(\hat{\underline{\beta}}_0 i\omega + \underline{\gamma}_0)[\underline{W}(i\omega) + \underline{\bar{F}}^{-1}] \\ &\quad - \tfrac{1}{2} i\omega(\hat{\underline{\beta}}_0 R - R^T \hat{\underline{\beta}}_0)\}. \end{aligned}$$

Thus, we note that taking a general matrix R prevents us from obtaining a result that is completely analogous to that of Chapter V, Section 1. Hence to complete this derivation we assume that $\hat{\underline{\beta}}_0 R = \underline{0}$,† so the frequency domain condition may be expressed as

$$\underline{H}_2(s) \triangleq (\hat{\underline{\beta}}_0 s + \underline{\gamma}_0)[\underline{W}(s) + \underline{\bar{F}}^{-1}] \in \{\underline{\text{MPR}}\}$$

provided $\hat{\beta}_{0i} \geqslant 0$. Rather than permitting some of the parameters $\hat{\beta}_{0i}$ to be negative, we use the alternative multiplier

$$\underline{Z}(s) \triangleq \text{diag}[Z_i^{\pm 1}(s) \in \{\underline{Z}_F(s)\}] \in \{\underline{\text{SPR}}\} \tag{9-32}$$

[where $\{Z_F(s)\}$ is defined in Eq. (5-11)] and require the condition

$$\underline{H}(s) \triangleq \underline{Z}(s)[\underline{W}(s) + \underline{\bar{F}}^{-1}] \in \{\underline{\text{MPR}}\}. \tag{9-33}$$

Multiple-Gain Stability Criterion 1b (MSC 1b). The system defined by Eq. (9-1) with

$$\begin{aligned} p &= -\underline{f}(r) = -F(r)r, \\ \underline{f}(r) &\in \{\underline{F}\}, \qquad F(r) \in [\underline{0}, \underline{\bar{F}}], \\ A &\in \{A_1\}, \qquad A_{\underline{F}} \in \{A_1\} \end{aligned} \tag{9-34}$$

is absolutely $\{\underline{F}\}$ stable if $\underline{Z}(s)$ [Eq. (9-32)] exists such that $\hat{\underline{\beta}}_0 R = \underline{0}$† and $\underline{H}(s)$ [Eq. (9-33)] satisfies $\underline{H}(s) \in \{\underline{\text{MPR}}\}$. □

Letov [1] and Sultanov [1] investigated the extension of the resolving equations of Lur'e to the absolute stability problem with multiple nonlinear gains. As in the scalar case the main contribution in deriving a frequency domain criterion was made by Popov [1]; there the result for $R = \underline{0}$ is given as He$\{\hat{\underline{H}}(i\omega)\} > \underline{0}$ [Eq. (9-31)]. A certain amount of duplication in the research effort in this area was caused by the inaccessibility of this original work of Popov.

Narendra and Goldwyn [2] gave the criterion for absolute stability as He $\{\hat{\underline{H}}(i\omega)\} \geqslant \underline{0}$, but failed to state all the conditions required for its validity. Ibrahim and Rekasius [1] obtained the less general result

$$\underline{\bar{F}}^{-1} + \text{He}\{(i\omega\hat{\underline{\beta}}_0 + I)C(i\omega I - A)^{-1}B\} > \text{He}\{\hat{\underline{\beta}}_0 CB\} \geqslant \underline{0}. \tag{9-35}$$

Results similar to those of Popov were presented by Tokumaru and Saito [1] and Jury and Lee [1].

Anderson [1] initially presented a restrictive form of the above result using a scalar $(\beta_0 s + \gamma_0)$ as the frequency domain multiplier; the more general result was given in a subsequent paper (Moore and Anderson [2]). As already indicated, it is his development relating the network theory concept of positive real functions to the control theory concept of minimal realizations that has been followed in this section.

† The significance of this condition is that the elements of $(\hat{\underline{\beta}}_0 s + \underline{\gamma}_0)\underline{W}(s)$ also must satisfy the condition that they have no more zeros than poles, that is, this matrix is proper. It could also be assumed that $\hat{\underline{\beta}}_0 R$ is symmetric, but this is somewhat artificial except when R is a diagonal matrix.

EXAMPLE 1. Jury and Lee [1] considered a general system with two nonlinearities. If we take $\underline{\beta}_0 R = \underline{0}$, the matrix $\underline{H}(s)$ has the form

$$\underline{H}(s) = \begin{bmatrix} (\beta_{01}s + \gamma_{01})^{\pm 1}[W_{11}(s) + (\bar{F}_1)^{-1}] & (\beta_{01}s + \gamma_{01})^{\pm 1}W_{12}(s) \\ (\beta_{02}s + \gamma_{02})^{\pm 1}W_{21}(s) & (\beta_{02}s + \gamma_{02})^{\pm 1}[W_{22}(s) + (\bar{F}_2)^{-1}] \end{bmatrix}$$

for the finite sector problem, $F(r) \in [\underline{0}, \underline{\bar{F}})$. The condition $\underline{J}(i\omega) \triangleq \text{He}\{\underline{H}(i\omega)\} \geqslant 0$ is satisfied if

$$\begin{aligned} &\text{(i)} \qquad J_{ii} \triangleq \text{Re}\{(\beta_{0i}i\omega + \gamma_{0i})^{\pm 1}[W_{ii}(i\omega) + (\bar{F}_i)^{-1}]\} \geqslant 0, \quad i = 1, 2, \\ &\text{(ii)} \quad J_{11}J_{22} \geqslant \tfrac{1}{4}|(\beta_{01}i\omega + \gamma_{01})^{\pm 1}W_{12}(i\omega) \\ &\qquad\qquad + (-\beta_{02}i\omega + \gamma_{02})^{\pm 1}W_{21}(-i\omega)|^2. \end{aligned} \tag{9-36}$$

Condition (i) indicates that the scalar finite sector Popov criterion (Chapter V, Section 1) must be satisfied by the diagonal terms. The second constraint takes into account the interaction between the two loops of this system.

EXAMPLE 2. A specific case in which a qualified verification of the Aizerman conjecture (Chapter IV, Section 1) may be obtained was first considered by Krasovskii [1] using the direct method of Lyapunov. The same result was subsequently rederived by Narendra and Neuman [3] using the frequency domain condition of MSC 1. Consider the following differential equation and corresponding transfer function matrix:

$$\dot{x} = \begin{bmatrix} -\alpha & 0 \\ 0 & -\beta \end{bmatrix} x - \begin{bmatrix} 1 & 0 \\ 0 & 1 \end{bmatrix} \begin{bmatrix} f_1(x_2) \\ f_2(x_1) \end{bmatrix}, \quad \underline{W}(s) = \begin{bmatrix} 0 & (s+\beta)^{-1} \\ (s+\alpha)^{-1} & 0 \end{bmatrix}. \tag{9-37}$$

By taking $(\underline{\beta}_0 s + \underline{\gamma}_0) = \text{diag}[(s + \beta_1), (s + \alpha_1)]$,

$$\underline{H}(s) = (\underline{\beta}_0 s + \underline{\gamma}_0)[\underline{W}(s) + \underline{\bar{F}}^{-1}] = \begin{bmatrix} \dfrac{s+\beta_1}{\bar{F}_1} & \dfrac{s+\beta_1}{s+\beta} \\ \dfrac{s+\alpha_1}{s+\alpha} & \dfrac{s+\alpha_1}{\bar{F}_2} \end{bmatrix}.$$

Writing $\underline{H}(s)$ in the form

$$\underline{H}(s) = s\underline{\bar{F}}^{-1} + \begin{bmatrix} \beta_1/\bar{F}_1 & 1 \\ 1 & \alpha_1/\bar{F}_2 \end{bmatrix} + \begin{bmatrix} 0 & (\beta_1 - \beta) \\ (\alpha_1 - \alpha) & 0 \end{bmatrix} (sI - A)^{-1}B,$$

we see that $\{C_1, A, B\}$ is a minimal realization of $[\underline{H}(s) - H_\infty - H_1 s]$ if $\alpha_1 \neq \alpha$ and $\beta_1 \neq \beta$. Further analysis reveals that choosing $\alpha_1 < \alpha$ and $\beta_1 < \beta$ ensures that $\underline{H}(s) \in \underline{\{\text{MPR}\}}$ if $\bar{F}_1\bar{F}_2 = \alpha_1\beta_1$. Hence by MSC 1b, the system is absolutely stable if

$$0 \leqslant \frac{f_1(x_2)}{x_2} \cdot \frac{f_2(x_1)}{x_1} \leqslant \alpha_1\beta_1 < \alpha\beta. \tag{9-38}$$

If $f_1(x_2) = \kappa_1 x_2$ and $f_2(x_1) = \kappa_2 x_1$, then the differential equation reduces to

$$\dot{x} = \begin{bmatrix} -\alpha & -\kappa_1 \\ -\kappa_2 & -\beta \end{bmatrix} x,$$

with the characteristic polynomial $|\lambda I - A_g| = \lambda^2 + (\alpha + \beta)\lambda + (\alpha\beta - \kappa_1\kappa_2)$, which demonstrates that $\kappa_1\kappa_2 < \alpha\beta$ is the necessary and sufficient condition to guarantee asymptotic stability. Thus, the result obtained from MSC lb is equivalent to the Hurwitz condition if it is stipulated that κ_1 and κ_2 are nonnegative.

EXAMPLE 3. A case in which it is possible to treat a system with two nonlinearities as two scalar systems is provided by the symmetric system considered by Lindgren and Pinkos [1], namely

$$\underline{W}(s) = \begin{bmatrix} W_1(s) & W_2(s) \\ W_2(s) & W_1(s) \end{bmatrix}; \qquad 0 \leqslant \frac{f_i(r_i)}{r_i} \leqslant \bar{F}, \qquad i = 1, 2.$$

We define two auxiliary functions for subsequent analysis:

$$\psi_1(\omega^2) \triangleq \operatorname{Re}\{(\beta_0 i\omega + \gamma_0)[W_1(i\omega) + \bar{F}^{-1}]\}$$
$$\psi_2(\omega^2) \triangleq \operatorname{Re}\{(\beta_0 i\omega + \gamma_0)W_2(i\omega)\}.$$

[The use of the scalar multiplier results in no loss in generality in this case.] For stability it is required that $\psi_1(\omega^2) \geqslant 0$ and $\psi_1{}^2 - \psi_2{}^2 \geqslant 0$; this in turn is equivalent to $\psi_1 + \psi_2 \geqslant 0$ and $\psi_1 - \psi_2 \geqslant 0$. Thus, the symmetric system is absolutely stable if

$$(\beta_0 s + \gamma_0)[W_1(s) + W_2(s) + \bar{F}^{-1}] \in \{\text{PR}\}$$
$$(\beta_0 s + \gamma_0)[W_1(s) - W_2(s) + \bar{F}^{-1}] \in \{\text{PR}\};$$

the graphical methods given in Chapter VII, Section 3 can then be used to check these conditions.

For completeness, it must be possible to obtain realizations $\{C, A, B, R\}$ of $\underline{W}(s)$ and $\{C_1, A, B, H_\infty, H_1\}$ of $\underline{H}(s)$ [Eq. (9-33)] such that $\{C_1, A, B\}$ is minimal in both Examples 1 and 3.

Examples for $m > 2$ are not presented. There is no difficulty encountered in generating higher order examples in SF1 that are easily treated using Lemma 9-2, while the direct application of MSC 1 becomes tedious.

B. Stability Criteria for Other Nonlinearity Classes

By following the developments presented in Section 5A, it is only an algebraic exercise to duplicate the extension of the stability criteria for all classes considered in Chapter V. The operations that have to be carried out in deriving these stability criteria are considerably more involved than in the scalar case. Further, since the positive realness of a matrix has to be established in all cases, the criteria are also significantly more complex to apply. With the possible exception of the generalized Popov criterion derived in Section 5A, it may thus safely be said that the criteria presented here are generally of little practical utility except for the case of systems in SF1.

In Chapter V it is shown that for all systems with a single nonlinearity in the feedback path the form of the frequency domain criterion remains essentially the same, that is, $Z(s)[W(s) + 1/\bar{F}_N] \in \{\text{PR}\}$. Different multiplier classes $\{Z_N(s)\}$ are shown to exist so that, given a function $f(\cdot) \in \{N\}$ in the feedback path, a multiplier $Z(s)$ such that $Z^{\pm 1}(s) \in \{Z_N(s)\}$ satisfying this condition must be selected to guarantee absolute stability. Virtually identical ideas carry over to the case when the system has multiple nonlinearities in the feedback path. In all cases the central condition in establishing the absolute stability of the system is the existence of a diagonal matrix $\underline{Z}(s)$ such that

$$\underline{H}(s) \triangleq \underline{Z}(s)[\underline{W}(s) + \underline{\bar{F}}_N^{-1}] \in \{\underline{\text{MPR}}\}. \tag{9-39}$$

When all the nonlinearities in the feedback path $f_i(\cdot)$ belong to the same class $\{N\}$, the corresponding elements of the multiplier matrix $\underline{Z}(s)$ in Eq. (9-39) are also found to belong to the same class $\{Z_N(s)\}$ or its inverse: $Z_{ii}(s) \in \{Z_N\}$, $i = 1, 2, \ldots, m$ or $Z_{ii}^{-1}(s) \in \{Z_N\}$, $i = 1, 2, \ldots, m$.

Without presenting a detailed derivation of all the frequency domain criteria we merely state below the stability criterion for a system with multiple monotonic nonlinearities in the feedback path.

The overall system is defined by Eq. (9-1) with

$$\begin{aligned} p &= -\underline{f}(r) = -F(r)r, \\ \underline{f}(r) &\in \{\underline{F}_m\}, \qquad F(r) \in [\underline{0}, \underline{\bar{M}}], \\ A &\in \{A_1\}, \qquad A_{\underline{\bar{M}}} \in \{A_1\}. \end{aligned} \tag{9-40}$$

For multiple monotonic nonlinear time-invariant gains, $\underline{f}(r) \in \{\underline{F}_m\}$, we have for any vectors r and q and any matrix $\underline{\gamma}_1 \triangleq \text{diag}[\gamma_{1i}] \geq \underline{0}$ that

$$(r - q)^{\mathrm{T}}\underline{\bar{M}}\underline{\gamma}_1(r - q) \geq [\underline{f}(r) - \underline{f}(q)]^{\mathrm{T}}\underline{\gamma}_1(r - q) \geq 0.$$

The multiplier class $\{\underline{Z}_{F_m}\}$ obtained using this relation is defined by

$$\{\underline{Z}_{F_m}(s)\} \triangleq \{\underline{Z}(s) = \text{diag}[Z_{\mathrm{RL}i}(s)]\} \tag{9-41}$$

where $Z_{\mathrm{RL}i}(s)$ are RL multipliers defined in Eq. (5-34). The problem of obtaining the inverse multiplier $\underline{Z}(s) = \text{diag}[Z_{\mathrm{RC}i}(s)]$ is not as straightforward as in the previous case but it can still be proved quite directly by using the range shifting transformation and inversion as in the scalar case [Eqs. (5-39) and (5-40)]; system inversion is guaranteed to be possible since $A_{\underline{\bar{M}}} \in \{A_1\}$ guarantees that $I + R\underline{\bar{M}}$ is nonsingular. We note that this procedure only permits the use of $\underline{Z}(s) = \text{diag}[Z_{RCi}(s)]$; it is not possible to obtain a diagonal multiplier composed of mixed RL and RC functions by this technique, in contrast with the extension of the Popov criterion (MSC 1b).

MULTIPLE-GAIN STABILITY CRITERION 2 (MSC 2). The system defined by Eqs. (9-1) and (9-40) is absolutely $\{\underline{F}_m\}$ stable if $\underline{Z}(s)$ exists such that $\underline{Z}^{\pm 1}(s) \in \{\underline{Z}_{F_m}(s)\}$ and

$$\underline{H}(s) = \underline{Z}(s)[\underline{W}(s) + \bar{\underline{M}}^{-1}] \in \{\underline{\text{MPR}}\}$$

where $\underline{Z}\underline{W}$ is proper if $\underline{Z} \in \{\underline{Z}_{F_m}\}$ or $\underline{Z}\underline{W}^{-1}$ is proper if $\underline{Z}^{-1} \in \{\underline{Z}_{F_m}\}$. □

The fact that $\underline{H}(s)$ must be achieved as a minimal realization using the given original system matrices (A, B) is in some cases a quite stringent condition. It may well preclude canceling poles of $\underline{W}(s)$ by zeros of $\underline{Z}(s)$, for example, although pole-zero cancellation is useful in that it simplifies checking the positive realness of $\underline{H}(s)$.

Many of the questions arising from the minimality condition may be clarified by considering an example in SF1: take

$$\underline{W}(s) = \hat{R} + \hat{C}(sI - \hat{A})^{-1}\hat{B} = \begin{bmatrix} \frac{(s+2)(s+3)}{(s+1)(s+2)} H_0(s) & \cdots \\ \frac{(s+1)(s+4)}{(s+1)(s+2)} H_0(s) & \end{bmatrix}$$

where $H_0(s)$ has no poles or zeros at $s = -1, -2, -3$ or -4, and it is the proper ratio of relatively prime polynomials.

(i) It is tempting to take $\underline{Z}_1(s) = \text{diag}[(s+1)/(s+3), (s+2)/(s+4)] \in \{\underline{Z}_{F_m}\}$ to obtain

$$\underline{H}_1(s) = \underline{Z}_1(s)\underline{W}(s) = H_0(s)\begin{bmatrix} 1 & 1 \\ 1 & 1 \end{bmatrix},$$

which would seem to imply absolute $\{\underline{F}_m\}$ stability if $H_0(s) \in \{\text{PR}\}$. However, by Lemmas 9-1 and 9-2 it is clear that the pair (A_1, B_1) in SF1 which provides a minimal realization of $\underline{H}_1(s)$ cannot be used to realize $\underline{W}(s)$.

(ii) We can, however, take $\underline{Z}_2(s)$ defined by $\underline{Z}_2^{-1}(s) = \text{diag}[(s+2)/(s+4), (s+1)/(s+3)] \in \{\underline{Z}_{F_m}\}$ to obtain

$$\underline{H}_2(s) = \underline{Z}_2(s)\underline{W}(s) = \frac{(s+3)(s+4)}{(s+1)(s+2)} H_0(s)\begin{bmatrix} 1 & 1 \\ 1 & 1 \end{bmatrix} \triangleq \hat{H}_0(s)\begin{bmatrix} 1 & 1 \\ 1 & 1 \end{bmatrix},$$

which by Lemma 9-2 guarantees absolute $\{\underline{F}_m\}$ stability if $\hat{H}_0(s) \in \{\text{PR}\}$, since from the given conditions $\hat{H}_0(s)$ is the ratio of relatively prime polynomials.

(iii) One important point to be stressed is that the condition $\underline{H}(s) \in \{\underline{\text{MPR}}\}$ does not preclude system augmentation to allow the use of more complicated multiplier matrices: take

$$\underline{W}_a(s) \triangleq \hat{R} + \hat{C}_a(sI - \hat{A}_a)^{-1}\hat{B}_a = \begin{bmatrix} \frac{(s+2)(s+3)(s+\lambda)}{(s+1)(s+2)(s+\lambda)} H_0(s) & \cdots \\ \frac{(s+1)(s+4)(s+\lambda)}{(s+1)(s+2)(s+\lambda)} H_0(s) & \end{bmatrix};$$

then if

$$\underline{Z}_3(s) = \operatorname{diag}\left[\frac{(s+4)(s+\eta)}{(s+2)(s+\lambda)}, \frac{(s+3)(s+\eta)}{(s+1)(s+\lambda)}\right],$$

we have

$$\underline{H}_3(s) = \underline{Z}_3(s)\underline{W}_a(s) = \frac{(s+3)(s+4)(s+\eta)}{(s+1)(s+2)(s+\lambda)} H_0(s) \begin{bmatrix} 1 & 1 \\ 1 & 1 \end{bmatrix} \triangleq \tilde{H}_0(s) \begin{bmatrix} 1 & 1 \\ 1 & 1 \end{bmatrix};$$

the scalars η and λ may be any real positive numbers chosen such that $Z_{ii}(s)$ are both either RL or RC functions and $\tilde{H}_0(s)$ is the ratio of relatively prime polynomials; absolute $\{\underline{F}_m\}$ stability is guaranteed if $\tilde{H}_0(s) \in \{\text{PR}\}$.

C. *Systems with Mixed Nonlinearities*

The criterion MSC 2 deals with systems with many nonlinearities, all of which belong to the same class $\{F_m\}$. The question naturally arises as to how the stability criterion may have to be modified when the nonlinearities in the feedback path belong to different classes, that is, $f_i(r_i) \in \{N_i\}$. The same approach used in earlier cases yields the result that elements of the diagonal matrix $\underline{Z}(s)$ in Eq. (9-39) may belong to different classes, depending upon the particular nonlinearity in the feedback path. If $f_i(r_i) \in \{N_i\}$, the element $Z_{ii}(s)$ of $\underline{Z}(s)$ would belong to the corresponding class $\{Z_{N_i}(s)\}$. The following example represents an interesting application of the criterion.

EXAMPLE 4. Consider the differential equation

$$\dot{x} = \begin{bmatrix} -\beta & \alpha \\ 0 & 0 \end{bmatrix} x - \begin{bmatrix} 0 & 0 \\ 1 & 1 \end{bmatrix} \begin{bmatrix} f_1(x_1) \\ f_2(x_2) \end{bmatrix}; \qquad \alpha > 0, \quad \beta > 0.$$

The transfer function is

$$\underline{W}(s) = (s(s+\beta))^{-1} \begin{bmatrix} \alpha & \cdots \\ (s+\beta) & \cdots \end{bmatrix} = \frac{\alpha}{s} \begin{bmatrix} (s+\beta)^{-1} & \cdots \\ \alpha^{-1} & \end{bmatrix}.$$

By an extension of MSC 1 to the particular case, it could be proven that the system is absolutely $\{\underline{F}\}$ stable in the infinite sector $(\underline{0}, \underline{\infty})$. However, the corresponding linear system

$$\dot{x} = \begin{bmatrix} -\beta & \alpha \\ -\kappa_1 & -\kappa_2 \end{bmatrix} x$$

has the characteristic polynomial $p(\lambda) = \lambda^2 + (\beta + \kappa_2)\lambda + (\alpha\kappa_1 + \beta\kappa_2)$, so the Hurwitz conditions are

(i) $\kappa_2 > -\beta$,

(ii) $\alpha\kappa_1 + \beta\kappa_2 > 0$,

which do not correspond to the condition $\underline{\kappa} \in (\underline{0}, \underline{\infty})$. Krasovskii [1] showed that the corresponding conditions $f_2(x_2)/x_2 > -\beta$ and $[\alpha(f_1(x_1)/x_1) + \beta(f_2(x_2)/x_2)] > 0$ do not guarantee equiasymptotic stability in the whole, even with a single nonlinearity. To see that this is so, refer to the differential equation (4-14).

If, however, we strengthen these constraints by imposing the conditions

(1) $\dfrac{f_2(x_2) - f_2(\hat{x}_2)}{x_2 - \hat{x}_2} \geqslant -(\beta - \varepsilon)$ for all x_2 and $\hat{x}_2 \neq x_2$, $\varepsilon > 0$,

(2) $\alpha f_1(x_1)/x_1 \geqslant \beta^2$,

we can guarantee absolute $\{F, F_m\}$ stability: Defining

$$h_1(x_1) \triangleq f_1(x_1) - (\beta^2/\alpha)x_1 \in \{F\},$$
$$h_2(x_2) \triangleq f_2(x_2) + (\beta - \varepsilon)x_2 \in \{F_m\},$$

the original differential equation transforms into the standard form with

$$\underline{W}(s) = (s^2 + \varepsilon s + \varepsilon\beta)^{-1}\begin{bmatrix} \alpha & \alpha \\ s + \beta & s + \beta \end{bmatrix}.$$

Using $\underline{Z}(s) \triangleq \operatorname{diag}[(s + \varepsilon)/\alpha, (s + \varepsilon)/(s + \beta)]$ yields

$$\underline{Z}(s)\underline{W}(s) = \frac{s + \varepsilon}{s^2 + \varepsilon s + \varepsilon\beta}\begin{bmatrix} 1 & 1 \\ 1 & 1 \end{bmatrix} \triangleq H_0(s)\begin{bmatrix} 1 & 1 \\ 1 & 1 \end{bmatrix}.$$

Given $\beta > \varepsilon > 0$, $H_0(s)$ is the positive real ratio of relatively prime polynomials. Since $((s + \varepsilon)/\alpha) \in \{Z_F\}$ and $((s + \varepsilon)/(s + \beta)) \in \{Z_{F_m}\}$, this condition suffices to guarantee absolute $\{F, F_m\}$ stability under constraints (1) and (2). To relate these restrictions to the Hurwitz conditions, we note that $\Delta f_2/\Delta x_2 \geqslant -(\beta - \varepsilon)$ directly corresponds in a strict sense to $\kappa_2 > -\beta$ and that this ensures that $f_2(x_2)/x_2 \geqslant -(\beta - \varepsilon)$. This latter condition in conjunction with constraint (2) then guarantees that

$$\left[\alpha\frac{f_1(x_1)}{x_1} + \beta\frac{f_2(x_2)}{x_2}\right] > 0,$$

which corresponds to (ii).

The stability criterion applied in Example 4 represents the most general criterion that may be derived for a system with multiple time-invariant nonlinearities using the technique discussed in this book. A formal statement of this criterion may be given as follows:

GENERAL MULTIPLE-GAIN STABILITY CRITERION 1 (GMSC 1). The system described by Eq. (9-1) with

$$\begin{aligned} p &= -\underline{f}(r) = -\underline{F}(r)r; \\ f_i(r_i) &\in \{N_i\}, \qquad \underline{F}(r) \in [\underline{0}, \bar{\underline{F}}_N]; \\ A &\in \{A_1\}, \qquad A_{\underline{F}_N} \in \{A_1\} \end{aligned} \tag{9-42}$$

is absolutely $\{\underline{N}\}$ stable if there exists a multiplier of the form $\underline{Z}(s) = \mathrm{diag}[Z_i(s) \in \{Z_{N_i}(s)\}]$ such that

$$\underline{H}(s) = \underline{Z}(s)[\underline{W}(s) + \bar{\underline{F}}_N^{-1}] \in \{\underline{\mathrm{MPR}}\}, \tag{9-43}$$

and $\underline{Z}(s)\underline{W}(s)$ is proper. □

The only major problem that arises in obtaining a general result with mixed NLTI gains is found in obtaining inverse multipliers. At the present stage of development, $\underline{Z}^{-1}(s)$ may be used only if all nonlinearities are at least monotonic.

As in all previous cases, this result is greatly simplified if the system is in SF1, and if the gains lie in the infinite sector. Then the function $\underline{H}(s)$ is in the form indicated in Lemma 9-2, and thus the condition for absolute stability is that

$$\underline{W}(s) = W_0(s) \begin{bmatrix} Z_1^{-1} & & \\ Z_2^{-1} & & \\ \cdot & & \cdots \\ \cdot & & \\ \cdot & & \\ Z_m^{-1} & & \end{bmatrix}$$

where $W_0(s)$ is a scalar transfer function that is the proper ratio of relatively prime polynominals and positive real, and $Z_i(s) \in \{Z_{N_i}(s)\}$, $i = 1, 2, \ldots, m$. Both of these results were obtained by Narendra and Neuman [3].

6. Stability of Nonlinear Time-Varying Systems

While all the criteria discussed in Section 5 can clearly be extended to the time-varying case, most of them have little practical utility. We consider briefly two generalizations of MSC 1 to NLTV systems primarily to indicate the form of the criteria in time-varying situations and the complexity involved in applying them. The frequency domain criteria (as in the scalar case) are identical to those derived in Section 5 with constraints imposed on the time variation of the feedback gains determined by the multipliers chosen. Both

Lyapunov's theorem and the theorem of Corduneanu are applied to obtain instantaneous and time-averaged constraints on the rate of time variation of the gains.

A. Instantaneous or Point Constraints on $(d/dt)K(t)$

The system model represented by Eqs. (9-1) and (9-23) is generalized by replacing $\underset{\sim}{f}(r)$ with $K(t)\underset{\sim}{f}(r)$.

$$
\begin{aligned}
&p = -K(t)\underset{\sim}{f}(r) = -K(t)F(r)r; \\
&f_i \in \{F\}, \qquad F(r) \in [\underline{0}, \underline{\bar{F}}], \\
&k_i \in \{K_1\}, \qquad K(t) \in [\underline{0}, \underline{\bar{K}}], \\
&A \in \{A_1\}, \qquad A_{\underline{G}} \in \{A_1\}, \qquad \underline{\bar{G}} = \underline{\bar{K}}\underline{\bar{F}},
\end{aligned} \tag{9-44}
$$

where $\bar{F}_i$ may be infinite but $\bar{K}_i$ may not be (Chapter II, Section 3).

The Lyapunov function candidate $v(x, t)$ for the system (9-44) has the same form as that used for the time-invariant case [Eq. (9-27)] with $\underset{\sim}{f}(r)$ being replaced by the time-varying nonlinearities $K(t)\underset{\sim}{f}(r)$. The time derivative of $v(x, t)$ is consequently similar to the expression in Eq. (9-28) but contains additional terms involving $\dot{K}(t)$. Using Lemma 10, it is then shown that if $\underline{\hat{H}}(s) \in \underline{\{\text{MSPR}\}}$, the first four terms of $\dot{v}$ are negative definite. The remaining terms of $\dot{v}$ are then shown to be nonpositive if

$$\underline{\Phi}_i k_i(t)[1 - (k_i(t)/\bar{K}_i)] \geqslant (\hat{\beta}_{0i}/\gamma_{0i})\,(dk_i/dt), \quad i = 1, 2, \ldots, m.$$

These conditions correspond exactly to those obtained in the scalar case; $\hat{\beta}_{0i}/\gamma_{0i}$ is determined by the scalar Popov multiplier chosen as the ith element of $\underline{Z}(s)$. The arguments presented previously allow $\hat{\beta}_{0i} < 0$ if desired. The frequency domain condition for those multiplier elements with $\hat{\beta}_{0i} < 0$ can be converted into the standard form by taking $(-\hat{\beta}_{0i}s + \gamma_{0i})^{-1} \triangleq (\beta_{0i}s + \gamma_{0i})^{-1} \in \underline{\{\text{SPR}\}}$, and constraining dk_i/dt to satisfy $dk_i/dt \geq -(\beta_{0i}/\gamma_{0i})\underline{\Phi}_i k_i[1 - k_i/\bar{K}_i]$ for those gains which correspond to the use of the inverse multiplier element in $\underline{Z}(s)$. A general stability criterion for systems described by Eqs. (9-1) and (9-44) may thus be stated as follows:

Multiple-gain Stability Criterion 3 (MSC 3). The system described by Eqs. (9-1) and (9-44) is absolutely $\{\underline{F}, \underline{K}_1\}$ stable if

(1) MSC 1b is strengthened by demanding that $\underline{H}(s)$ [Eq. (9-33) with $\underline{\bar{F}}$ replaced by $\underline{\bar{G}}$] satisfy $\underline{H}(s) \in \underline{\{\text{MSPR}\}}$,
(2) $(d/dt)K(t)$ is restricted by

$$\underline{\Delta}\dot{K} \leqslant \underline{\bar{\Lambda}}\underline{\underline{\Phi}}K(I - K\underline{\bar{K}}^{-1}) \tag{9-45}$$

where $\bar{\underline{\Lambda}}$ and $\underline{\Delta}$ are defined as follows:

$$\underline{\Delta} \triangleq \text{diag}[+1 \quad \text{if } Z_i(s) \in \{Z_{N_i}(s)\}, \quad -1 \quad \text{if } Z_i^{-1}(s) \in \{Z_{N_i}(s)\}] \tag{9-46}$$

$$\bar{\underline{\Lambda}} \triangleq \max\{\underline{\Delta}\colon \underline{Z}_{\lambda}(s) \triangleq \text{diag}[Z_i^{\pm 1}(s - \lambda_i)] \in \{\underline{Z}_N(s)\}, \quad \underline{0} \leqslant \underline{\lambda} < \bar{\underline{\Lambda}}\}. \tag{9-47}$$

□

The interpretation of the time domain constraint on $K(t)$ as given in Eq. (9-45) remains unchanged from the scalar case.

If the frequency domain condition of MSC 3 is satisfied for $\beta_{0i} = 0$, then we obtain a constraint on $k_i(t)$ which may be viewed as the extension of the circle criterion for $[0, \bar{G}_i]$. Usually no geometrical interpretation can be attempted. An exception is again provided by the symmetric system (see Example 3): For the system described by

$$\underline{W}(s) = \begin{bmatrix} W_1(s) & W_2(s) \\ W_2(s) & W_1(s) \end{bmatrix}; \; 0 \leqslant \frac{g_i(r_i, t)}{r_i} \leqslant \bar{G}, \quad i = 1, 2,$$

then $G_1(s) \triangleq [W_1(s) + W_2(s)]$ and $G_2(s) \triangleq [W_1(s) - W_2(s)]$ must each satisfy the scalar circle criterion (Chapter VII, Section 2).

EXAMPLE 6. An interesting general example has been provided by Moore and Anderson [1]. While the time invariant case is considered in this paper, we treat the corresponding NLTV case. In the classical optimal control problem dealing with a linear system $\dot{x} = Ax + Bp$ an input function $p(t)$ which minimizes the quadratic index of performance

$$V(x_0, p) = \int_{t_0}^{\infty} [p^{\mathrm{T}}p + x^{\mathrm{T}}Qx]\, dt; \qquad Q = Q^{\mathrm{T}} > 0$$

is to be determined. This is found to be provided by state variable feedback, that is,

$$p = -Mx,$$

where M satisfies $M^{\mathrm{T}} = PB$ and P is the unique real positive definite symmetric matrix satisfying

$$A^{\mathrm{T}}P + PA + Q = PBB^{\mathrm{T}}P = M^{\mathrm{T}}M. \tag{9-48}$$

Assuming that certain nonlinearity and time variation is unavoidable in implementing the optimal control law, we are interested in determining how much this relation can deviate from linear control while still guaranteeing absolute stability. If we define $r \triangleq Mx$ as the nominal plant output and $p = -g(r, t) = -G(r, t)r$, this scheme corresponds to $\underline{W}(s) = M(sI - A)^{-1}B$ in Fig. 9-1 where the nominal optimal feedback control law is given by $p = -r$, or $G(r, t) = I$.

It can be shown in a straightforward manner that the conditions $g_i(r_i, t)/r_i \in [\frac{1}{2}, \infty)$ and $\{M, \hat{A}, B\}$ minimal where $\hat{A} \triangleq A - \frac{1}{2}BM$ guarantee absolute stability. Transforming the problem to the standard infinite sector case, we obtain

$$\hat{g}(r,t) \triangleq [g(r,t) - \tfrac{1}{2}r] \triangleq \hat{G}(r,t)r$$
$$\hat{W}(s) \triangleq W(I + \tfrac{1}{2}W)^{-1} = M[sI - A + \tfrac{1}{2}BM]^{-1}B;$$

absolute stability for $\hat{G} \in [0, \infty)$ is guaranteed if $\hat{W}(s) \in \{\underline{\text{MSPR}}\}$ by the extended circle criterion.

Since $Q = Q^{\mathrm{T}} > 0$, we can always find some $\varepsilon > 0$ and some real matrix Q_1 such that $Q = 2\varepsilon P + Q_1 Q_1^{\mathrm{T}}$; thus Eq. (9-48) may be put in the form

$$(A - \tfrac{1}{2}BM)^{\mathrm{T}}P + P(A - \tfrac{1}{2}BM) = -Q_1 Q_1^{\mathrm{T}} - 2\varepsilon P,$$

which together with $PB = M^{\mathrm{T}}$, demonstrates that $\hat{W}(s) \in \{\underline{\text{MSPR}}\}$ as required (refer to Lemma 10 and Definition PR6).

Another example of the application of the extended circle criterion may be found in Sandberg [4], where the stability of networks with time-varying capacitors is considered. For other classes of nonlinear gains and for LTV systems, similar results can be obtained for the finite sector case essentially by inspection. The ability to mix classes of NLTV gains is again permitted, as in the NLTI case. These comments motivate the following generalization of General Stability Criterion 2 (Chapter VI, Section 4).

General Multiple-gain Stability Criterion 2 (GMSC 2). The system described by Eq. (9-1) with

$$\begin{aligned} p &= -K(t)f(r) = -K(t)F(r)r \\ f_i(r_i) &\in \{N_i\}, \quad F(r) \in [0, \bar{F}_N], \\ k_i(t) &\in \{K_i\},\dagger \quad K(t) \in [0, \bar{K}], \qquad \bar{K} < \infty \\ A &\in \{A_1\}, \quad A_{\bar{G}_N} \in \{A_1\}, \qquad \bar{G}_N = \bar{F}_N \bar{K}, \end{aligned} \tag{9-49}$$

is absolutely $\{N_i, K_i\}$ stable if

(1) GMSC 1 is strengthened by demanding that $H(s)$ [Eq. (9-43) where $\bar{F}_N$ is replaced by $\bar{G}_N$] satisfy $H(s) \in \{\underline{\text{MSPR}}\}$,
(2) condition (2) of MSC 3 is satisfied with $\Delta = I$. □

It is more difficult to give meaningful examples for the application of stability criteria to systems with multiple NLTV gains since well-known problems (for instance, the Mathieu equation) do not exist.

B. *Time-Averaged or Integral Constraints on $(d/dt)K(t)$*

Based on the previous derivations for systems with multiple NLTI and NLTV gains, the stability conditions resulting from an application of the

† $K_i = K_1$ unless the frequency domain multiplier element is $Z_i = \gamma_{0i}$, which corresponds to the circle criterion, in which case $k_i(t) \in \{K_0\}$ is allowed.

theorem of Corduneanu (Chapter III, Section 3) and Lemma 10 (Section 3) may be stated directly.

GENERAL MULTIPLE-GAIN STABILITY CRITERION 3 (GMSC 3). The system described by Eq. (9-1) with

$$p = -K(t)\underset{\sim}{f}(r) = -K(t)F(r)\mathrm{r};$$

$$f_i(r_i) \in \{N_i\}, \qquad F(r) \in [\underset{\sim}{0}, \underset{\sim}{\infty}),$$

$$k_i(t) \in \{K_i\}^\dagger, \qquad K(t) \in (\underset{\sim}{0}, \underset{\sim}{\bar{K}} < \underset{\sim}{\infty}),$$

$$(A + \mu I) \in \{A_1\}, \qquad \mu > 0$$

is absolutely $\{N_i, K_i\}$ stable if

(1) $\underset{\sim}{Z}(s)$ exists such that
  (i) $\underset{\sim}{Z}_\lambda(s) \triangleq \mathrm{diag}[Z_i(s - \lambda_i)]$ satisfies $\underset{\sim}{Z}_\lambda(s) \in \{\underset{\sim}{Z}_N(s)\}$, $\underset{\sim}{0} \leqslant \underset{\sim}{\lambda} < \underset{\sim}{\bar{\Lambda}}$;
  (ii) $\underset{\sim}{H}_\mu(s) \triangleq \underset{\sim}{Z}(s - \mu)\underset{\sim}{W}(s - \mu) \in \{\underline{\mathrm{MPR}}\}$ is proper.

(2) $\hat{p}(t) \triangleq \sup\{-2\mu; -[\underset{\sim}{\Phi}_i\bar{\Lambda}_i - (k_i)^{-1}dk_i/dt], i = 1, 2, \ldots, m\}$ satisfies

$$\lim_{t\to\infty} \int_{t_0}^{t} \hat{p}(\tau)\, d\tau = -\infty \tag{6-37a}$$

uniformly with respect to t_0 if $\hat{p}(t)$ is aperiodic, or

$$\int_0^T \hat{p}(t)\, dt < 0 \tag{6-37b}$$

if $\hat{p}(t) = \hat{p}(t + T)$ for all t. □

The complex interrelations implicit in condition (2) make the criterion primarily of academic interest.

It is indicated throughout this chapter that the absolute stability problem for multivariable systems becomes quite unwieldy when compared with the analogous problems arising in scalar situations. This is particularly evident in the difficulty encountered in applying the resulting criteria, as well as in the number of details that must be considered before any derivation can be carried out rigorously.

While the results presented are tentative and incomplete in many respects, Lemmas 7–10 are interesting in their own right, relating as they do the concept of positive realness in network theory to the canonical representation of dynamical systems and in turn to the stability of such systems in the sense of Lyapunov. We can anticipate further modifications and extensions of these lemmas in the future which will lead to a better understanding of the close interrelationships in the three areas.

† $K_i = K_2$ unless the frequency domain multiplier element is $Z_j = \gamma_{0j}$, in which case $k_j(t) \in \{K_0\}$ is permitted since $v(x, t)$ and $\hat{p}(t)$ are independent of $k_j(t)$ for that particular gain.

APPENDIX†

MATRIX VERSION OF THE KALMAN–YAKUBOVICH LEMMA

The proofs of Lemmas 7–9, which are extensions of the Kalman–Yakubovich Lemma to the matrix case, depend on the important result that if $\underline{H}(s)$ is a positive real matrix with a minimal realization $\{C_1, A_1, B_1\}$ then there exist two minimal realizations of the matrix $\underline{Y}(s)$ defined by $\underline{Y}(s) \triangleq \underline{H}(s) + \underline{H}^{\mathrm{T}}(-s)$ given by:

$$\left\{\begin{bmatrix} C_1^{\mathrm{T}} \\ -B_1 \end{bmatrix}^{\mathrm{T}}, \begin{bmatrix} A_1 & \underline{0} \\ \underline{0} & -A_1^{\mathrm{T}} \end{bmatrix}, \begin{bmatrix} B_1 \\ C_1^{\mathrm{T}} \end{bmatrix}\right\} \quad \text{and} \quad \left\{\begin{bmatrix} PB_1 \\ -B_1 \end{bmatrix}^{\mathrm{T}}, \begin{bmatrix} A_1 & \underline{0} \\ \underline{0} & -A_1^{\mathrm{T}} \end{bmatrix}, \begin{bmatrix} B_1 \\ PB_1 \end{bmatrix}\right\}, \tag{A-1}$$

where P is a positive definite matrix. This result in turn depends on a lemma on spectral factorization due to Youla [1] and the properties of minimal realizations for dynamical systems. We first briefly present the preliminary result given in Eq. (A-1) before proceeding to prove Lemmas 7–9.

If $\{C_1, A_1, B_1\}$ is a minimal realization of a positive real matrix $\underline{H}(s)$, by direct calculation we have one realization of $\underline{Y}(s) = \underline{H}(s) + \underline{H}^{\mathrm{T}}(-s)$:

$$\{C_2, A_2, B_2\} = \left\{\begin{bmatrix} C_1^{\mathrm{T}} \\ -B_1 \end{bmatrix}^{\mathrm{T}}, \begin{bmatrix} A_1 & \underline{0} \\ \underline{0} & -A_1^{\mathrm{T}} \end{bmatrix}, \begin{bmatrix} B_1 \\ C_1^{\mathrm{T}} \end{bmatrix}\right\}, \tag{A-2}$$

where A_2 is a $(2n \times 2n)$ matrix. If $\underline{H}(s)$ has only poles in Re $s < 0$ and thus

† The proofs presented here follow closely those presented in Anderson [2].

$\underline{H}^{\mathrm{T}}(-s)$ in Re $s > 0$, the dimension of the minimal realization of $\underline{Y}(s)$ is $2n$, so $\{C_2, A_2, B_2\}$ is minimal.

By the spectral factorization lemma of Youla, if $\underline{H}(s)$ is positive real and $\underline{H}(s) + \underline{H}^{\mathrm{T}}(-s)$ has rank r almost everywhere, then there exists an $(r \times m)$ matrix $\underline{G}(s)$ such that

$$\underline{Y}(s) = \underline{H}(s) + \underline{H}^{\mathrm{T}}(-s) = \underline{G}^{\mathrm{T}}(-s)\underline{G}(s), \tag{A-3}$$

and $\underline{G}(s)$ is analytic in Re $s > 0$. If $\{Q_1{}^{\mathrm{T}}, R_1, S_1\}$ is a minimal realization of $\underline{G}(s)$, by direct calculation it is seen that

$$\{C_3, A_3, B_3\} = \left\{\begin{bmatrix} \underline{0} \\ -S_1 \end{bmatrix}^{\mathrm{T}}, \begin{bmatrix} R_1 & \underline{0} \\ Q_1Q_1{}^{\mathrm{T}} & -R_1{}^{\mathrm{T}} \end{bmatrix}, \begin{bmatrix} S_1 \\ \underline{0} \end{bmatrix}\right\} \tag{A-4}$$

is also a minimal realization of $\underline{Y}(s)$. Using a nonsingular transformation

$$T = \begin{bmatrix} I & \underline{0} \\ P & I \end{bmatrix}, \qquad \text{where} \quad PR_1 + R_1{}^{\mathrm{T}}P = -Q_1Q_1{}^{\mathrm{T}},$$

an alternative minimal realization of $\underline{Y}(s)$ is obtained:

$$\{C_4, A_4, B_4\} = \left\{\begin{bmatrix} PS_1 \\ -S_1 \end{bmatrix}^{\mathrm{T}}, \begin{bmatrix} R_1 & \underline{0} \\ \underline{0} & -R_1{}^{\mathrm{T}} \end{bmatrix}, \begin{bmatrix} S_1 \\ PS_1 \end{bmatrix}\right\}. \tag{A-5}$$

Since $\{C_2, A_2, B_2\}$ and $\{C_4, A_4, B_4\}$ are minimal realizations of the same matrix $\underline{Y}(s)$ and A_1 and R_1 have strictly negative eigenvalues, then A_1 and R_1 are similar. This in turn is found to imply that a minimal realization of the form $\{Q^{\mathrm{T}}, A_1, B_1\}$ exists for $\underline{G}(s)$. Using this minimal realization we obtain relation (A-1) (substituting $S_1 = B_1$, $R_1 = A_1$ in Eq. (A-5)) where

$$A_1{}^{\mathrm{T}}P + PA_1 = -QQ^{\mathrm{T}}. \tag{A-6}$$

Proof of Lemma 7:

Sufficiency. Of the three conditions for positive realness in Definition PR4, we need to verify only (iii): consider $\underline{J}(s) \triangleq \mathrm{He}\{\underline{H}(s)\}$:

$$2\underline{J}(s) = \underline{H}^*(s) + \underline{H}(s) = B_1{}^{\mathrm{T}}(s^*I - A_1{}^{\mathrm{T}})^{-1}C_1{}^{\mathrm{T}} + C_1(sI - A_1)^{-1}B_1.$$

Substituting $C_1{}^{\mathrm{T}} = PB_1$,

$$\begin{aligned} 2\underline{J}(s) &= B_1{}^{\mathrm{T}}\{(s^*I - A_1{}^{\mathrm{T}})^{-1}P + P(sI - A_1)^{-1}\}B_1 \\ &= B_1{}^{\mathrm{T}}(s^*I - A_1{}^{\mathrm{T}})^{-1}P(sI - A_1)^{-1}B_1(s + s^*) \\ &\quad + B_1{}^{\mathrm{T}}(s^*I - A_1{}^{\mathrm{T}})^{-1}QQ^{\mathrm{T}}(sI - A_1)^{-1}B_1. \end{aligned} \tag{A-7}$$

The right-hand side is clearly nonnegative for Re $s > 0$.

Necessity. Since the two minimal realizations given in Eq. (A-1) are for the same matrix $\underline{Y}(s)$, a nonsingular matrix T must exist such that

$$T\begin{bmatrix} A_1 & \underline{0} \\ \underline{0} & -A_1{}^{\mathrm{T}} \end{bmatrix} = \begin{bmatrix} A_1 & \underline{0} \\ \underline{0} & -A_1{}^{\mathrm{T}} \end{bmatrix}T$$

and

$$T\begin{bmatrix} B_1 \\ C_1{}^{\mathrm{T}} \end{bmatrix} = \begin{bmatrix} B_1 \\ PB_1 \end{bmatrix} \quad \text{and} \quad (T^{-1})^{\mathrm{T}}\begin{bmatrix} C_1{}^{\mathrm{T}} \\ -B_1 \end{bmatrix} = \begin{bmatrix} PB_1 \\ -B_1 \end{bmatrix}. \tag{A-8}$$

Such a matrix T must have the form

$$T = \begin{bmatrix} T_1 & \underset{\sim}{0} \\ \underset{\sim}{0} & T_2 \end{bmatrix},$$

where T_1 and $T_2{}^{\mathrm{T}}$ commute with A_1. Hence, from Eq. (A-8) we have $T_1B_1 = B_1$ and $(T_1^{-1})^{\mathrm{T}}C_1{}^{\mathrm{T}} = PB_1$. Since $T_1B_1 = B_1$, we use the commutative property $T_1A_1 = A_1T_1$ to write the controllability matrix:

$$[B_1, A_1B_1, \ldots] = [T_1B_1, A_1T_1B_1, \ldots] = T_1[B_1, A_1B_1, \ldots]. \tag{A-9}$$

Since $\{C_1, A_1, B_1\}$ is a minimal realization of $\underset{\sim}{H}(s)$, (A_1, B_1) is completely controllable and thus the controllability matrix must have rank n. Equation (A-9) can then only be satisfied if $T_1 = I$. Thus completes the proof by showing that $PB_1 = C_1{}^{\mathrm{T}}$. □

Proof of Lemma 8: We first consider a matrix $\underset{\sim}{H}_1(s)$ defined by

$$\underset{\sim}{H}_1(s) = \frac{aa^*}{s - i\omega_0} + \frac{(aa^*)^{\mathrm{T}}}{s + i\omega_0}, \tag{A-10}$$

where a is an n-dimensional complex vector. If $b \triangleq (a + a^{*\mathrm{T}})/\sqrt{2}$ and $c \triangleq i(a - a^{*\mathrm{T}})/\sqrt{2}$, such a matrix can be expressed as

$$\underset{\sim}{H}_1(s) = [b, c](s^2 + \omega_0{}^2)^{-1}\begin{bmatrix} s & -\omega_0 \\ \omega_0 & s \end{bmatrix}\begin{bmatrix} b^{\mathrm{T}} \\ c^{\mathrm{T}} \end{bmatrix}. \tag{A-11}$$

Hence

$$\{\hat{C}, \hat{A}, \hat{B}\} = \left\{\begin{bmatrix} b^{\mathrm{T}} \\ c^{\mathrm{T}} \end{bmatrix}^{\mathrm{T}}, \begin{bmatrix} 0 & -\omega_0 \\ \omega_0 & 0 \end{bmatrix}, \begin{bmatrix} b^{\mathrm{T}} \\ c^{\mathrm{T}} \end{bmatrix}\right\} \tag{A-12}$$

is a minimal realization of $\underset{\sim}{H}_1(s)$. If P is the unit matrix, P satisfies the equations

$$P\hat{A} + \hat{A}^{\mathrm{T}}P = \underset{\sim}{0}; \qquad P\hat{B} = \hat{C}^{\mathrm{T}}. \tag{A-13}$$

If $\underset{\sim}{H}(s)$ is a positive real matrix with poles only on the imaginary axis, it has been shown [Newcomb (1)] that $\underset{\sim}{H}(s)$ may be expressed as

$$\underset{\sim}{H}(s) = \sum_j \frac{E_js + F_j}{s^2 + \omega_j{}^2}, \tag{A-14}$$

where E_j and F_j satisfy certain requirements and the frequencies ω_j are distinct. Each term in the summation can in turn be expressed as

$$\frac{Es + F}{s^2 + \omega^2} = \sum_{i=1}^{k}\left[\frac{a_ia_i^*}{s - i\omega} + \frac{(a_ia_i^*)^{\mathrm{T}}}{s + i\omega}\right] \tag{A-15}$$

if $\underline{H}(s)$ is of degree $2k$. Since the minimal realization of $\underline{H}_1(s)$ is given by Eq. (A-12), the minimal realization of each term of Eq. (A-14) can be directly computed as

$$\{\hat{C}_j, \hat{A}_j, \hat{B}_j\} = \left\{\begin{bmatrix} d_j^{\mathrm{T}} \\ e_j^{\mathrm{T}} \end{bmatrix}^{\mathrm{T}}, \begin{bmatrix} 0 & -\omega_{0j} \\ \omega_{0j} & 0 \end{bmatrix}, \begin{bmatrix} d_j^{\mathrm{T}} \\ e_j^{\mathrm{T}} \end{bmatrix}\right\}. \tag{A-16}$$

The minimal realization $\{C_1, A_1, B_1\}$ of $\underline{H}(s)$ in Eq. (A-14) and the corresponding P satisfying Eq. (A-13) are obtained as

$$\begin{aligned} A_1 &= \bigoplus_i \hat{A}_i \qquad \text{(where } \oplus \text{ denotes direct sum)}, \\ B_1^{\mathrm{T}} &= [\hat{B}_1^{\mathrm{T}}, \hat{B}_2^{\mathrm{T}}, \ldots], \\ C_1 &= [\hat{C}_1, \hat{C}_2, \ldots], \\ P &= \bigoplus_i P_i. \end{aligned} \tag{A-17}$$

The above developments indicate that for one minimal realization of a positive real matrix with poles on the imaginary axis a matrix P can be found satisfying Eq. (A-13). For any other realization $\{C_1T^{-1}, TA_1T^{-1}, TB_1\}$ the matrix $\hat{P} = [T^{\mathrm{T}}]^{-1}PT^{-1}$ is seen to satisfy relations (A-13). Hence it follows that a positive definite matrix P exists for every minimal realization. □

Proof of Lemma 9:

Sufficiency. Again it is only necessary to verify the positive realness of $\underline{H}^*(s) + \underline{H}(s)$ in $\operatorname{Re} s > 0$; consider

$$\begin{aligned} 2\underline{J}(s) &= H_\infty + H_\infty^{\mathrm{T}} + B_1^{\mathrm{T}}(s^*I - A_1^{\mathrm{T}})^{-1}C_1^{\mathrm{T}} \\ &\quad + C_1(sI - A_1)^{-1}B_1 \\ &= D^{\mathrm{T}}D + B_1^{\mathrm{T}}\{(s^*I - A_1^{\mathrm{T}})^{-1}P + P(sI - A_1)^{-1}\}B_1 \\ &\quad + B_1^{\mathrm{T}}(s^*I - A_1^{\mathrm{T}})^{-1}QD + D^{\mathrm{T}}Q^{\mathrm{T}}(sI - A_1)^{-1}B_1 \\ &= \{D^{\mathrm{T}} + B_1^{\mathrm{T}}(s^*I - A_1^{\mathrm{T}})^{-1}Q\}\{D + Q^{\mathrm{T}}(sI - A_1)^{-1}B_1\} \\ &\quad + B_1^{\mathrm{T}}(s^*I - A_1^{\mathrm{T}})^{-1}P(sI - A_1)^{-1}B_1(s + s^*), \end{aligned} \tag{A-18}$$

which is positive semidefinite for $\operatorname{Re} s > 0$.

Necessity. The proof of necessity for the case when $\underline{H}(s)$ has poles only in $\operatorname{Re} s < 0$ follows exactly along the lines indicated for Lemma 7, with PB_1 replaced by $PB_1 + QD$ in Eq. (A-1). For the case when $\underline{H}(s)$ has poles on the imaginary axis also, $\underline{H}(s)$ is expressed as

$$\underline{H}(s) = \underline{H}_a(s) + \underline{H}_b(s)$$

where $\underline{H}_a(s)$ has poles only on the imaginary axis and $\underline{H}_b(s)$ has poles only in $\operatorname{Re} s < 0$.

If the minimal realization of $\underline{H}_a(s)$ is $\{C_a, A_a, B_a\}$, a positive definite matrix P_a exists by Lemma 8 such that $P_aA_a + A_a^{\mathrm{T}}P_a = \underline{0}$, $P_aB_a = C_a^{\mathrm{T}}$. Similarly, if $\underline{H}_b(s)$ has a minimal realization $\{C_b, A_b, B_b, H_{b\infty}\}$, matrices $P_b > \underline{0}$, Q_b, and D exist so that $P_b A_b + A_b{}^T P_b = -Q_b Q_b{}^T$, $PB_b = C_b^T - Q_b D$ and $D^T D = H_{b\infty} + H_{b\infty}^T$. By taking direct sums $P = P_a \oplus P_b$; $A_1 = A_a \oplus A_b$, and defining $B_1{}^T = [B_a{}^T, B_b{}^T]$, $C_1 = [C_a, C_b]$, and $Q^T = [\underline{0}, Q_b{}^T]$, the lemma is established.

Note that these matrices clearly show that $\{Q^T, A_1, B_1\}$ is not a minimal realization (that is, (Q^T, A_1) is not completely observable) unless $\underline{H}(s)$ has poles only in $\mathrm{Re}\, s < 0$, in accord with the Corollary to Lemma 9. □

REFERENCES

Aizerman, M. A.

(1) On a problem relating to the global stability of dynamic systems, *Uspehi Mat. Nauk* **4**, No. 4 (1949).

Aizerman, M. A., and Gantmacher, F. R.

(1) "Absolute Stability of Regulator Systems." English ed. Holden-Day, San Francisco, California 1964 (Russian ed., 1963).

Anderson, B. D. O.

(1) Stability of control systems with multiple nonlinearities, *J. Franklin Inst.* **282**, No. 3 (1966).

(2) A system theory criterion for positive real matrices, *SIAM J. Control*, **5**, No. 2 (1967).

Anderson, B. D. O., and Moore, J. B.

(1) Algebraic structure of generalized positive real matrices, *SIAM J. Control* **6**, No. 4 (1968).

Antosiewicz, H.

(1) A survey of Lyapunov's second method, *in* "Contributions to Nonlinear Oscillations" (Ann. Math. Study, Vol. IV). Princeton Univ. Press, Princeton, New Jersey 1958.

Baker, R. A., and Desoer, C. A.

(1) Asymptotic stability in the large of a class of single-loop feedback systems, *SIAM J. Control* **6**, No. 1 (1968).

Bergen, A. R., and Sapiro, M. A.

(1) The parabola test for absolute stability, *IEEE Trans. Automatic Control* **AC-12**, No. 3 (1967).

Bergen, A. R., and Willems, I. J.
(1) Verification of Aizerman's conjecture for a class of third-order systems, *IRE Trans. Automatic Control* **7**, No. 3 (1962).

Bongiorno, J. J. Jr.
(1) An extension of the Nyquist–Barkhausen stability criterion to linear lumped-parameter systems with time-varying elements, *IEEE Trans. Automatic Control* **AC-8**, No. 2 (1963).
(2) Real frequency stability criteria for linear time-varying systems, *Proc. IEEE* **52**, No. 7 (1964).

Brockett, R. W.
(1) Variational methods for the stability of periodic equations, *in* "Differential Equations and Dynamical Systems" (J. Hale and J. P. LaSalle eds.). Academic Press, New York 1966.

Brockett, R. W., and Forys, L. J.
(1) On the stability of systems containing a time-varying gain, *Proc. 2nd Allerton Conf. Circuit and System Theory, Urbana, Illinois* (1964).

Brockett, R. W. and Willems, J. L.
(1) Frequency domain stability criteria, *IEEE Trans. Automatic Control* **AC-10**, pt. I: No. 3, pt. II: No. 4 (1965).

Chen, C-T.
(1) Stability of linear multivariable feedback systems, *Proc. IEEE* **46**, No. 5 (1968).

Cho, Y. S., and Narendra, K. S.
(1) Stability of nonlinear time-varying feedback systems, *Automatica* **4**, Nos. 5, 6 (1968).
(2) An off-axis circle criterion for the stability of feedback systems with a monotonic nonlinearity, *IEEE Trans. Automatic Control* **AC-13**, No. 4 (1968).

Coddington, E. A., and Levinson, N.
(1) "Theory of Ordinary Differential Equations." McGraw-Hill, New York 1955.

Corduneanu, C.
(1) The application of differential inequalities to the theory of stability, *An. Sti. Univ. Al. I. Cuza, Iasi, Sect. I a Mat.* (N. S.) **6**, pp. 47–60 (1960); (see G. Sansone and R. Conti, "Nonlinear Differential Equations." Pergamon Press, New York 1964).

Cunningham, W. J.
(1) "Nonlinear Analysis." McGraw-Hill, New York 1958.

Desoer, C. A.
(1) A generalization of the Popov criterion, *IEEE Trans. Automatic Control* **AC-10**, No. 2 (1965).

Desoer, C. A., and Wu, M. Y.
(1) Stability of linear time-invariant systems, *IEEE Trans. Circuit Theory* **CT-15**, No. 3 (1968).
(2) Stability of multiple-loop feedback linear time-invariant systems, *J. Math. Anal. Appl.* **23**, No. 1 (1968).

Dewey, A. G.
(1) On the stability of feedback systems with one differentiable nonlinear element, *IEEE Trans. Automatic Control* **AC-11**, No. 3 (1966).

Dewey, A. G., and Jury, E. I.
(1) A note on Aizerman's conjecture, *IEEE Trans. Automatic Control* **AC-10**, No. 4 (1965).

(2) A stability inequality for a class of nonlinear feedback systems, *IEEE Trans. Automatic Control* **AC-11**, No. 1 (1966).

Erugin, N. P.

(1) A problem in the theory of stability of servomechanisms, *Prikl. Mat. Meh.* **16**, No. 5 (1952).

Falb, P. L., and Zames, G.

(1) On cross-correlation bounds and the positivity of certain nonlinear operators, *IEEE Trans. Automatic Control* **AC-12**, No. 2 (1967).

(2) Multipliers with real poles and zeros: an application of a theorem on stability conditions, *IEEE Trans. Automatic Control* **AC-13**, No. 1 (1968).

Fitts, R. E.

(1) Two counter-examples to Aizerman's conjecture, *IEEE Trans. Automatic Control* **AC-11**, No. 3 (1966).

Freedman, M., and Zames, G.

(1) Logarithmic variation criteria for the stability of systems with time-varying gains, *SIAM J. Control* **6**, No. 3 (1968).

Gilbert, E. G.

(1) Controllability and observability in multivariable control systems, *SIAM J. Control* **1**, No. 1 (1963).

Graham, D.

(1) Discussion on the paper of Bongiorno (1), *IEEE Trans. Automatic Control* **AC-8**, No. 2 (1963).

Gruber, M., and Willems, J. L.

(1) On a generalization of the circle criterion, *Proc. 4th Allerton Conf. Circuit and System Theory, Urbana, Illinois* (1966).

Guillemin, E. A.

(1) "Synthesis of Passive Networks." Wiley, New York 1957.

Hahn, W.

(1) "Stability of Motion." Springer-Verlag, Berlin and New York 1967.

Hsu, C-H., and Chen, C-T.

(1) A proof of the stability of multivariable feedback systems, *Proc. IEEE* **56**, No. 11 (1968).

Ibrahim, E. S., and Rekasius, Z. V.

(1) A stability criterion for nonlinear feedback systems, *IEEE Trans. Automatic Control* **AC-9**, No. 2 (1964).

Johnson, C. D., and Wonham, W. M.

(1) A note on the transformation to canonical (phase-variable) form, *IEEE Trans. Automatic Control* **AC-9**, No. 3 (1964).

Jury, E. I., and Lee, B. W.

(1) The absolute stability of systems with many nonlinearities, *Avtomat. i Telemeh.* **26**, No. 6 (1965).

Kalman, R. E.

(1) On physical and mathematical mechanisms of instability in nonlinear automatic control systems, *J. Appl. Mech. Trans. ASME* **79**, No. 3 (1957).

(2) Lyapunov functions for the problem of Lur'e in automatic control, *Proc. Nat. Acad. Sci.* (U.S.A.) **49**, No. 2 (1963).

(3) Mathematical description of linear dynamical systems, *SIAM J. Control* **1**, No. 1 (1963).

Kalman, R. E., and Bertram, J. E.
(1) Control system design via the second method of Lyapunov (parts I and II), *J. Basic Engrg. Trans. ASME* **82**, No. 2 (1960).

Kalman, R. E., Ho, Y. C., and Narendra, K. S.
(1) Controllability of linear dynamical systems, *in* "Contributions to Differential Equations" (Ann. Math. Study No. 2, Vol. I). Princeton Univ. Press, Princeton, New Jersey 1963.

Krasovskii, N. N.
(1) "Stability of Motion." English ed. Stanford Univ. Press, Stanford, California 1963 (Russian ed., 1959).

Kudrewicz, J.
(1) Stability of nonlinear systems with feedback, *Avtomat. i Telemeh.* **25**, No. 8 (1964).

LaSalle, J. P.
(1) Asymptotic stability criteria, *Proc. Symp. Appl. Math.* **13** (*Hydrodynamics Instability*) Amer. Math. Soc., Providence, Rhode Island (1962).

LaSalle, J. P., and Lefschetz, S.
(1) "Stability by Lyapunov's Direct Method with Applications." Academic Press, New York 1961.

Lefschetz, S.
(1) "Stability of Nonlinear Control Systems." Academic Press, New York 1965.

Lesfchetz, S., Meyer, K. R., and Wonham, W. M.
(1) A correction to the propagating error in the Lur'e problem, *J. Differential Equations* **3**, No. 3 (1967).

Letov, A. M.
(1) "Stability in Nonlinear Control Systems." English ed. Princeton Univ. Press, Princeton, N. J. 1961 (Russian ed., 1955).

Lindgren, A. G., and Pinkos, R. F.
(1) Stability of symmetric nonlinear multivariable systems, *J. Franklin Inst.* **282**, No. 2 (1966).

Lur'e, A. I.
(1) "On Some Nonlinear Problems in the Theory of Automatic Control." English ed. HM Stationery Office, London 1957 (Russian ed., 1951).

Lur'e, A. I., and Postnikov, V. N.
(1) On the theory of stability of control systems, *Prikl. Mat. Meh.* **8**, No. 3 (1944).

Malkin, I. G.
(1) "Theory of Stability of Motion." English ed. U. S. Atomic Energy Comm., Tr. 3352, Dept. of Commerce 1958 (Russian ed., 1952).

Massera, J. L.
(1) Contributions to stability theory, *Ann. of Math.* **64**, No. 1 (1956).

McLachlan, N. W.
(1) "Theory and Application of Mathieu Functions." Oxford Univ. Press (Clarendon), London and New York 1951.

Meyer, K. R.
(1) On the existence of Lyapunov functions for the problem of Lur'e, *SIAM J. Control* **3**, No. 3 (1966).

Moore, J. B., and Anderson, B. D. O.
(1) Applications of the multivariable Popov criterion, *Internat. J. Control* **5**, No. 4 (1967).
(2) A generalization of the Popov criterion, *J. Franklin Inst.* **285**, No. 6 (1968).

Narendra, K. S., and Cho, Y. S.
(1) Stability of feedback systems containing a single odd monotonic nonlinearity, *IEEE Trans. Automatic Control* **AC-12**, No. 4 (1967).

Narendra, K. S., and Goldwyn, R. M.
(1) A geometrical criterion for the stability of certain nonlinear nonautonomous systems, *IEEE Trans. Circuit Theory* **CT-11**, No. 3 (1964).
(2) Existence of quadratic type Lyapunov functions for a class of nonlinear systems, *Internat. J. Engrg. Sci.* **2**, No. 4 (1964).
(3) Generation of quadratic-type Liapunov functions for linear time-varying systems, Cruft Laboratory, Harvard University, Cambridge, Massachusetts, Technical Report No. 421 (November 1963).

Narendra, K. S. and Neuman, C. P.
(1) Stability of a class of differential equations with a single monotonic nonlinearity, *SIAM J. Control* **4**, No. 2 (1966).
(2) A conjecture on the existence of a quadratic Lyapunov function for linear time-invariant feedback systems, *Proc. 4th Allerton Conf. Circuit and System Theory, Urbana, Illinois* (1966).
(3) Stability of continuous-time dynamical systems with *m*-feedback nonlinearities, *AIAA J.* **5**, No. 11 (1967).

Narendra, K. S., and Taylor, J. H.
(1) Stability of nonlinear time-varying systems, *IEEE Trans. Automatic Control* **AC-12**, No. 5 (1967).
(2) Lyapunov functions for nonlinear time-varying systems, *Information and Control* **12**, Nos. 5–6 (1968).

Newcomb, R. W.
(1) "Linear Multiport Synthesis." McGraw-Hill, New York 1966.

O'Shea, R. P.
(1) A combined frequency-time domain stability criterion for autonomous linear systems, *IEEE Trans. Automatic Control* **AC-11,** No. 3 (1966).
(2) An improved frequency time domain stability criterion for autonomous continuous systems, *IEEE Trans. Automatic Control* **AC-12**, No. 6 (1967).

Parks, P. C.
(1) A new proof of the Routh–Hurwitz stability criterion using the second method of Lyapunov, *Proc. Cambridge Philos. Soc.* **58**, No. 4 (1962).

Persidskii, S. K.
(1) On Liapunov's second method, *Prikl. Mat. Meh.* **25**, No. 1 (1961).

Pliss, V. A.
(1) "Certain Problems in the Theory of Stability of Motion." Russian ed. Leningrad Univ. Press, Leningrad 1958.

Popov, V. M.
(1) Nouveaux criteriums de stabilité pour les systemès automatiques non-linéaries, *Revue d'Electrotechnique et d'Energetique, Acad. de la Rep. Populaire Romaine* **5**, No. 1 (1960).

(2) Absolute stability of nonlinear systems of automatic control, *Avtomat. i Telemeh.* **22**, No. 8 (1961).
(3) New graphical criteria for the stability of the steady state of nonlinear control systems, *Revue d'Electrotech. et d'Energ., Acad. Rep. Pop. Romaine* **6**, No. 1 (1961).
(4) Hyperstability and optimality of automatic systems with several control functions, *Revue d'Electrotech. et d'Energ., Acad. Rep. Pop. Romaine* **9** (1964).

Rekasius, Z. V., and Gibson, J. E.
(1) Stability analysis of nonlinear control systems by the second method of Lyapunov, *IRE Trans. Automatic Control* **AC-7**, No. 1 (1962).

Rekasius, Z. V., and Rowland, J. R.
(1) A stability criterion for feedback systems containing a single time-varying nonlinear element, *IEEE Trans. Automatic Control* **AC-10**, No. 3 (1965).

Rosenbrock, H. H.
(1) The stability of multivariable systems, *IEEE Trans. Automatic Control* **AC-17**, No. 1 (1972).

Rozenvasser, E. N.
(1) The absolute stability of nonlinear systems, *Avtomat. i Telemeh.* **24**, No. 3 (1963).

Sandberg, I. W.
(1) On the theory of linear multiloop feedback systems, *Bell System Tech. J.* **42**, No. 2 (1963).
(2a) On the L_2-boundedness of solutions of nonlinear functional equations, *Bell System Tech. J.* **43**, No. 4 (1964).
(2b) A frequency domain condition for the stability of feedback systems containing a single time-varying nonlinear element, *Bell System Tech. J.* **43**, No. 4 (1964).
(3) Some results on the theory of physical systems governed by nonlinear functional equations, *Bell System Tech. J.* **44**, No. 5 (1965).
(4) A stability criterion for linear networks containing time-varying capacitors, *IEEE Trans. Circuit Theory* **CT-12**, No. 1 (1965).
(5) On generalizations and extensions of the Popov criterion, *IEEE Trans. Circuit Theory* **CT-13**, No. 1 (1966).

Schwarz, H. R.
(1) A method for determining stability of matrix differential equations, *Z. Angew. Math. Phys.* **7**, pp. 473–500 (1956).

Smirnov, Yu. N.
(1) Iso-μ curves in the zones of instability of Mathieu's equation, *Dokl. Akad. Nauk USSR* **178**, No. 3 (1968).

Srinath, M. D., and Thathachar, M. A. L.
(1) A stability criterion for systems with a single slope-restricted monotone nonlinearity, *Proc. Asilomar Conf. Systems and Circuits* (November 1967).

Srinath, M. D., Thathachar, M. A. L., and Ramapriyan, H. K.
(1) Stability of a class of nonlinear time-varying systems, *Internat. J. Control* **7**, No. 2 (1968).
(2) Absolute stability of systems with multiple nonlinearities, *Internat. J. Control* **7**, No. 4 (1968).

Stoker, J. J.
(1) "Nonlinear Vibrations." Wiley (Interscience), New York 1950.

Sultanov, I. A.

(1) Resolving equations for the investigation of absolute stability of control systems with many control elements, *Avtomat. i Telemeh.* **25**, No. 2 (1964).

Taylor, J. H.

(1) Strictly positive real functions and the Lefschetz–Kalman–Yakubovich lemma, Tech. Rep. EE 22/72, Dept. Electrical Eng., Indian Institute of Science, Bangalore, India (August, 1972).

(2) An all-geometric absolute stability criterion for monotonic nonlinear time-varying systems, Tech. Rep. EE 23/72, Dept. Electrical Eng., Indian Institute of Science, Bangalore, India (September, 1972).

Taylor, J. H., and Narendra, K. S.

(1) Stability regions for the damped Mathieu equation, *SIAM J. Appl. Math.* **17**, No. 2 (1969).

(2) The Corduneanu–Popov approach to the stability of nonlinear time-varying systems, *SIAM J. Appl. Math.* **18**, No. 2 (1970).

Thathachar, M. A. L.

(1) Stability of systems with power-law nonlinearities, *Automatica* **6**, No. 5 (1970).

Thathachar, M. A. L., and Srinath, M. D.

(1) Stability of linear time-invariant systems, *IEEE Trans. Automatic Control* **AC-12**, No. 3 (1967).

(2) Some aspects of the Lur'e problem, *IEEE Trans. Automatic Control* **AC-12**, No. 4 (1967).

(3) An improved stability criterion for a system with a nonmonotonic nonlinearity, *Internat. J. Control* **12**, No. 1 (1970).

Thathachar, M. A. L., Srinath, M. D., and Krishna, G.

(1) Stability with nonlinearity in a sector, *IEEE Trans. Automatic Control* **AC-11**, No. 2 (1966).

Thathachar, M. A. L., Srinath, M. D., and Ramapriyan, H. K.

(1) On a modified Lur'e problem, *IEEE Trans. Automatic Control* **AC-12**, No. 6 (1967).

Tokumaru, H., and Saito, N.

(1) On the absolute stability of automatic control systems with many nonlinear characteristics, *Mem. Fac. Engrg. Kyoto Univ.* **27**, No. 3 (1965).

Willems, J. L.

(1) Comments on "A stability inequality for a class of nonlinear feedback systems" (Dewey and Jury, (2)), *IEEE Trans. Automatic Control* **AC-12**, No. 2 (1967).

Wolovich, W. A., and Falb, P. L.

(1) On the structure of multivariable systems, *SIAM J. Control* **7**, No. 3 (1969).

Yakubovich, V. A.

(1) Solution of certain matrix inequalities occurring in the theory of automatic controls, *Dokl. Akad. Nauk USSR* **143**, No. 1 (1962).

(2) The method of matrix inequalities in the stability theory of nonlinear control systems: I. Absolute stability of forced vibrations; II. Absolute stability for a class of nonlinearities with a condition on the derivative, *Avtomat. i Telemeh.*; I: **25**, No. 7 (1965); II: **26**, No. 4 (1966).

(3) Absolute stability of nonlinear control systems, *Avtomat. i Telemeh.*; I: **31**, No. 12 (1970); II: **32**, No. 6 (1971).

Yoshizawa, T.

(1) Lyapunov's function and boundedness of solutions, *Funkcial. Ekvac.* **2**, No. 2 (1959).

Youla, D. C.

(1) On the factorization of rational matrices, *IEEE Trans. Information Theory* **IT-7**, No. 4 (1961).

Zames, G.

(1) Functional analysis applied to nonlinear feedback systems, *IEEE Trans. Circuit Theory* **CT-10**, No. 3 (1963).

(2) On the stability of nonlinear, time-varying feedback systems, *Proc. Natl. Electronics Conf.* **20**, pp. 725–730 (Oct. 1964).

(3) Nonlinear time-varying feedback systems—conditions for L_∞-boundedness derived using conic operators on exponentially weighted spaces, *Proc. 3rd Allerton Conf. Circuit and System Theory, Urbana, Illinois* (1965).

(4) On the input–output stability of time-varying nonlinear feedback systems, *IEEE Trans. Automatic Control* **AC-11**, No. 2 (Part I) and No. 3 (Part II) (1966).

Zames, G., and Falb, P. L.

(1) On the stability of systems with monotonic and odd monotonic nonlinearities, *IEEE Trans. Automatic Control* **AC-12**, No. 2 (1967).

(2) Stability conditions for systems with monotonic and slope-restricted nonlinearities, *SIAM J. Control* **6**, No. 1 (1968).

INDEX

A

A matrix
  LTI, 4, 18ff
  LTV, 4
  stability properties of, 25ff, 208
A_κ matrix, 66, 91
Aizerman conjecture, 13, 69ff, 219
  counterexamples to, 70ff
Antisymmetric matrix, 45
Attractivity, 7
  uniform, 34
  uniform in the whole, 36
Augmentation, 101ff
Augmented system, stability of, 105

C

Canonical decomposition, 23
Canonical form
  phase-variable, 19
  Schwarz, 83
Characteristic equation, 67, 81
Characteristic polynomial, 67, 81
  Hurwitz, 67, 81
    conditions for, 81
Circle criterion, 79, 123ff, 155ff, 227
  Off-axis, 166ff
    failure of, 175
Common quadratic Lyapunov function, 75ff, 123ff
Comparison functions of Hahn, 8, 35ff
Compensation to achieve stability, 149
Conjecture
  of Aizerman, 13, 69ff, 219
  concerning the $\underline{\kappa}$-dependent Lyapunov function, 214
  of Kalman, 13, 72
  of Narendra and Neuman, 86ff
Controllability (complete), 22ff, 206
  condition for, 24, 206
Controller (NLTV), 5, 18ff, 203
Corduneanu, stability theorem of, 10, 43ff, 138, 229
Corduneanu function, absolute $\{G_i[N, T]\}$, 44

D

Decrescent function, 41ff
Describing function method, 72ff
  counterexamples to its use in stability analysis, 75
Direct control, equation of, 30
Driving point impedance, 57, 60, 100

E

Eigenvalues, 67

Equilibrium, 2
  origin as the sole, 32ff, 208
Existence of solutions, 2ff
Existence theorems for stability, 39, 61ff

F

Finite sector, 28ff
Finite sector–infinite sector transformation, 108, 132
First and third quadrant function, 14, 27
Floquet theory, 186, 189ff
Frequency domain compensation, 149
Frequency domain multiplier, 99
Frequency response, 84ff, 149ff
Function class
  $\{F\}$, 14, 27, 92ff
  $\{F_m\}$, 28, 100ff
  $\{F_{mo}\}$, 28, 116ff
  $\{K\}$, 8, 35, 41
  $\{K_i\}$, 27
  $\{L\}$, 8, 35
  $\{\underline{\mathrm{MPR}}\}$, 212
  $\{\underline{\mathrm{MSPR}}\}$, 213
  $\{PR\}$, 57ff
  $\{\underline{PR}\}$, 209
  $\{S\}$, 4
  $\{SPR\}$, 59
  $\{SPR_0\}$, 60
  $\{Z_F\}$, 99
  $\{\underline{Z}_F\}$, 218
  $\{Z_{F_m}\}$, 107
  $\{\underline{Z}_{F_m}\}$, 221
  $\{Z_{F_{mo}}\}$, 117
  $\{Z_L\}$, 120
  $\{Z_{LC}\}$, 86
  $\{\underline{Z}_{N_i}\}$, 225
  $\{Z_{RC}\}$, 109
  $\{Z_{RL}\}$, 107
  $\{Z_\alpha\}$, 113, 118
  $\{Z_\theta\}$, 167
Functional analysis, 9, 99, 118

G

General finite sector–finite sector transformation, 119
Generalized circle $C[\underline{G}, \bar{G}]$, 156
  closed interior $I[\underline{G}, \bar{G}]$ of, 156
Generalized off-axis circle $C_\mu[\underline{G}, \bar{G}]$, 170
  closed interior $I_\mu[\underline{G}, \bar{G}]$ of, 170

H

Hermetian part of a matrix, 209ff
Hill's equation, 185
Hurwitz polynomial, 67
  conditions for, 81, 214
Hurwitz range, 80

I

Impulse response, 154
Index of nonlinearity $\underline{\Phi}$, 29, 127, 136, 208
Indirect control, equation of, 30
Infinite sector case, 27ff
Integral condition for absolute stability, 138ff, 228ff
Inverse multiplier, 109ff, 133, 221
Inversion of the system, 109ff, 221
Iso-μ curve, 186

K

Kalman conjecture, 13, 72
  counterexample to, 72
Kalman–Yakubovich lemma, 48ff, 209ff
  Anderson form, 210
  Lefschetz form, 49
  Meyer form, 50
  Rekasius–Rowland form, 124
κ-dependent Lyapunov function, 79ff, 84ff, 214

L

Laplace transform, 19
LC function, 86
  shifted, 88, 113
Leading principal minor, 46
Limit cycle, 74
Linearization
  total, 5
  partial, 5, 14
  quasi-, 72ff
Lipschitz condition
  global uniform, 3, 37
  local, 3, 61, 68
  uniform, 3, 37, 62
LTI plant, 5, 18, 203
  unstable, 151ff
LTI system, 4
  stability of, 81ff, 113, 213
LTV system, 4
  stability of, 135, 142ff, 181ff

Lur'e and Postnikov, problem of, 14
  Lyapunov function of, 14, 80, 92
    multivariable generalization of, 216
Lyapunov, direct method of, 7ff, 39ff
Lyapunov function, 9
  absolute $\{G_i[N, T]\}$, 44
  candidate, 39
    absolute $\{G_i[N, T]\}$, 44ff
    global, 42
  common (quadratic), 75ff, 123ff
  existence of, 10, 42, 61ff
  global, 42
  κ-dependent, 79, 214
  quadratic, 10, 45ff
  time derivative along trajectories (total time derivative), 10, 41, 47

M

Mathieu's equation, 185ff
Minimal positive real matrix, 212
Minimal strictly positive real matrix, 213
Monotonic gain function, 28, 100ff
  separate sector and slope restriction, 112
Monotonic odd gain function, 28, 116ff
Motion, 2
  stability of, 6
Motion space, 2
Multiplier, frequency domain, 99
  classes of, 99, 120
Multiplier margin $\bar{\Lambda}$, 128, 136, 227

N

NLTI system, 4
NLTV system, 5
NLTV gain (or element), 5
  nonlinear behavior of, 27ff
  separability of, 26, 207
  time behavior of, 26
Nonlinearity, index of $\underline{\Phi}$, 29, 127, 208
Nyquist criterion
  generalized to multivariable systems, 214ff
  for single-variable stable plants, 84ff, 150ff
  for single-variable unstable plants, 151

O

Observability (complete), 22ff, 206
  condition for, 24, 206
Odd monotonic gain function, 28, 116ff
Off-axis circle criterion, 166ff
  failure of, 175

P

Parabola criterion, 163
Particular case, 26, 30, 97, 130, 138
Periodic gains
  LTV, 144ff
  NLTV, 137ff
Perturbation analysis, 187ff
Phase plane, 175ff, 192
Phase variable canonical form, 19
Plant (LTI), 5, 18, 203
Point condition for absolute stability, 126ff, 226ff
Pole-zero cancellation, 33, 101
Popov
  criterion of, 15, 91ff, 158ff
  system model of, 91
Popov line, 159ff
Positive definite function, 41
Positive definite matrix, 45ff
Positive real function, 57ff
  conditions for, 59
Positive real matrix, 209
Positive semi-definite matrix, 46, 64
Principal case, 25, 30
Principal minor, 46
Proper transfer function, 66
  matrix, 218

Q

Quadratic form, 45
Quadratic Lyapunov function, existence of, 63ff

R

Radially unbounded function, 41
Range of gain, 26ff, 208
Realization, 206
Region
  $\mathfrak{F}$(s-plane), 170
  $\mathfrak{F}_\lambda$(s-plane), 181
  $\mathfrak{R}$(s-plane), 151
  $\mathfrak{R}_\lambda$(s-plane), 181
RL function, 107ff, 167, 221
Root locus technique, 153, 182

S

Scalar differential equation formulation, 20, 204
Schwarz canonical form, 83
Separability of NLTV gain, 26, 207
Solutions $x(t; x_0, t_0)$, 2, 21
  existence of, 2ff
Stability, 7
  absolute, 14, 38
    algebraic conditions for, 158
    geometric conditions for, 159ff
      failure of, 161
    parabola criterion for, 163
  absolute $\{G_i[N, T]\}$, 38
  asymptotic, 7
  equiasymptotic, 35
  in the first approximation, 67
  infinitesimal, 68
  in the limit, 97
  of motion, 6
  regions
    in function space, 12
    in parameter space, 11, 185ff
    in state space, 11, 68
  with specified inputs, 7
  theorems
    of Corduneanu, 10, 43ff, 138, 229
    existence, 61ff
    of Lyapunov, 10, 40ff
      (LaSalle), 41, 137
    sufficiency, 40ff
  uniform, 34
    asymptotic, 35
    asymptotic in the whole, 37
State
  space, 1
  variables, 1
State vector, 1
  formulation, 18, 21ff, 203
Strictly positive real function, 59
Strictly positive real matrix, 213
Symmetric matrix, 45

T

Total time derivative, 39, 41, 47
Trajectory, 2
Transfer function formulation, 19, 203
Transition matrix, 21, 62, 64
Transition points, 186

ELECTRICAL SCIENCE

A Series of Monographs and Texts

Editors

Henry G. Booker
UNIVERSITY OF CALIFORNIA AT SAN DIEGO
LA JOLLA, CALIFORNIA

Nicholas DeClaris
UNIVERSITY OF MARYLAND
COLLEGE PARK, MARYLAND

Joseph E. Rowe. Nonlinear Electron-Wave Interaction Phenomena. 1965

Max J. O. Strutt. Semiconductor Devices: Volume I.
Semiconductors and Semiconductor Diodes. 1966

Austin Blaquiere. Nonlinear System Analysis. 1966

Victor Rumsey. Frequency Independent Antennas. 1966

Charles K. Birdsall and William B. Bridges. Electron Dynamics of Diode Regions. 1966

A. D. Kuz'min and A. E. Salomonovich. Radioastronomical Methods of Antenna Measurements. 1966

Charles Cook and Marvin Bernfeld. Radar Signals: An Introduction to Theory and Application. 1967

J. W. Crispin, Jr., and K. M. Siegel (eds.). Methods of Radar Cross Section Analysis. 1968

Giuseppe Biorci (ed.). Network and Switching Theory. 1968

Ernest C. Okress (ed.). Microwave Power Engineering:
Volume 1. Generation, Transmission, Rectification. 1968
Volume 2. Applications. 1968

T. R. Bashkow (ed.). Engineering Applications of Digital Computers. 1968

Julius T. Tou (ed.). Applied Automata Theory. 1968

Robert Lyon-Caen. Diodes, Transistors, and Integrated Circuits for Switching Systems. 1969

M. Ronald Wohlers. Lumped and Distributed Passive Networks. 1969

Michel Cuenod and Allen E. Durling. A Discrete-Time Approach for System Analysis. 1969

K. Kurokawa. An Introduction to the Theory of Microwave Circuits. 1969

H. K. Messerle. Energy Conversion Statics. 1969

George Tyras. Radiation and Propagation of Electromagnetic Waves. 1969

Georges Metzger and Jean-Paul Vabre. Transmission Lines with Pulse Excitation. 1969

C. L. Sheng. Threshold Logic. 1969

Dale M. Grimes. Electromagnetism and Quantum Theory. 1969

Robert O. Harger. Synthetic Aperture Radar Systems: Theory and Design. 1970

M. A. Lampert and P. Mark. Current Injection in Solids. 1970

W. V. T. Rusch and P. D. Potter. Analysis of Reflector Antennas. 1970

Amar Mukhopadhyay. Recent Developments in Switching Theory. 1971

A. D. Whalen. Detection of Signals in Noise. 1971

J. E. Rubio. The Theory of Linear Systems. 1971

Keinosuke Fukunaga. Introduction To Statistical Pattern Recognition. 1972

Jacob Klapper and John T. Frankle. Phase-Locked and Frequency-Feedback Systems: Principles and Techniques. 1972

Kumpati S. Narendra and James H. Taylor. Frequency Domain Criteria for Absolute Stability. 1973

In Preparation

Daniel P. Meyer and Herbert A. Mayer. Radar Target Detection: Handbook of Theory and Practice

T. R. N. Rao. Error Coding for Arithmetic Processors